Vrikshayurvedic Farming

Prof. C. Swaminathan
Dr. M.R. Nandhakumar

2022
Daya Publishing House®
A Division of
Astral International Pvt. Ltd.
New Delhi – 110 002

Publisher's Note:

Every possible effort has been made to ensure that the information contained in this book is accurate at the time of going to press, and the publisher and author cannot accept responsibility for any errors or omissions, however caused. No responsibility for loss or damage occasioned to any person acting, or refraining from action, as a result of the material in this publication can be accepted by the editor, the publisher or the author. The Publisher is not associated with any product or vendor mentioned in the book. The contents of this work are intended to further general scientific research, understanding and discussion only. Readers should consult with a specialist where appropriate.

Every effort has been made to trace the owners of copyright material used in this book, if any. The author and the publisher will be grateful for any omission brought to their notice for acknowledgement in the future editions of the book.

Cataloging in Publication Data--DK
Courtesy: D.K. Agencies (P) Ltd. <docinfo@dkagencies.com>

Swaminathan, C., 1964- **author.**
Vrikshayurvedic farming / Prof. C. Swaminathan, Dr. M.R. Nandhakumar.
pages cm
ISBN 978-93-5124-735-7 (Hardbound)
ISBN 978-93-86071-16-3 (International Edition)

1. Trees--India. I. Nandhakumar, M. R., author. II. Title.

SD391.S93 2016 DDC 634.90954 23

Published by : **Daya Publishing House®**
A Division of
Astral International Pvt. Ltd.
– ISO 9001:2015 Certified Company –
4736/23, Ansari Road, Darya Ganj
New Delhi-110 002
Ph. 011-43549197, 23278134
E-mail: info@astralint.com
Website: www.astralint.com

Acknowledgement

I depict my profoundest feelings of gratitude for the people who have rendered invaluable help for this book in the form of words.

I thank my teacher and respected and revered Prof. Dr. K. Ramasamy, Vice-Chancellor, Tamil Nadu Agricultural University and authorities for giving consent to write the book on ***Vrikshayurvedic Farming.***

I show gratitude to all my post graduate scholars M.R. Nandhakumar, A. Sugana, A. Sakthivel, B. Rajalakshmi, V. Chelladurai and J. Anbarasi and Foreign student, Rajasekaran Murugan, Germany who are the front runners in executing my concept at field and gave scientific validity.

I thank my young and energetic colleagues Dr. P. Kannan, Dr. M. Sivaprakash, Dr. P. Arunachalam, Dr. P. Kathirvelan, Dr. R. Sasikala, Dr. E. Tamilselvi and Dr. M. Sudha of Dryland Agricultural Research Station, TNAU, Chettinad, Sivaganga district, Tamil Nadu for their consistent support during preparation of the texts and writing.

I thank entomologists Dr. K. Premalatha, Dr. S. Raguraman for their involvement during the early stages of research work and participation; Dr. N. Muthukrishnan, Dr. J. Jeyarai of Department of Agricultural Entomology; Dr. V. Swaminathan, Dr. A. Sathasakthi of Department of Horticulture, Agricultural College and Research Institute, Madurai for their technical support in executing research on onion and bhendi.

I also thank my coauthor cum my student Dr. M.R. Nandhakumar for all his contributions.

I place on record for ever the thanks and gratitude to my academic *Guru* Dr. V. Veerabadran, Chief Scientist and Professor of Agronomy (Retd.) and Late Dr. R. Stephen Vinaya Rai, Professor of Forestry to my disciple Dr. Rajasekaran

Murugan of UniKassel, Germany for improvising this book at every stage and their forewords.

No words to thank my wife and only son; family members for their continued encouragement and support to put my work in writing.

I am grateful to my *Guruji*, Divine Mother Aurobindo Mirra for their continued blessings and energy in all walks of my life.

I dedicate this book to my academic *Gurus*

1. Dr. V. Veerabadran, Professor of Agronomy (Retd.), Architect of Modern Agronomy and crop production and
2. Late Dr. R. Stephen Vinayaraj, Professor of Forestry, Architect of Forestry Education
3. All my graduates and post graduates students of Mettupalayam, Coimbatore, Trichy, Killikulam, Periakulam and Madurai campuses of TNAU.

Prof. C. Swaminathan

'A' Note

The plant products are naturally evolved ingredients on the biosphere; they not only have an edge over the synthetic alien molecules but also rate preference and acceptance from the viewpoint of environmental safety and eco-friendly. *Vrikshayurveda* (Indian plant science) provides ample scope to produce quality food products by adopting traditional methods and by utilizing plant, including trees, extracts and by products. The three major ancient texts *inter alia* are 1) Varahamitra, 505 AD., Birhat Samhita, *Vrikshayurveda* Part I, Chapter 55 (edited by M. Ramakrishna Bhat, Motilal Banarasidas, Bangalore, India 1950) 2) Chaundarya, 1025 A.D. Lokopakaran, *Vrikshayurveda*, chapter 6 (edited by H. Shesa Iyengar, Government central Manuscripts library, Chennai, India1950) and 3) Sarangadhara, 1363 A.D. *Vrikshayurveda* (edited by S.K.Ramachandra rao, Kalpatharu Research Academy, P.O. Box 1857, Bangalore, 1993, India). The most detailed of all the three seems to be that of Chaundarya, 1025 A.D. (Shamarao Jahadirgar *et al.*, 2003). The use of leaves of leguminous trees as green manures, because of its high nutrient content and faster decomposition rate and the role of plant extracts in growth promotion, pest and disease management has been studied by me for a decade. Simultaneously the allelopthic proclivities of extracts of trees and their parts have been studied by me for more than two decades and had been published in peer reviewed international journals like International Tree crop journal, Allelopathy journal, Journal of Tropical Forest science, Biological agriculture and horticulture, Journal of sustainable agriculture. My first work on *Vrikshayurvedic* farming started during 1997-99 wherein the leaf extracts of tree species (*Alangium salvifolium*, *Aegle marmellos* and *Annona squamosa* and *Azdirachta indica*) were found to have effects on growth and development of cowpea and greengram. Then I decided that it's the right time to explore the information available in the ancient texts and find a new way out and accordingly a decade of research work done on these lines. Later, I continued my research work on these lines for more than a decade and the outcome of the work is brought as a nice compilation in this book.

I put the methodologies and practices adopted for different crops under this research work comprehensively and coined a new terminology Vrikshayurvedic Farming.

Vrikshayurvedic farming is scientific reorientation of eco-friendly ancient system of Indian agriculture by looking back the traditional and natural ways of food production and adopting traditional and indigenous practices and methods for cultivation of crops; by utilizing trees, plants and animal products, by products, extracts and other means with the objective of enhancing the quality of food produce (Swaminathan, 2012). This is the definition coined by me for the first time in the world.

The entire research work done by me alongwith my post graduate scholars are presented in this compilation for further use by the aspirants, agricultural researchers, students and carry forward the work in newer dimensions in the years to come.

I appreciate and wish the Astral International, leader in academic publishing to reach new heights besides placing it on record my gratitude for bringing out this nice compilation. I am indebted.

Enjoy reading the texts, contemplate and innovate

Thanks for reading the original book.

Prof. C. Swaminathan

U N I K A S S E L
V E R S I T Ä T

Organic
Agricultural Sciences

University of Kassel, FB 11, BBP

Dr. Rajasekaran Murugan
Junior research group leader
Department of Soil Biology and Plant Nutrition
University of Kassel, Nordbahnhof straße 1a
37213 Witzenhausen, Germnay
murugan@uni-kassel.de
tel +49 5542 98 1532
fax +49 5542 98 1596

Foreword-I

It is fascinating that ancient India not only had a medical science for the humans (Ayurveda) but also a one for the plants, called Vrikshayurvedha. Vrikshayurvedha – a Sanskrit word meaning *the science of plant life* (Vriksha – plants; ayur – life and vedha – science or knowledge). In the recent past a plant cell structure is studied with the help of microscope by botanists and taxonomists. What is astonishing is that about 2000 years ago prior to that date, Sage Parashara (presumably around BC 100) had clearly explained the structure of a plant cell and the phenomenon of photosynthesis (process of self-nourishment in the plants) in the fourth chapter (Vriksha sharira dharma sastram) of the classical text "Vrikshayurvedha".

This book is an enthusiastic celebration of many agricultural practices, especially those which are of ancient origin and biological relevance. It is also a unique tribute to the ancient texts and sages who passed their traditional knowledge on beneficial effects of applying various plant decoctions to seeds, diseased plants and soil. My guru, Prof. C. Swaminathan should be acknowledged for his work to bring this ancient knowledge for the benefit of agricultural ecosystem. On top of that, there are innumerable historical facts on agricultural sciences were proved through his experiments in a very appealing way. Although these represent only a small sample of the world of agricultural sciences, they amply illustrate importance of this field of science to humankind and contain much that will be of interest to those outside - indeed to anyone with a fascination with the world of Ayurvedha.

I think that the authors can be confident that there will be many grateful readers who will have gained a broader perspective of the disciplines of Vrikshayurvedha and agricultural sustainability as a result of their efforts. Indigenous knowledge on plant life is the basis for self-sufficiency and recognised to be more eco-friendly and sustainable. The authors of *Vrikshayurvedhic farming* demonstrate clearly the impact we can have on our own destiny through plants. Their book is destined to play a major role in exciting, motivating and educating the next generation of scientists and farmers from all over the world to plunge into ancient texts like Vrikshayurvedha to overcome new hurdles in ever expanding fields of science.

Dr. Rajasekaran Murugan

Vrikshayurvedic Farming

THE AUTHORS

Prof. C. Swaminathan (born on July 19th 1964, grandson of a freedom fighter in Kanyakumari district, Tamil Nadu), with a first class academic career, had his bachelors and master in Agriculture from the reputed Tamil Nadu Agricultural University. He is one among the top ten in the country to receive Doctorate in Forestry from Forest College and Research Institute, TNAU as early as in 1996. He is merited with many awards and recognitions *viz.,* Dr. A. Venkataraman Medal for the best scholar and best research program in the doctoral degree; Nanda Memorial Award for the best National Young Forestry Scientist; His biography is included in the Asia pacific Who's who in 2000; FAO Fellowship in 2001 on Wood Science for post doctoral research at University of Melbourne and he had visited Australia and Malaysia. Further he has been appreciated for his Book on Brown Revolution (Revolution on/of woods) and received Letter of Appreciation from His Excellency Dr.A.P.J. Abdul Kalam, President of India in 2002. In 2004, He received Best Paper Award in National Wasteland Development seminar. During 2005-06, he is awarded BOLT award for the best Teacher in Thoothukudi district of Tamil Nadu. During 2009 he has been awarded with IASc- INSA- NASI Fellowship for forestry at the Institute of Wood Sciences, Bengaluru. He is adjudged as the BEST TEACHER by TNAU in 2010. He has also been awarded as the 'Best Teacher with Social concern' in South Tamil Nadu.He has been serving TNAU in various capacities at different Colleges and Research Stations since 1988. Presently he is holding the position of Professor and Head, Department of Agronomy, Agricultural College and Research Institute, Tamil Nadu Agricultural University, Madurai, Tamil Nadu. As a flourishing and dedicated worker of great prominence he has utilized his diverse academic and practical pursuits in the development of tree sciences. The wide arrays of research dimensions focused by him are agronomy, dryland farming, organic farming, silviculture, Agro forestry, Nursery Technology of Trees Afforestation and Environmental Protection. He has to to his credit about 35 research articles published in international and 50 articles in national journals of repute.

Dr. M.R. Nandhakumar, Senior Research Fellow (Agronomy) is working in department of Crop Physiology, Tamil Nadu Agricultural University, Coimbatore. He has completed his UG at Annamalai University during 2004-2008. He has obtained M.Sc. (Ag.) and Ph.D with specialization in Agronomy from TNAU, Coimbatore. He has received best Rice Thesis Awad in Agronomy from TNAU, oral and poster presentation awards in National level conferences. He has published 1 book, 9 research papers and 3 popular articles.

Foreword-II

Tradition has its own value in any field of human activity, Agriculture is no exception. The major concern today is how to revitalize the natural resources damaged by misuse and abuse of many modern agricultural concepts and technologies through ignorance and indifference. Traditional agricultural practices evolved over hundreds of years based on the experience of farmers are now being promoted as an answer for most the ills affecting agriculture today. Sustainable agriculture is a keyword today in agricultural is a keyword today in agricultural development. Many traditional agricultural concepts and practices have the needed ingredients for sustainable agriculture. Documentation of traditional agriculture is always a felt need and welcome attempt in the promotion of sustainable agriculture. *Vrikshayurveda*, Indian plant science as defined by the authors is yet another collection of traditional agricultural practices.

The essential feature of *Vrikshayurveda* is the reliance on plant products for all aspects of crop production. This book presents an extensive review of the use of various plant and tree products for seed treatment, growth promotion, crop nutritional, soil fertility maintenance, pest and disease control and yield enhancement with no adverse affect on soil, water and air. This projected ad the most prominent benefit of *Vrikshayurveda*. Most of the problems faced in agriculture today are attributed to the long term indiscriminate use of agricultural chemicals. *Vrikshayurveda* offers scope for substitution of these agrochemicals by plant and tree products. Exlusion of inorganic chemicals and inclusion of organic products of biological origin constitute the basic concept of organic farming for conservation of natural resources and sustainable agriculture. *Vrikshayurveda* is proposed by the authors as another format of organic farming with dependence on plant and tree products. It is a kin to the switching over from allopathy to Indian systems of medicine (Siddha, Ayurvedha, Unani) for healthcare. *Vrikshayurveda* is literally difined as "Ayurveda of plants".

The efforts of the authors to collect, review and present information available through *Vrikshayurveda* deserve commendation. This book on *Vrikshayurveda* will be an excellent source of reference to those who believe in a shift to traditional agriculture for future sustenance of agriculture.

Dr. V. Veerabadran

Professor of Agronomy (Retd.)
Tamil Nadu Agricultural University
TNAU Avenue,
Y.Othakkadai, Madurai – 625 107

Contents

Chapter 1

Introduction, Meaning and History

Introduction

India over several millennia has been the treasure land of biological wealth, intellectual knowledge and spiritual wisdom. However, the influence of modernization and changes in lifestyle has resulted in a decline in traditional health and also shift in agricultural practices. Moreover, many modern practices show indifferences or negative attitudes towards traditional practices. Nevertheless, a sector of farm scientists are now focusing on evolving ecologically sound, biologically sustainable and socio economically viable technologies. Scientific evaluation of practices followed by farmers, establishment of cropping systems with organic farming practices and quantifying the responses and identifying plants suitable for pest, diseases and weed control are the major issues requiring consideration for the prospects of growing crops organically.

Meanwhile, India has also been blessed with largest collections of ancient manuscpripts in the world. While there has never been a precise account, estimates suggest there may be as many as 300 million texts. These old Indian scripts paid considerable attention to agriculture, livestock, rain harvests, philosophy, religion and health care. These provide amble scope and support to the researchers. For example, Ethano farming, one of the manuscripts referring agriculture to a "Farming systems adopted by a particular ethnic group of farmers in a society to maintain their livelihood on sustainable basis with least possible damage to the nature" and is still being practiced by the farmers in India. Apart from manuscripts, the Indigenous Technical Knowledge (ITK) possessed by different sector of peasants in India has been, over years passed on to generation to generations. It is held in different brains, languages and as skills in a many groups, cultures and environments.

In the past decades, agricultural development focused on short-term productivity, ignoring the maintenance of soil health, based on external inputs resulting in neglect and improper use of local resources. This has resulted in damage to the environmental resources and disappearance of indigenous knowledge and hence, the agriculture can hardly be perceived as sustainable. In the present scenario, across the world, efforts are being made to work towards ecologically friendly agriculture. One such option is organic farming as it is a holistic approach which takes care of all the components of the system. It is nature based, environment friendly and sustainable ensuring not only the requirement of the present but also ensures the conservation of resources for future.

In India, several instances showed that farmers' practices coincide with prescriptions found in *vrikshayurveda* which are systematizations of the practices the farmers followed at field level, placed in a theoretical framework. Many references regarding the relationship of plants with other plants and animal, soil characteristics, soil fertility, nature of plant life, evolution of plants are available in ancient Indian texts. In general the word *'vrikshayurveda'* literally meaning ayurveda for plants, is used to denote knowledge of plant life in all its varied aspects. It is of great relevance to organic agriculture because it incorporates many aspects of organic agriculture. **Vrikshyurveda farming** is a traditional Indian system of farming aiming at ecologically friendly agriculture and conservation of natural resources (Swaminathan, 2011). Here, tree leaves are used as green leaf manure and leaf extracts as foliar spray. The tree leaf extracts offer potential in development of new bio-formulations, which would be used as growth stimulants; pest and diseases control agents as well as growth tonics and providing an environmentally compatible crop production. These tree products offer excellent potential to keep off the inorganic chemicals and fertilizers, to certain extent.

Trees and plants are vital for ecosystem and are essential part of human life. Since the start of civilization during the Vedic period, greater importance was given to trees as a source of almost all the essentials in life particularly for agriculture. Enormous and elaborate literatures are found in Indian plant science (*Vrikshayurveda*) such practices are in vogue in certain areas, though not in a large scale. The recent awareness on harmful and everlasting ill effects on human beings due to indiscriminative usage of chemical fertilizers in crop production among food producers and consumers have paved way for going back to human friendly, eco-friendly and traditional agriculture. The use of green leaves of leguminous trees as green leaf manures, because of its high nutrient content and faster decomposition rate has been very well known. The leaves of *Albizia lebbek, Delonix regia, Leucaena, Azadirachta indica* and many other species are commonly used green leaf manure trees. The role of plant extracts as growth stimulant, pest and diseases control are also very well documented. The leaf extracts from Bael, Notchi and *Albizia amara* tree species are having insecticidal, fungicidal properties. Similarly some tree species (*Alangium salvifolium, Aegle marmellos, Annona squamosa* and neem) are found to have effects on growth and development of crops. (Swaminathan, 2004). Incorporation of leaves *Gliricidia sepium* before sowing of crop and followed by sprayings of leaf extracts of *Aegle marmellos* recorded higher yields (Swaminathan *et al.*, 2008).

Trees have been so far viewed as material for construction, furniture making and a provider of shade. The indirect roles that tree species perform are being ignored by the people. These trees played a major role in food production in yester years. These tree leaves extracts offer potential for development of new bio-formulations, which would be used as growth stimulants, pest and diseases control agents as well as growth tonics for providing an environmentally compatible crop production. A scientific method of discovering and developing new tree based bio formulation with help of indigenous knowledge is easier and less expensive. The use of natural tree leaf extracts as crop health tonics are attractive, due to the increasing need for, eco-friendly safe and selective bio formulations, for total health care in crop production.

Chapter 2

Heritage of Indian System of Agriculture

The Indus civilization was thought to have been confined to the valley of the river Indus; hence the name given to it was Indus Valley civilization. This civilization was a highly developed urban one and two of its towns, Mohenjo-Daro and Harappa, represent the important settlements. Subsequent archaeological excavations established that this civilization were not restricted to the Indus valley but spread to a wide area in northwestern and western India. Thus this civilization is now better known as the Harappa civilization and the principal sites in India include Ropar in Punjab, Lothal in Gujarat and Kalibangan in Rajasthan. The Indus Valley Civilization stretched across the whole of Sindh, Baluchistan, Punjab, Northern Rajasthan, Kathiawar and Gujarat.

Developments in Agriculture and Trade

The economy of Indus valley civilization appears to have depended significantly on trade, which was facilitated by major advances in transport technology. These advances included bullock-driven carts that are identical to those seen throughout South Asia today, as well as boats.

Agriculture during Indus valley civilization must have been highly productive; as it was capable of generating surpluses sufficient to support tens of thousands of urban residents who were not primarily engaged in agriculture. It relied on the considerable technological achievements of the pre-Harappan culture, including the plough. It should be noted that people of Indus valley civilization built their lives around the monsoon, a weather pattern in which the bulk of a year's rainfall occurs in a four-month period.

During this period the people identified the importance of ploughing for the proper sowing of crop (*i.e.*) soil has to be stirred and seed has to be covered. Ox-drawn wheel cart was used for transport. The people cultivated wheat, barley, gram, peas, sesame and rape. They also cultivated cotton and also devised methods of ginning, spinning and weaving. Animal husbandry was also given more importance during this period. They domesticated buffalo, cattle, camel, horse, elephant, ass and birds. They utilized them in agriculture and also for transport.

The most remarkable discovery in Harappa is the Great Granary used for storing food grain. These granaries, each 50x20 feet overall, are arranged with central passage 23 feet wide. From the size of the granary it can be concluded that the peasants paid their dues to the Government in kind, used the kinds in granary for payments to employees. The artisans, carpenters and others received their wages in kind from the farmers.

In the Chalcolithic period, Harappans had reached a high state of culture. They wore cotton garments and used ivory combs and copper mirrors. The women wore ornaments of bronze and gold. They used implements such as sickles, saws, knife blades, spears, axes, arrowheads and daggers made of bronze and copper fish-hooks. Harappans cultivated bread wheat, barley, sesame, pea, melon, date palm and *Brassica* spp. Cotton was an important crop, the centre of its origin the Indus valley. The rice cultivated in Harappa had long seeded grain and perhaps was the ancestor of the fragrant basmati rice. Wheat and jowar were the other food crops. Specialized occupation besides agriculture developed and these articles were produced by skilled craftsman such as coppersmiths, carpenters, jewelers, goldsmiths, stone cutters and potters. Trade with other countries especially with Mesopotamia flourished and some items such as metals timber and precious stones were imported.

The Vedic Civilization

The word "Veda" is derived from "Vid" which means "Knowledge" Veda is the only literary source from which we know about the land of Aryans called as land of seven rivers *i.e.* (Sutlej, Beas, Ravi, Chenab, Jhelum, Indus and Saraswathi). The Rig-Veda was the oldest book, which was compiled around 3700 BC. At the beginning of the cropping season, the ploughing was done with great fanfare associated with several rituals. There are several hymns address to Shuna, Sita and Shunshira. Sita has been referred to as the goddess of the early and also the share of the plough. Barley, sesame and sugarcane were the main crops.

Status of Agriculture in the Vedic Period (1600 BC–1000 BC)

Early Vedic people were primarily pastoral and settled in Indus valley. They cut jungles, built their villages, grazed their cattle in jungles and planted barley in the land close to their habitation. They were accustomed to cows, horses; buffalo was a new animal which they called gouri or govala which appears to be an extension of the word gau (cow). Indus valley is the land of seven rivers was called '*Saptasindhavah*'. Cattle owners apparently knew the benefits of providing clean safe water from ponds. Dogs were used for managing herds of cows and for recovering stolen cows. Burning of dried cow dung is practiced as fuel for fire. Killing of cows was clearly discouraged not only because it played an important part in human subsistence, but also for the cow's innocence.

Rig-Veda clarified that there are six seasons in a year; namely Grishma (May – June), Varsha (July-August), Hemant (September-October), Sharad (November-December), Shishir (January-February) and Vasant (March-April). The beginning of the rainy season is after 21 June when the sun starts 'moving south'. There were dams on the seven rives. Constructing dams on rivers must have meant cutting off water to Vedic people to irrigate lands and to provide water to people and animals after the rains for increasing the food production.

During Rig-Veda: Soil preparation was done through repeated ploughings. A bamboo stick of a specific size was used for measuring land. Bullock power was used for ploughing and for pulling bullock carts. Other farm operations included bird scaring, harvesting with sickle, threshing, winnowing, storing grains in storage bins and burning of trash/wastes. Barley was ratooned on residual moisture possibly for fodder after harvest of grain crop. Vedic people realized the importance of off-season ploughing and they started ploughing as and when the rain was received. The first ploughing of the season was inaugurated amidst much ritual. The plough used was large and heavy, but a yoke does not seem to have been used. Bullocks and ox were used for ploughing. With regard to irrigation, channels were formed the rivers. Wells were in use for supply of drinking water and irrigation. There were kucha wells, which were just holes, dug in the ground. Even now such wells are in use in the river rain areas of northern India.

End of the Indus Civilization

After 1750 BC, the Mohenjo-Daro and Harappan culture slowly declined and gradually faded out. The cause or causes of the end of the Indus civilization are not easy to determine. Some ascribe this to the decreasing fertility of the soil on account of the increasing salinity, caused by the expansion of the neighboring desert. Others attribute it to some kind of depression in the land, which caused floods. Even though there are various theories for the downfall of this civilization, there is no clear picture as to how or why it came to an end.

Status of Farmers in the Society

The Indus Civilization had the first farming cultures in South Asia, which emerged in the hills of what is now called Baluchistan, to the west of the Indus Valley. The farmers took part in the so-called Neolithic Revolution, which took place 9000 to 6000 BCE. These early farmers domesticated wheat and a variety of animals, including cattle. In the "Era" terminology, the Neolithic is known as the "Early Food Producing Era". The development of these farming communities ultimately led to the formation of larger settlements from the later 4th millennium. Indus valley civilization was composite product of different races that lived and worked together in a particular environment. Mohenjo-Daro had easy land and water communication; it was the meeting ground of people for different parts of Asia. Farmers had, by this time, domesticated numerous crops, including peas, sesame seeds, dates and cotton, as well as a wide range of domestic animals, including the water buffalo.

Sangam Period

During Sangam literature, agriculture was the main occupation and hence the position of the farmers in the society was also high during this period. Agriculture Sangam was developed in Madurai. The farmers are called uzhavar (plough man) and also they are called as Kalmar. The farmers' status was mainly determined by the holding of land and animal population.

Thirukural Period

Thiruvalluvar mentioned about importance of farmers in the society. In his statement,

> "Farmers alone live an independent life. Others worship them and are second to them"
>
> "If farmers stop cultivation, even Rishis (sages) can not survive".

Water Harvesting and Irrigation Developments during different Periods

The need for continuous supply of water for irrigation whether from canal, well, pond or lake is realized as the most important for agriculture in ancient period. It should be preserved by all sorts of efforts for the benefit of agriculture. The different irrigation principles adopted in ancient period are:

- ✰ Construction of large mud embankment on a stone foundation for diverting flood water from the river.
- ✰ Building of small tanks.
- ✰ It is also indicated that severe penalty was imposed when water is let out other than sluice gate.
- ✰ Extensive tank irrigation systems were adopted in South India. It was practiced that not even a drop of rain water should go to sea without benefiting human populations.
- ✰ The topography of Telengana region of Andhra Pradesh and Karnataka is ideally suited for the construction of tanks. A special feature of tanks in Telengana tank construction in series, by bunding the same valley at several points and surplus water from lower elevation and so on. Even now the tanks constructed by Chola king in the same way exist today in Tamil Nadu.
- ✰ Every farmer emphasizes the efficient and skillful use of water.
- ✰ It is also suggested that preference of the use of water should be in the order of food crop, vegetables and flowers.
- ✰ For the maintenance of tanks, a committee of villages called 'eri-variyam' was appointed. The committee ensured repairs and de-silting tanks and distribution of irrigation from wells.

In Tamil Nadu, the ancient people stored rainwater in public, placed separately one for drinking purposes and another for bathing and other domestic purposes and called them as Ooranies. They also formed percolation tanks or ponds, for the purpose of recharging irrigation or domestic wells. They periodically clean the water ways so as to get clean water throughout the year. These are instances in the history that people constructed crude rubble bunds across river courses either for diversion of water or for augmenting the ground water. The various methods of rainwater harvesting are classified below under two category, Traditional and Modern methods.

Traditional rainwater harvesting, which is still prevalent in rural areas, was done in surface storage bodies like lakes, ponds, irrigation tanks, temple tanks etc. In urban areas, due to shrinking of open spaces, rainwater will have to necessarily be harvested as ground water, Hence harvesting in such places will depend very much on the nature of the soil *viz.*, clayey, sandy etc. The below listed are the various kinds of traditional rainwater harvesting methods.

The Modern methods of rainwater harvesting are categorized under two; they are Artificial Recharging and Rainwater harvesting. The former is classified into Absorption Pit Method, Absorption Well Method, Well cum Bore Method and Recharge trench cum injection well. The later is categorized into Individual Houses and Grouped Houses which are further classified into Percolation Pit Method, Bore Well with Settlement Tank, Open Well Method with filter bed Sump and percolation Pit with Bore Method.

Plant Protection, Harvesting, Threshing and Storage in Ancient Period

In the past when Indians were gaining knowledge on the prediction of rainfall, management of agriculture, farm operations, harvesting and storage, nothing was known about plant protection. The only methods to protect the crop were prayers and mantras. It was believed that the crop is protected if the mantra was written with red lac-dye and tied to the crop. But it cannot be said that the people of that time were unaware of insects and other pests and their damage. Besides these pests, goats, rats, wild boars, pigs, deer, parrots and sparrows were mentioned as destroyers of crops. In fact when the damage to crops due to different pests reached the economic injury level, they might have started thinking about plant protection and diverted their efforts to develop protection technology. It is significant that people at that time considered that plants and human beings have similar physiology. Therefore, they divided the diseases of plants into two categories: (1) internal; and (2) external. The internal diseases were those which were caused by "vata", "pitta" and "kapha" and external diseases were those which were caused by insects, birds and weather. These categories can be attributed today to fungi, bacteria, viruses and nematodes-internal diseases; and insects, non-insect pests, frost, water logging and drought-external diseases.

Today integrated pest management (IPM) is considered a recent approach for plant protection but the so called recent approach was conceptualized and practiced centuries ago in India. Some of the practices adopted in those days are given below.

Table 2.1: Information Contained in Surapala's *Vrikshayurveda*, Related to Kinds of Internal Disorders Observed in Trees, Causes and Symptoms Attributed and Remedies Suggested

Cause given	*Symptoms*	*Cause Elaborated*	*Possible Causes*
Vata	Trunk slender and crooked; knots on trunk or leaves; hard fruits (less juicy and sweet); gradual defoliation; flower and fruit drop; generally yellowing of leaves and fruits.	Arid land on account of excessive supply of dry and pungent matters.	Underground mechanical barrier, leaf-galling insects; root-infecting fungi or nematodes, viruses, saline or alkaline soils.
Pitta	Leaf yellowing; premature drop; decay of flowers and fruits	Occur at the end of summer if trees are excessively watered with bitter, sour, salty and strong materials	Viral disease; salinity in irrigation water; predisposal to blossom blight; fruit decays due to fungal/bacterial infections
Kapha	Fruit-bearing delayed and fruits are tasteless and ripen prematurely; oozing without wounds	Appear in winter and spring if trees are excessively watered with sweet, oily, sour or cold materials	Fungal gummosis/rot, nutrient deficiencies or toxicities, excessive watering

Seed Treatment

Seed treatment which is considered an important component of IPM to ensure better germination was given a lot of importance in ancient times. The seed was treated with milk, mustard, sesame-ash and cow dung for better germination and protection against insect pests.

Fumigation

Fumigation methods were not as developed in those days as they are today but the concept of fumigation was prevalent. For example, diseases of cucurbits were controlled by smoking the bones of cow and dog mixed with the excreta of cat.

Field Application

The plant protection appliances of the present era were not developed at that time; sprinkling of aqueous suspension and hand-dusting of various materials were used. For the control of insect pests several ancient recommendations are available. Some of them are as follows:

- ✰ Insects infesting trees can be removed by smoking a mixture of white mustard, black pepper, asafoetida, vidanga (*Embelia ribes*), vaca (*Zingiber zerumbet*) and water mixed with beef, horn of a buffalo, flesh of pigeon and the powder of bhillata (*Semecarpus anacardium*).
- ✰ Insects infesting creepers can be controlled by sprinkling water mixed oilcake.
- ✰ Leaf-eating insects can be destroyed by dusting cow dung-ash and brick-dust.
- ✰ Trees are watered with cold water for 7 days to remove insects from the roots and branches.
- ✰ A wound caused by insects is healed if sprinkled with milk after being anointed with mixture of vidanga, sesame, cow's urine, ghee and mustard.

A new term, eco-friendly pesticides, has been coined recently. In IPM more emphasis is laid on this term and botanicals are being used instead of chemical pesticides. In fact this is not new. Years ago several botanicals and other materials which have biocide properties were identified and recommended by Surapala to control plant diseases. The famous "panchamula" (roots of five plants) which was commonly used at that time has antifungal, antiviral, antibacterial and antifeeding properties. Likewise, mustard had been used for all kinds of diseases caused by "kapha". We now know that mustard causes antibiosis in insects; in addition it is antifungal and has nematicidal activity.

Materials and Practices that Need our Early Attention

Milk and milk products: Milk and ghee have been used for centuries. Even buttermilk was found useful. About 40 per cent of total amino acids in milk are glutamate, leucine and proline. Milk is reported to contain plant growth promoters. A recent report claimed that milk sprays induced systemically acquired resistance in

Table 2.2: Some Important Products Used in Pest Management during Ancient and Medieval Periods in India

Material	*Author/Period*	*Properties*
Root of vasika (*Justicia adhatodaa*)	Varahamihira (505-587 AD)	Soothing effect, insecticidal, antifungal, antibacterial, anthelmintic.
Branches and leaves of atimuktaka (*Hiptage enghalensis*)	Varahamihira (505-587 AD)	Leaf juice insecticidal; bark contains glucoside (hiptagin) and tannins ‘
Mustard(*Sinapis alba = Brassica alba*)	Surapala (1000AD)	Insect antixenosis and antibiosis; acaricidal; nematicidal; antifungal
Bidanga (vidanga) (*Embelia ribes*)	Surapala (1000AD) Someshwara Deva (1126 AD)	Anthemintic; antibacterial insecticidal
Ash	Someshwara Deva (1126 AD)	Desiccates insect eggs on seed; speeds up germination by softening seed coat through mild alkalinity; provides micronutrients
Sesame (*Sesamum indicum*)	Surapala (1000 AD)	Allelopathic to rice; insect repellent; insecticidal
Mahua (*Madhuca* spp.)	Surapala (1000 AD)	Insecticidal oil; piscicidal; antibacterial
Kusta (costus) (*Saussurea lappa*)	Surapala (1000 AD)	Insecticidal (repellents, anti-feedant); antiseptic
Bhillata (Bhallataka) (*Semecarpus anarcardium*)	Surapala (1000 AD)	Insecticidal, antiseptic, termite-repellent, anthelmintic, antibacterial

chilli against leaf curl, a viral disease. Milk (10 per cent aqueous suspension) also has been effectively used for controlling powdery mildews. Besides, milk has excellent sticker-spreader properties. The amino acid proline has been found to systemically induce resistance in plants. It stimulates production of antimicrobial phenolics. High amounts of endogenous proline increase contents of cytokinin and auxins. Besides milk, proline is present in the connective tissues of animals including fish.

Application of Cow Dung

Use of cow dung for dressing seeds, plastering cut ends of vegetative propagating units such as sugarcane setts, dressing wounds, sprinkling diluted suspension on plants and applying to soil has been indicated since the time of Kautilya (c.300 BC). Indian farmers continue to use cow dung in various ways, but the agricultural scientists have ignored its use for other purposes except manure. Briefly speaking cow dung from the cattle shed is a mixture of dung and urine, generally in a ratio of 3:1. Cow dung consists of crude fiber, crude protein and materials that can be obtained in nitrogen-free extracts and ether extracts. Cellulose along with lignin makes up most of the crude fiber; hemicelluloses and pentosans (poly saccharides based on pentose sugars) are also present. Micronutrients too are present in cow dung. The urine portion of cow dung contains nitrogen, potash and sulphur and only traces of phosphorus. The nitrogenous compounds excreted in fecal matter consist in part undigested or unabsorbed food nitrogen and in part another fraction called metabolic nitrogen. The metabolic fraction comprises substances originating in the body such as residues of the bile and other digestive juices, epithelial cells from the alimentary tract and the bacterial residues. In short, fecal residues comprise undigested fiber, debris from sloughed-off intestinal epithelium, some excreted products derived from bile (*e.g.* pigments), intestinal bacteria and mucus. There are more than 60 species of bacteria and over 100 species of protozoa encountered in the rumen of a cow. A majority of the bacteria are cellulose, hemicelluloses and pectin fermenters. The bile constituents are bile salts, bile acids and bile pigments. Bile salts confer hydrophilic coat to otherwise hydrophobic droplets, thus acting as emulsifying agents. No bile salt is supposed to be present in the dung because these are reabsorbed through the intestine and are put back in the bile. Bile salts have antiseptic properties. Two chief bile pigments are bilirubin (reddish/golden yellow) and biliverdin (green). It is the biliverdin ($C_{33}H_{36}N_4O_8$) which is chiefly present in herbivorous animals and gives greenish color to the dung.

Harvesting and Threshing

Kautilya Arthasasthra states "Grains and other crops shall be collected as often as they are harvested. The threshing of different fields shall be in close proximity. In Sangam literature it is mentioned that paddy was removed from the stalks by beating them on ground or by making the bullocks to tread on them. Cleaned paddy was collected, measured and stored in proper places. Sickles and swords were used for harvesting millet. For threshing, buffaloes were made to tread or men are used to tread ears with feet. Black gram was threshed with sticks. Women considerably contributed to threshing and cleaning. A common vessel for measuring grain was referred as "ambanam". Festivals were celebrated before the commencement of harvest

Table 2.3 Materials Recommended by Surapala to Control Tree Disorders and their Currently known Properties

Materials	*Properties*
	Plant Species
Acorus calamus L.	Antibacterial
Brassica alba (L.) Rabenh/ *Sinapis alba* L. (white mustard)	Insectantixenosis; antifungal; acaricidal; nematicidal; glucosinolate sinalbin "anti-insect" and "anti-nematode" allyl isothiosinate antifungal
Curcuma longa Koenig non L. *Curcuma domestica* Val. (turmeric)	Antioxidative curcuminoids; antimicrobial
Embelia ribes Burm. F	Anthelmintic; antibacterial; insecticidal
Emblica officinalis Gaertn. (triphala)	Anthelmintic with other two species of triphala.
Ficus benghalensis L. (banyan)	Latex with good sealing property; tannin
Ficus glomerata Roxb.	Latex; bark 14 per cent tannin; some *Ficus* spp. are antibacterial.
Piper nigrum L. (black pepper)	Oleoresin antibacterial/antifungal; alkaloid piperin is insecticidal.
Sesamum indicum L. (sesame)	Insecticidal and repellent; oil synergistic to pyrethrums; antioxidative lignins in seed; 17 per cent protein; 800 mg per 100g calcium, phosphorous and potassium; 14 per cent iron (ash) - highest.
Solanum indicum L.	Fruits/leaves antifungal/antibacterial; glyco-alkaloid solasonine present.
	Animal Products and other Materials
Ash	Particles hygroscopic; absorb moisture from insect eggs and spores; interfere with insect feeding; ash potassium interacts with surface fats.
Cow dung	With urine it is antiseptic; rich in bacteria which compete with pathogens; good medium for biocontrol agents; beneficial to Rhizobium and Azotobacter.
Fish meal	Rich in protein; releases aminoacids including proline.
Ghee	Same as animal fat
Honey	Antimicrobial; protects wounds in plants/animals; proline present; honeybee peptide apidaecin is antibacterial.
Liquid manure (kunapa)	Effects would include : healthy crop/tree; crop tolerance to abiotic stresses such as frost, heat, etc. as well as to insect pests and disease; high yields; high quality produce.

and during the time of harvest. For threshing, Parashara mentioned a levelled threshing "pit" and installation of threshing pillar called "medhi" are mentioned. The wood for the pillar was obtained from a tree that produces milky sap, preferably by silk cotton, *Ficus bengalensis*, *F. glomerata*.

Tree Worship and Trees and Environment

Tree motifs have been found in the art of Indus Valley, Mauryan ring stones and gateways and railings of stupas at Bharhut, Bodhgaya, Sanchi, Amaravati and

Nagajunakonda, Mauryan relief sculptures, particularly on the Ashokan Pillar Capitals and Rampura Bull Capital of Ashoka. Old sculptures and architecture of Mathura (Kanishka period, AD 78-101) and Ajanta frescoes (AD 100-600) also bear testimony to the importance of plants and flowers in ancient India. Trees and flowers have been also delineated in ancient coins found at the pre-Mauryan site, Sugh, Taxila, Ayodhya during Mitra Kings, Kausambi and Mathura and also of the Andhra dynasty and Pandyan territory. The ancient Sanskrit and other literature and texts, mythological epics and legends, paintings, cave murals and frescoes, sculptures, architecture, folklores and tribal arts and crafts provide evidences of the kind of plants and trees and flowers growing in the forests and gardens. The science of plant life, (*Vrikshayurveda*) and arbori-horticulture and usefulness of forests and gardens were well-known in ancient India. The utilitarian qualities of trees and plants for food, medicine, shelter, shadow and fuel and the relationship of trees with fertility were also known to ancient Indians. They were concerned with the conservation of trees and biodiversity in nature and ecological balance in environment. Their concept of identifying trees with gods and goddesses and threats and punishments against the destruction of useful trees helped to save the trees and flora, which is a remarkable contribution of our ancient people.

Lokopakarum

It is a text consisting of ten chapters. It deals with subjects such as *Vrikshyurveda* (treatment of plants), *Udarkargalam* (water divining and sources of underground water), *Vasthusilpa* (architecture) and *Naravaidyadi Prakaranam* (human medicine). To quote stanzas 6 and 7 of the Vrksayurveda chapter on fruit seed selection, treatment and sowing: "Select seeds from fallen, well developed, ripened fruit. Treat with cow dung and dry for five days. Soak seeds in milk for a week and mix them with the juice of *Solanum indicum* and salt water. Incense the seeds with *Embelia ribes* and ghee. Sow the seeds at the right time with joy and enthusiasm. Fresh water should be sprinkled for good growth. Shravana, Magha, Mula, Chitra, Vishaka, Mrugashira, Rohini and Hastha are good astral times for sowing".

In an experiment by Karnataka Forest department, 10 kilos of leaves of three different species, Ankole (*Alangium salvifolium*), Sabbasige (*Anethum sowa*) and Fenugreek (*Trigonella foenum-graecum*) were added to the soil around a single Phyllanthus tree. Then fresh fox meat was added and the soil was irrigated daily. *Vetivera* and *Withania somnifera* roots were used to cover the stem portion and water was sprinkled over this. One kilo of honey was diluted and applied once a month for three months. The treatment resulted in the tree growing very fast. It yielded 10 times more biomass than surrounding plants.

(*Courtesy/Source* : KPP, Malnad, Karnataka).

Kunapajala

([illegible])

Chapter 3
Indigenous Technical Knowledge in Farming

Introduction

The indigenous technological knowledge possessed by farmers has great potential in solving their problems. The traditional knowledge is the sum total of practices based on the peoples' experience in dealing with situation and problems in various walks of life. In agriculture, this pertains to crop production, plant protection, animal husbandry and fisheries development. Though the agricultural; operations in rural areas are performed by people having little exposures to the outside and who are devoid of advance education and training, these farmers do not give up their creative urge to experiment and innovate to develop local solutions without the encouragement or recognition they deserve.

As a result, a number of indigenous farm management practices have been evolved by the farmers. The indigenous technical knowledge, possessed by the farmers, has emerged out of field experiences, passed from one generation to another orally. The wisdom gained and the practices adopted by the ancient farmers was passed down through generations and got ingrained in the cultural outlook of the society. These indigenous systems though low-yielding are accepted to be more dynamic as well as stable in nature. These are eco-friendly and the practices were found to be more economical and effective over a long period of time.

Indigenous agricultural and environmental knowledge gained global recognition though the United Nations Conference on Environment and Development in 1992 and documents such as the World Conservation Strategy (International Union for the Conservation of Nature and Natural Resources (IUCN) 1980 and Brundtland commission's our common future (World Commission on Environment and Development (WCED) 1987. Indigenous knowledge is an immensely valuable resource that provides humankind with insights on how communities' have interacted with their changing environment.

As a result, a number of indigenous farm management practices have been evolved by the farmers. For instance, choice of indigenous varieties, seed treatments, storage and preservation, water harvesting, soil and water conservation, maintaining soil fertility, disease and pest management, farm implements and animal health.

What is Indigenous Knowledge?

Indigenous knowledge is the local knowledge that is unique to a community or society. It is the basic knowledge for local-level and on the spot decision making in agriculture and allied activities like animal husbandry, tree culture *etc* and local natural resources management. It is the information base for a society, which facilities communication and decision-making in a dynamic way which is influenced by local people's creativity as well as by external systems present in the said environment. Indigenous knowledge is the knowledge that people in a given community have developed over time and continue to develop. It is based on experience, often tested over centuries of use, adapted to local culture and environment. It is also referred to as traditional or local knowledge, refers to the large body of knowledge and skills that has been developed outside the formal educational system. It is embedded in culture and is unique to a given location or society. Indigenous knowledge is an important part of the lives of the rural poor.

It can be abstracted as follows:

- Orally transmitted knowledge and generally not documented
- Non-formal knowledge
- Locally bound, indigenous to a specific region
- Context and culture specific, Dynamic and adaptive
- Holistic in nature
- Intimately related to survival and subsistence for people

Characteristics of ITK

- ITK can show an understanding of the complex relationship between these individual components and the dynamic ecosystems within they act.
- ITK is frequently linked with the sustainable use of local resources. Since people depend on the resource, survival requires that it is maintained.
- ITK often focuses on particularly important local attributes, *e.g.* beluga whales, polar bears, fish resources.
- ITK is based on experience, it is local knowledge acquired from observations over time-it is argued that it may be most useful for local scale decision – making.
- ITK often describes these symbiotic relationships and provides that basis for life-sustaining decisions about how to relate to the environment.

- ☆ Sustainability, cultural survival and ITK are related through social organization, human interaction, institutional arrangements, values and moral codes of conduct.
- ☆ Dependence on the health of the local environment, wildlife, etc., promotes consideration of the relationships between human and biological systems.
- ☆ This is reflected on the belief of many indigenous people that they were put on the earth to take care of the land; if they destroy it then they themselves are destroyed.

Application

- ☆ Subsistence/traditional pursuits
- ☆ Basis for local decision – making
- ☆ Cultural preservation
- ☆ Resource management
- ☆ Land-use planning
- ☆ Development assessment
- ☆ Environment assessment
- ☆ Conservation
- ☆ Co-management
- ☆ Ecological consciousness

Traditional ecological knowledge is another term used to describe those aspects of an IKS that are directly related to management and conservation of the environment. Johnson (1992) defines TEK as a 'body of knowledge built by a groups of people through generations living in close contact with nature.

Why is Indigenous Knowledge Important?

In the emerging global knowledge economy a country's ability to build and mobilize knowledge capita, is equally essential for sustainable development as the availability of physical and financial capital (World Bank, 1997). The bases component of any country's knowledge system is its indigenous knowledge. It encompasses the skills, experiences and insights of people, applied to maintain or improve their livelihood.

Significant contribution to global knowledge have originated form indigenous people for instance in medicine and veterinary medicine with their intimate understanding of their environments. Indigenous knowledge is also the social capital of the poor, their asset to invest in the struggle for survival.

Accordingly, indigenous knowledge is of great relevance for the development process in the following sectors:

- ☆ Agriculture
- ☆ Animals husbandry and ethnic veterinary medicine

- Use and management of natural resources
- Primary health care (PHC), preventive medicine and psychological care
- Spacing and lending
- Community development
- Poverty alleviation

Sources of ITK

The following sources of ITK may be explored

- Farmers, community members, especially elders, are the best sources of ITK. But, since ITK is unevenly distributed in communities, it is important to find out who know what in order to tap the right sources. For example, asking men about garden plants when are in charge of homes gardens, might lead you to conclude that villagers know little about gardening.
- Folklore, songs, poetry and theatre can reveal a great deal about people's values, history and practices. These are often not written down and need to be recorded.
- Community recorded – although ITK is mostly transmitted by world of mouth, some indigenous forms or recorded keeping exist. These include writings, paintings and carving and many other forms.
- People working with communities, such as extensionists, can be valuable sources of ITK. Other resource persons are local school headmaster, credit cooperative society officials, village milk co-operative members, men and women labourers and village Panchayat Sarpanch.
- Secondary sources include published and unpublished documents, databases, videos, photos, museums and exhibit.

IK, IKS and ITK

Indigenous knowledge is not confined to tribal groups or the original inhabitants of an area as is commonly thought of. It is not even confined to only rural people. Rather, any community possesses indigenous knowledge-rural and urban, settled and nomadic, original inhabitants and migrants. Other names of an indigenous knowledge or "local knowledge", "indigenous technical knowledge" and traditional knowledge".

Indigenous knowledge is the participants' knowledge of the temporal and social space and refers not only to the knowledge of indigenous peoples, but to that of any other defined community.

The concept of indigenous knowledge systems delineates a cognitive structure in which theories and perceptions of nature and culture are conceptualized. Thus, it includes definitions, classifications and concepts of the physical, natural, social and economical environments.

Indigenous technology knowledge is of a practical nature, concerned with operational zed local thinking in such fields as agriculture, fisheries, health, horticulture and forestry.

Underlying Principles behind Indigenous Technologies

Field studies elsewhere have clearly brought out that indigenous technologies are based upon different principles and perceptions in order to meet additional requirement, which are specific to small holding situation. Some of the major points on these aspects are:

1. Enhancement of Productivity besides Conservation of Land Resources

This is the crucial distinguishing feature of the indigenous technologies for management of land resources. The existing available exogenous technologies focus more on in-situ conservation of land resource with a view to achieve long-term sustainability in production while. The indigenous technologies focus on short term as well as long term increase in productivity.

2. Smaller Size of Groups for Conservation of Land Resources

Group activity is a pre-requisite for implementation of soil conservation measures. The facilitation process of meetings this requirement becomes more difficulty as a size of social action increase. The exogenous technology of contour/grade bunding requires a large size of group action among participating farmers before its implementation could start.

3. Creation of Microenvironments

Indigenous technologies work with the concept of concentration of an resource (at appropriate place) rather than in-situ conservation of this resource an original place. These technologies are evolved to make the best use to micro level variability. The microenvironment thus created through indigenous treatments may have small but significant contribution towards sustainability of overall production over the vagaries of seasonal rainfall.

Local Knowledge and Beliefs

A number of traditional knowledge in agriculture is always meant for productivity, sustainability, stability and equitability. The examples are showing in Table 3.1

Some of the Indigenous Technical Knowledge Possessed and Used by the Farmers in India

Genetic Resource Management

The management is based non conventional technology and included day to day techniques *viz.*, traditional farmers usually grow several varieties of a crop to compensate loss due to any single factor and to conserve or maintain gene pool resistant to several biotic and abiotic factors; farmers identify rows where plant growth is better than the surrounding area. These plants are harvested separately and the produce is used as seed; and people of Uttarakhand grow about 12 kinds of crops and their varieties differing in growing behavior, maturity, rooting behavior drought tolerance and other characteristics.

Table 3.1: Traditional Knowledge in Agriculture

What	*Why*	*Where*
A. General agriculture		
Mixing sand with seeds of sesame and tobacco while sowing	Ensures uniform stand	Andhra Pradesh and Karnataka
Covering of curds of cauliflower with the top leaves	Prevents direct hating insect attack and produces curds of high quality	Rajasthan
B. Sowing/Transplanting		
Sowing of paddy must be completed in Mrigasira Karthi (1st week of June)	A good harvest is ensured	Andhra Pradesh
Sowing of mustard in the end of autumn	Avoidance of aphid attack	West Bengal
Paddy transplanting during mid June and mid July	For photo and thermo sensitivity it increases straw yield and grain yield	West Bengal
C. Plant protection		
Growing of castor (in stray lines) in the field of chillies	Control leaf curl virus in chillies	Andhra Pradesh and Karnataka
Broadcasting of ash in kitchen garden crops particularly in cucurbits and brinjal	Reduces the pest population by causing damage/ irritation to be cuticle of insects	Rajasthan, Tamil Nadu and West Bengal
Mixing of neem leaves with stored products	An insurance against the attack of the pest	All over India
D. Rainfall		
Fluttering of feathers by cockeral	Indication of rain	South Indian states
Thunder occurring in eastern skies	Bound to give rain	South India states
Rainbow in the east sky	Heavy rain	West Bengal

Crop Production

- ✰ Besides sprinkling mixture of powdered dung of goats and sheep, wheat particles, beef and water kept for seven nights for increasing flowers and fruits of trees, creepers and shrubs.
- ✰ Minimizing premature flower and fruit drops of trees by irrigating with milk that has been cooled after being boiled with horse gram, black gram, green gram, sesamum and barley.
- ✰ Treating cotton seed with red lac juice in a special manner to get red tinged cotton;
- ✰ And planting of deep rooted plants which serve as natural aerators and are able to extract water from deeper layers of soil along with shallow rooted plants.

Water Management

Historical records show the existence of a number of irrigation works in different parts of the country. In the Himalayas, the perennial river Ganges made it relatively easy to diverts its flow through inundation channels. In the south, where rainfall is scanty, the practice of trapping rainwater in large tanks and ponds for agricultural purposes is widely adopted. From time immemorial, surface irrigation methods have been followed. The most effective irrigation method for a particular area depends on the slope of the land, the nature of the soil, the soil type of the crop and availability of funds.

Indigenous technical knowledge of farmers of Southern Meghalaya use intricate networks of bamboos pipelines to deliver water to betel leaf plantations in rocky areas where channels cannot be built. Bamboo culm is used for controlling gully erosion, the fibrous root system of the bamboo acts as an excellent soil binder.

A big stream in Tamil Nadu is often diverted to feed a chain of tanks in sequence; ponds are constructed by Knkuma tribe in such a manner that surplus water from one pond flows down to the pond below: tribal people in the hills place large boulders where water is falling from considerable height. Stones are also bound with ropes or wire mesh for the same purpose.

Rooftop harvesting of rainwater is practiced in Mizoram. Water is harvested has it falls from the rooftop.

The Madhya Pradesh farmers store rainwater in the agricultural fields itself. The collected water seeps into the sol and gives enough moisture for growing crops in succeeding dry period.

Disease Control

The ancient Indian literature documents use of plants products, animal's products and wastes for the curing the diseases of human beings and plants. Plant diseases are a major source of field reduction. There are about 1000 diseases, which affect economic crop plants causing significant damages. Majority of plant diseases are caused by fungi followed by viruses, bacteria, nematodes and a number of other plant pathogens. Imbalance of nutrients and extreme variation in environmental factors also leads to serious plant diseases. Integrated management of these diseases needs to synergy with the natural environment. Traditional agricultural practices were tailor made to maintain this synergy and from the basis for plant disease management.

During Ancient Times Plant Diseases were Controlled by Conventional Technology. This Indigenous Technology Includes

- ☆ Decoction of *Asafoetida indica* for seed treatment of sorghum seeds against smut disease;
- ☆ Soaking of sorghum seeds in cow urine for 2-3 m, drying them immediately under sun before sowing for protection against grain smut.
- ☆ Seed treatment with cow dung paste to control seed borne diseases;
- ☆ Soaking of wheat seeds in milk before sowing to avoid rust disease; and soaking the seeds in milk, smearing with ghee and rolling them many times in cow dung for protection against diseases cereals.
- ☆ Plants attacked by pests are to be treated with mixture of fresh cow urine, clarified butter, *Embelia ribes*, mustard and sesamum, applied to trunk and then are to water with milk and water.
- ☆ Trees which attacked by any type of pests are to be treated with the paste from bark of *Pongamia glabra, Cassia fistula, Embelia ribes* and *Cyperus esculentus* in cow's urine.
- ☆ Exudation of trees can be stopped with the application of bark paste of *Setaria italica* and *Terminalia arjuna* in boiled milk.

Some Classical Examples of Local Practices Adopted for Plant Disease Management

- ☆ Shade regulation for coffee leaf rust and blister blight of tea
- ☆ Growing wind breaks like silver oak, Casuarina, jack, etc., to avoid sun scorching of young shoots of plantation crops.
- ☆ Tieing of arecanut seedlings with coconut and arecanut fronds to protect them from western sun scorching.
- ☆ Pasting of lime on areca trees to avoid ill effect due to sun scorching.
- ☆ Watering nursery beds during early morning for higher seedlings vigour and stand particularly followed in chilli and brinjal.
- ☆ Burning nursery with leaf litter and farm waste to overcome certain soil borne pathogens.
- ☆ Raised beds, fields and ridges used to mange some soil borne pathogens.
- ☆ Flooding with water to overcome the problem of soil borne pathogens by creating anaerobic conditions.
- ☆ Mixed cropping of jowar with Tur to prevent movement of mites which transmit sterility mosaic of pigeon pea and to minimized wilt.

☆ Salt-water treatment for wheat to over come the problem of seed borne diseases. *e.g.* Bunt and seed gall in wheat.

Methods of Irrigation

Indigenous Drip Irrigation

The practice of using pitcher water as a source of irrigation on new fruit plantation in sandy loam/loamy sand soils, in areas of scantly rainfall is prevalent in temperate districts of Himachal Pradesh. The pitcher is placed in soil and the new plant is planted close to it. The pitcher is filled with water during summer months (April – June) and stone/slate lid is placed on the top. The roots draw moisture/water from pitcher, which in turn reduces the mortality. The pitcher once filled, supply sufficient moisture for atleast two weeks and then again it is filled with water.

Bamboo Drip Irrigation System

In this system of irrigation bamboo channels (open) are used for irrigating the fields. This system is common in North – East regions of India. Small holes are made at the internodes of open bamboo channels, from where water gets trickled down in the field. These channels are placed along the natural gradients. In these channels, no uniform head for water trickling is maintained.

Harvesting of Rainwater

In the hills, rains are erratic and torrential. Relatively high percentage of rainwater goes as run-off and stream flow. It carries fertile soil and plant nutrients, which makes the soil degraded and barren. In some areas this excess water is stored directly in the farm ponds, depression or stream flow or is diverted to safer points where it is stored.

The stored water in ponds and depressions is used for irrigational purposes, as a life saver or for supplementary irrigation during lean periods. It is also stored in dugout structures. In some areas during summer, it is used as drinking water human, livestock and for other domestic purposes.

Soil Management

In West Himalayan region, in the month of March/April, when snow melts and weather condition improves, the bunds and comers of the fields are dug-out and weeds and grasses are removed with the help of spade and clods. The grasses or weeds are beaten up and then soil is separated from these clods and collected in lower fields.

This practice of removing weeds and grasses from bunds and corners by digging helps in weed control in the cultivated fields. Secondly area under crops remains the same as that of previous crop *i.e.*, area is not wasted for weeds and grasses. Thirdly the soil added in lower fields from the bunds of upper field is rich in nutrients and it improves the soil fertility.

Traditional Rainfed Farming

In the hilly areas, most of the area is rainfed except for a few pockets in valleys where irrigation facilities are existing. The choice of crop and rotation, completely depends on crops which require less volume of water. For rotation, legumes, are important as mixed crop. During rotation, when rainy season erosion permitting crops are grown, such as cowpea, which form integral part of the mixed cropping system. The crops are chosen as per their nutrient *e.g.* from old ages protein rich pulses are part of cropping pattern. The coarse grains like 'phaphra', 'chulai' are also grown very commonly which are very rich in nutrition.

Within the premises of the house it is mandatory to have fruit plants such as citrus, mango, anar which provide seasonal fruits rich in vitamin C, carbohydrates etc. The fields are well protected with biofence of thorny shrubs or their cut pieces. The traditional rainfed farming is done irrespective of land with respect to slope and other characteristics. There are chances of sheet erosion but with traditional knowledge, crop rotation is adopted in such a way that during peak runoff periods sowing of close growing crops provide protection to the soil.

Terracing

From old times, land in the hills has been put under cultivation on scientific lines as cultivation is done up to 25-100 degree slope, where there are many chances of landslips, sheet and gully erosion. But with 'bench terracing practices' the menace of soil erosion is controlled and is very common in hill fanning.

The terraces are constructed across the slope *i.e.* along the contour. The size of the terrace is decided by the prevailing degree of slope. The terraces are supported by risers of suitable heights and width. The height of riser is again decided by the degree of slope. The risers are sometime made of loose boulders supported by grasses. The roots of grasses help in binding and keeping the boulders intact at a place. The roots of grasses help in drainage and keeping the boulders intact at a place. The roots of grasses help in drainage of excess water. With the traditional knowledge, farmers are keeping the risers toward inner slopes. In paddy growing areas the risers are erected to facilitate the pounding of water in the field. This type of bunding and terracing is continuing from centuries and terraces are still intact. The bunds are again used for growing palatable grasses which is used as fodder for livestock and trees are meant for fuel, fodder and fibre. The example are beul, shisham, mango etc. In lower areas the bench terraces are known as "khet". Sometime on the risers contour hedge of grass like khus, local grasses also established.

Use of Ash

i) Nutrient Recycling

The farm use cattle dung, shrubs and bushes as the main source of fuel. Ashes available, there upon, are mixed either with household waste or human excreta. Sometimes ashes are also broadcasted in the fields.

Mixing of ash with household waste and human excreta aids in nutrient availability and recycling. Ash primarily meets the deficiency of potash. Availability of phosphorus is also ensured.

ii) Softening of Hard Soils

Hard soils are softened by putting ash obtained from cowdung, sheep/goat manure, fuelwood etc. Through this practice upper layers of soils are not only softened but their fertility status is also improved, as ash contains phosphorus.

iii) Poultry Manure and Ash for Increased Vegetable Production

This specific technology is used only in case of tomato, brinjal, capsicum and cauliflower, Kitchen ash and poultry manure mix enhances vegetable production levels.

Sprawling of Ash Dust in Vegetables Crops

In some parts of the country, ash dust is a product obtained after the combustion of fuelwood. It has been observed that dusting of material in the fields enhance early maturity and high yield of vegetable crops.

The reason for the early maturity of cucurbits and vegetable crops is due to the fact that ash dust contains sufficient quantity of phosphorus in available form to the plants. Secondly, in cucurbits the ash dust has been used to repel the insect pest of the crops. Thirdly, amendments of ash dust in the soil, improves soil structure and fertility. Ash dust is also useful in enhancing the maturity of bulb crops which normally takes 6-7 months for obtaining economic yield.

Green Manuring

The practice of green manuring is localized in many villages (paddy growing) in India. Leaves and twigs of trees and bushes are used.

Use of Goat Manure

Goat manure improves not only the millet production but also its taste. According to farmers vegetables grown in goat manure have longer keeping quality. It is easy to plough fields manured with goat excreta. Actually with the addition of goat excreta, there is improvement in the physical properties like soil structure, water holding capacity and porosity. There is also an improvement in soil fertility as it contains 3 per cent nitrogen, 1 per cent phosphorus and 2 per cent potassium.

Biofencing

This practice is prevalent in Tamil Nadu and other states. There is a common practice to provide biofencing with *Commiphora* sp., *Jatropha* spp., etc.

The biofence of with thorny species protects crop from stray animals. Its multipurpose utility as a nitrogen fixer, checks against soil erosion, conservation of soil and moisture, source of fuelwood and indigenous drug (rich source of vitamin C) makes it a promising plants for eco-economic rehabilitation of the region.

Weather Forecasting

Weather is certainly the most important factor determining the success or the failure of agricultural enterprises. It manifests itself through its effects on soil, plant growth as well as on every phase of animal growth and development. A greater proportion of the total annual crop loss results from aberrant weather. Also crop and animal disease are greatly influenced by weather. In all, weather accounts for approximately three fourth of the annual loss in farm production both directly and indirectly.

However, the crop losses can be reduced substantially by affecting adjustments through timely and accurate weather forecasts. Such weather forecast support and also provides guidelines for long range or seasonal planning and selection of crops best suited to the anticipated climatic conditions.

Weather forecasts for agriculture can be grouped into short range forecast (upto 48 hours), long range forecast (310 days) and medium range forecast (one week to entire season). Each plays an important role in farm operations and planning of agriculture activities. Most important effects since time immemorial have been on rain making and weather for casting.

From the food gathering stage to the last quarter of the twentieth century, man's need for water has been increasing. However, the unpredictable nature of rainfall has not changed. Adequate availability of water through rainfall has, therefore, always been the major worry of man, leaving him susceptible to any activity that could bring rain. Just as present day heads of government are concerned and incorporate plans for the adequate supply of water for their people, similarly primitive tribal chiefs were also responsible for bringing rain to their people. Like the politicians to today, the tribal chiefs, delegated the responsibility of rain making to someone else, obviously as a hedge against failure. If there was no rain, it was the luckless assistant who was punished rather than the chief.

The early rainmakers, fore-runners of the present day meteorologists, were perhaps among the most intelligent of the primitive men who had to guess the scientific reasons for the phenomenon of rain and the fact that these gentlemen often succeeded in duping their chiefs to preserve their own lives is in it self a tribute to their intelligence. Again, it was these very people who set the stage 'for scientific enquiry into the causes of weather and rainfall.

It was believed that prayer itself could bring rain. But as an insurance, non living gifts were presented to the rain God during the prayers. Men prayed for rain to the dead and then to the Gods. Praying for rain is firmly routed in Indian culture even today.

In the Himalayas, people sing the traditional song *Singaitoo* up to mid night during the prolonged hot dry summer. They then organize a feast (*Havan*) from the collections made, in honour of God. People claim that in this way the rain Gods are appeased and within 2-3 days of the feast rainfall is assured. A similar type of feast (varisty Puja) is organized by villagers collectively during summers in the even of a long dry spell in Punjab. All cry for rain and the children sing a song. The essence of the song is:

"When God listen to children's voices there is rainfall" in the next stage of its evolution rain making was centered around the bestowing of gifts trough sacrificial offerings.

Weather Prediction

Out of various the factors which control agricultural production, weather is the only factor over which man has no control and hence it has an overwhelming dominance over the success or failure of agricultural enterprise. It is an accepted fact that food production is inextricably linked with climate and weather. It is also reported that weather induced variability of food production is more than 10 per cent. This variability can be as high as 50 per cent of the normal production in respect of smaller areas situated in arid and semi-arid regions. In order to reduce risks of loss in food production due to the vagaries of weather, weather *per se*, should be taken into account as one of the major inputs in agricultural planning. That is why forecast of weather parameters play a vital role in agricultural production. It also aids in minimize crop losses to a considerable extent. Thus development and refinement of the art of weather prediction has been essential since time immemorial.

In present times we have many improved technologies for making weather forecasts as well as for their dissemination. Previously when there was no such technology available farmers based their prediction on many natural, cultural and social phenomena. Some of these are given below:

Cloud and Wind Direction

If there is an accumulation of clouds in the South-East direction in a layered form accompanied by winds blowing from the southern direction then it is claimed that there will be rainfall within a day or two.

Farmers also predict weather by observing closely the different activities of various birds, animals etc. The following are some indigenous beliefs;

During the rainy season farmers observe the House swift and they predict heavy rainfall if the bird flies high in the sky if the Maina bird bathes in the water it indicates that there will be rainfall within one or two days.

A group of sparrows frolicking in the sand indicates that there will be rainfall that day or the next day and if they are observed to be playing in water then it is believed that the weather will be dry for some days to come.

If the Lapwing lays her eggs on the higher portion of the field then heavy rainfall is predicted during the coming rainy season but if the eggs are laid in the lower portion of the field then a drought is predicted. These birds never construct a nest but lay their eggs on bare soil. Further it is also believed that if a single egg is laid, then there will be rainfall only for one month out of four months of the rainy season. If two eggs are laid then rainfall will occur for two months and similarly four eggs indicate there will be rainfall during all the four months of the rainy season.

If there is a swelling on the lower portion of the camel's legs then rainfall is predicted by the farmers. The swellings are probably caused due to higher relative humidity.

If the Dragon fly, which appears generally in the rainy season, are observed to swarm in a large group over a water surface predicted by the farmers.

If the colour of the clouds is similar to the colour of the wings of the Partridge *i.e.* grey or black-grey and strong eastern winds are also blowing then assured rainfall is predicted by the farmers. The clouds of a colour similar to that of the said bird are rain bearing clouds *i.e.* of cumulonimbus type.

When spider nets are plentiful on grasses, sticks of tomato crop and on trench bean crop then it is estimated that the rainy season is over.

Bio-indicators of Rain

- ✰ Ants shifting their eggs without any apparent cause and coming out of their holes digging the ground, funning aimlessly.
- ✰ If neem tree produces plenty of fruits and the bava tree produces plenty of pods, the total rainfall on the monsoon will be high.
- ✰ Rain is indicated by musical sound of cuckoo and jumping of fishes on the surfaces of ponds and rivers.

Frequent coughing of cow, crying of jackle and jumping of fishers in the ponds and frequent dips of crow in water are indicators of advent of rain (Parasar, 1994).

Social and Cultural Beliefs

Many cultural, social and religious beliefs and activities superstitious pertaining to the prediction of future weather prevail since generations. From time immemorial farmers have predicted the weather on the basis of these beliefs/activities. The following are some examples from the western Himalayana region.

If the first 10-15 of the month "Jeth" (May-June) are very hot then good rainfall/ monsoon is predicted during the ensuing rainy season. This results probably from the low pressure zone in north-west India that is generated due to the high temperatures.

If the "Khejri" tree bears good fruit in a particular year then farmers predict good rainfall during the next rainy season and vice versa less rain is predicted in the even of a poor fruit crop.

Preservation and Storage

In India, large quantity of food grain is stored in villages in different traditional storage structures and containers, it is estimated that sixty to seventy per cent of food grains produced in the country is stored at home level in indigenous storage structures, the storage methods range from mud structures to modern bins, the containers are made from variety of locality available materials used in design, shape, size and functions, the materials used include paddy straw, wheat straw, wood, bamboo, reeds, mud, bricks and cow dung etc.

Central India especially Madhya Pradesh, has a large ratio of tribal population. In general, it is believed that the tribal farmers are unskilled and have primordial knowledge of agriculture. This 'myth' has also perhaps hastened disappearance of indigenous wisdom. The tribal farmers, such as those of central India, hence been the custodians of practical knowledge based on their experiences and continued experimentation.

Rise of Storage Structures

Growing crops is a daily routine; but, when the crop matures, farmers become alert for safe harvesting of the crop and its storage. Inaccessibility to developed market centers and the imperfect pricing of commodities in local rural markets have created the necessity to restore the agricultural produce. This necessity of storing the produce gave rise to innovative indigenous methods of post harvest storage.

Grains are mechanically taken out from dry plants of many crops like wheat, paddy and coarser millets. Such grains are stored in round, rectangular or square shaped containers made up of earth or in bamboo baskets woven in the shape of earthen containers. These containers are lined with neem leaves or with the leaves of any wild plants having preserving properties. The bamboo baskets are lined with or layered outside with the cow dung, which protect from all insects, fungi and pests to protect the stored plant material from perpetual rains, the hanging kernels are wrapped by teak leaves or gunny bags, or otherwise removed from the poles and stored in a shed. Moreover, the fungi, moss and microorganisms also do not invade the grains inside. Villagers claim that there method of storing the maize and pearl millet far superior than other chemical intensive methods.

Indigenous Postharvest Management Technologies

The hilly communities of Kalvarayan hills in Tamil Nadu, India consists of tribal population, who depend largely on agriculture and livestock. In this area, the tribal women adopt many indigenous cost-effective post-harvest techniques. Some of them are mentioned below.

- ☆ The harvested millet seeds are mixed with wet ash. After mixing thoroughly, the seeds are dried. This is done to control the storage pest.
- ☆ Dried chillies are crushed and mixed with cut pieces of lemon and ash. This mixtures is mixed with millets seeds during storage. The pungent smell of chillies and the acidity of lemon prevent attack by storage pests.
- ☆ Cereal grains are stored in tall mud pots or bins, which are popularly called as "Kudhiris". These mud pots are made up of clay soils and plant fibre. The height ranges from 10 to 3 metres. And opening at the base, helps to draw grains from kudhir, whenever needed. There are normally 1 to 3 kudhris in each house depending on the requirements.
- ☆ Neem leaves and neem tips are kept along with cereal grains like paddy, ragi, etc. the pungent smell of neem leaves prevents attack by storage pests.
- ☆ Nochi (*Vitex negundo*) leaves are kept along with the cumbu seeds during storage, to prevent the occurrence of pests. The pungent smell of nochi leaves resists the pest attack.
- ☆ Paddy seeds are mixed with the *Ipomea* leaves, which acts as pest repellent.
- ☆ Pulse seeds are stored in salted gunny bag's to resist the storage pest.
- ☆ Red soil in mixed with pulse seed and kept in a separate place for storage, it helps to prevent attack by pests.

- For quick ripening of jackfruit, a small piece of neem stick is inserted in the top portion of the fruit.
- After harvesting, the tomato fruits are given a coating with charcoal, for longer period.

Conclusion

By and large the traditional technologies for development and conservation of land resource are found to be highly relevant particularly for small holding farmers. This is evident due to the fact that these measures are being implemented and maintained by farmers at their own cost. The percentage of their adoption is however limited; and the reason for non-adoption vary from farmers to farmer.

Lack of technological knowledge is not the main reason for non-adoption of soil conservation measures. Other reasons like limitation of finance, shortage of labour availability, lack of proper demarcation of filed boundary, difficulty in facilitation group action among neighboring farmers, low motivation towards long term measures, low priority to rainfed agriculture, etc. are major reasons which are adversely affected the adoption of these technologies. Likewise for harvesting runoff water the indigenous technologies are found to be sound but they are gradually becoming non-functional due to land of proper mechanism for their maintenance. The original institutional base in the rural areas is becoming weaker. Although currently large investments are being made on construction of new water harvesting structures at the micro level, this investment is rarely made for construction of indigenous type of structure or for facilitation for maintenance of such structure. Construction as well as maintenance of water harvesting structure requires regular group action among its users. In the watershed area under study, farmers have shown keen interest towards maintenance of old structure (Closing the breached tank, desilting of tank bed etc.) and also converting the irrigation tank into percolation tank. However, lack of finance and inability of facilitate group action are found to be the main constraints in adoption of these technologies/measures.

Indigenous Technical knowledge provides food tree from pesticidal residue and knowledge of Genetic Resource Management, Crop Production, Pest Management, Disease Control, Preservation and Storage, Water Management, Farm Implements and Animal Health". It also helps in explains the Vrikshayurvedic farming studies.

Chapter 4

Vrikshayurveda Farming: A Revisit and Review*

Indian System of Agriculture *viz., Vrikshayurveda* is an ancient Indian farming and traditional ways for food production in concomitant with Mother Nature but now-a-days it is a remedial measure to manoeuvre the ill effects of modern chemical agriculture. Based on the authors experiences in *vrikshayurveda* with crops like green gram, onion, sorghum, maize, black gram, cow pea and bhendi, the available literature pertaining to this traditional knowledge is reviewed, discussed and presented.

Review

Our Indian knowledge system is a treasure trove of information and traditional agriculture is usually associated with pre-industrial peasant agriculture. Indigenous knowledge is the "largest single knowledge source not yet mobilized in development enterprises". India has a treasure of indigenous knowledge concerning plant health, which was developed and documented several centuries ago (Sharmarao Jahagirdar *et al.*, 2003). This traditional knowledge, has been subjected to a process of refinement through generations of experience and now is receiving recognition. The scientific basis for such indigenous technology is evaluated and perfected.

Scope of *Vrikshayurveda*

Trees and plants have an essential part on human life, since the start of civilization. During Vedic period, greater importance was given to trees as a source of almost all the essentials in life, particularly agriculture. India had a rich agricultural heritage, since the time of Rig Veda (3700 BC), it is time now to reorient Indian System of Agriculture by looking back into traditional ways of food production. *Vrikshayurveda* (Indian plant science) provides ample scope to produce quality food products by

* *Courtesy*: Article published in Madras Agric. J., 98 (4-6): 99-108, June 2011

adopting traditional methods by utilizing plants including tree extracts and by products. Enormous and elaborate literature is found in *Vrikshayurveda* on procedures and practices for raising a crop, pest and disease control, seed treatment for increasing crop growth and its yield (Vijayalakshmi and Shyamsundar, 1994). In modern day scientific agriculture, the importance of understanding *Vrikshayurveda* needs to be brought out because food consumers have got awareness on environmentally disastrous and harmful effects of various agrochemicals used for pest control, disease management, nutrition of crops, growth regulation and promotion. A detailed understanding of *Vrikshayurveda* along with knowledge on trees and plant species widely utilized and methods adopted for soil health building, crop growth enhancement, pest and disease control and management, enhancing shelf life of food grains, etc. would be highly beneficial to the agricultural scientists of different disciplines.

Meaning and History of *Vrikshayurveda*

Vrikshayurveda is an ancient system exploited in Indian System of Agriculture. The three major ancient texts *inter alia* are : i) Varahamihira, 505 AD, *Birhat samhita Vrikshayurveda* part I (chapter (55) edited by M. Ramakrishna Bhatt, Bangalore, 1950) ii) Chaundarya, 1025 AD, Lokopakaran *Vrkshay urveda* a (chapter (6) edited by H. Shesa Iyengar, Government Central Manuscript Library, Chennai, 1950) and iii) Sarangadhara 1363 AD, *Vrikshayurveda* (edited by S.K. Ramachandra rao, Kalpatharu Research Academy, P.O. Box 1857, Bangalore (1993) India. The most detailed of all three seems to be that of Chaundarya 1025, AD (Sharmarao Jahadirager *et al.*, 2003). '*Vrikshayurveda*' of Chaundaraya (Lokopakarum) (1025 AD), Kanada, which deals with agriculture and botany, describes the use of milk that changed the flower colour and enhances the fruit taste (Upendra Shenoy *et al.*, 2000). It deals with various species of trees and their healthy growth and productivity of the crops, suraphala *Vrikshayurveda* text mentioned glorification of trees, composition of planting and preparation of agriculture, horticultural crops and all the cultivation aspects of plant and animal production (Vijayalakshmi, 1993a and Nen, 2004). In general, the word *Vrikshayurveda* (literally meaning the "*ayurveda for plants*") is used to denote knowledge of plant life (Vijayalakshmi, 1993a). The 100 texts on *Vrikshayurveda* are systematizations of the practices that the farmers followed at field level, placed in a theoretical frame work and it defined certain plant growth stimulants. *Vrikshayurveda* aims at producing quality food products by adopting traditional methods, by utilizing plant and/or tree extracts and by their by-products. Plant products are naturally evolved ingredients on the biosphere, they not only have an edge over the synthetic alien molecules but also merit reference and acceptance from the viewpoint of environmental safety and eco-friendliness (Swaminathan *et al.*, 2007b).

Vrikshayurveda on Growth of Crops

Vijayalakshmi (1993b) reported that application of *Ferula asafoetida* leaves in ash gourd improved vegetative growth, flowering and yield while leguminous tree leaf extracts at 3 per cent increased the vegetative growth and flowering in balsam plant (Vaidya *et al.*,1994). Devarani and Rangasamy (1998) found that 2 per cent *Morinda tinctoria* leaf extract treated with sorghum seeds enhanced germination, plant height and root length. In ragi, Palanisamy and Punithavathi (1998) reported that application of *Prosopsis* and pungam leaf extract 2 per cent as seed treatment increased root length, dry-matter production (DMP), shoot length and also enhanced seed germination by 90 per cent. Some tree species (*Alangium salvifolium, Aegle marmelos, Annona squamosa* and *Azadirachta indica*) were found to influence growth and development of crops (Swaminathan, 2000). Tripathi *et al.* (2000) reported that the *Dalbergia sissoo* leaf extract at 10 per cent enhanced the seed germination compared to water spray alone. Prasad (2002) reported that neem leaf extract treated turnip seed enhanced the seed weight and germination up to 40 per cent. Manikandan (2002) reported that application of FYM @ 1 kg mixed with 100 g of asafoetida powder around the plants at a depth of half a foot in bottle gourd promoted flowering and also fruiting. In green gram basal application of *Gliricidia* leaves at 10 t ha^{-1} and two sprays of leaf extract of *Aegle marmellos* at 5 per cent during 30 and 45 days after sowing recorded a grain yield of 2.14 t ha^{-1} (Swaminathan *et al.*, 2007b). Zodape (2010) found that foliar application of *Kappaphycus alvarezii* extract increased the yield and nutrition of green gram due to the presence of micro elements and growth regulators. Kavitha *et al.* (2005) reported that application of panchagavya along with moringa leaf extract increased the number of leaves per plant (116.4) and shoot length in amaranthus. The foliar spraying of moringa leaf extract five per cent produced second best result, next to panchagavya, in terms of dry matter production and single plant yield in *Solanum nigrum* (Sritharan *et al.*, 2010). Nandhakumar (2010) observed higher germination in maize at 5 DAS with *Albizia lebbek* as green leaf manure and seed soaking with *vitex negundo* leaf extract. This might be due to the presence of secondary metabolites in the leaf extracts, which would have promoted the germination in maize

Vrikshayurveda on Crop Yields

Rice

Incorporation of *Gliricidia, Leucaena* and *Sesbania bispinosa* as green leaf manure significantly increased rice yield up to 70 per cent (Brewbaker and Glover, 1988). However, Karim and Savill (1991) reported that alley cropping and incorporation of *Gliricidia* leaves at 10 t ha^{-1} increased the yield of rice to 30 per cent. According to Prakash and Bhusan (2003) 100 per cent substitution of nitrogen through the green leaves of *Leucaena* improved yield (2055 kg ha^{-1}) compared to 25 per cent nitrogen substitution (1987 kg ha^{-1}) in rice. In rice field, application of *Gliricidia* leaves recorded 0.46 t ha^{-1} higher straw yield over control (Pal *et al.*, 2005).

Vegetables

Antonysamy (2004) found that aggregatum onion treated with three per cent of *panchagavya* along with Notchi (*Vitex negundo*), Calotropis (*Calotropis gigantea*), Nerium (*Nerium oleander*), Aloe (*Aloe vera*) and Pungam (*Pongamia pinnata*) in cow's urine for seven days increased the yield of bulbs by 17.5 t ha^{-1}. While in chillies the studies of Meera *et al.* (2004) revealed that the extract of *Allium sativum* sprayed at 20 per cent concentration recorded maximum seed germination (80 per cent) and higher yield and fruit length and fruit weight. Christopher Lourduraj *et al.* (2005) opined that application of neem cake + *panchagavya* in bhendi, increased the number of fruits per plant (10.66), fruit girth (6.20 cm), fruit length (16.06 cm), fruit weight (16.06 g) and fruit yield (9.25 t ha^{-1}). Chelladurai (2006) reported that the yield parameters were greater in the treatment with *Gliricidia sepium* as green leaf manure, bulb soaking and foliar sprays of *Aegle marmelos* on onion (Table 4.1). Swaminathan *et al.* (2008) revealed that application of leaves of *Gliricidia sepium* @ 10 t ha^{-1} at 45 days before sowing of onion and followed by two sprayings of leaf extracts of *Aegle marmelos* @ 5 per cent during 30 and 45 days after sowing recorded higher yield.

Pulses and Oilseeds

In castor, alley cropping of *Dalbergia sissoo* and *Leucaena* and incorporation of leaves significantly increased yield by 79 per cent (Vani and Bheemaiah, 101 2003). In cluster bean, incorporation of *Gliricidia* leaves alone doubled the yield compared to control (Patel *et al.*, 2003). Singh and Attrey (2002) reported that in case of crop grown from the roots and seedlings the best yield was observed in beet leaf, when cake manure (Neem + Madhuca + Mustard cake in equal quantity) was applied @ 30 t ha^{-1}. Swaminathan and Gururajan (2005) found that application of *Gliricidia* leaves @ 10 t ha^{-1} in combination with five per cent of *Aegle marmelos* spraying after 30 and 45 days in cowpea and green gram increased yield by 2.14 t ha^{-1} (Table 4.2). It was followed by *Delonix regia leaves* 10 t ha^{-1} with combination of *Morinda tinctoria* 5 per cent spraying, which recorded yield of 2.12 t ha^{-1} compared to control (1.19 t ha^{-1}).

Maize

Nandhakumar (2010) reported that the yield of maize was higher in combination *Albizia lebbek* as green leaf manure and foliar spray of *Annona squamosa* leaf extract at 5 per cent concentration. The combination had higher grain number and grain row, cob weight and also the test grain weight (Table 4.3).

Vrikshayurveda for Pest Management

Hazara *et al.* (1999) reported that neem leaf extract and Neem Seed Kernel Extract (NSKE) at 5 per cent effectively controlled onion thrips. Pawar *et al.* (2003) found that plant extracts of *Mentha arvensis* and *Nerium oleander* at five per cent concentration were effective in reducing the same pest in onion. For chilli thrips, the plant extracts of *Azadirachta indica*, *Pongamia pinnata* and *Vitex negundo* @ 5 per cent proved effective control with 50 per cent mortality for each treatment compared to application of dimethoate which had only 29.58 per cent mortality (Thambidurai and Jayaraj, 2000). Stoll (2000) was of the opinion that *Gliricida sepium* leaf extract at two per cent was effective in controlling cabbage army worm and tobacco bud worm. Neem oil in 50 EC one per cent, neem oil three per cent, Vitexol 50 EC one per cent, neem seed kernel extract five per cent and neem oil one per cent were also effective in reducing the population leafhopper on egg-plant (Ramamurthy and Rajaram, 2001). Aqueous leaf and seed extracts of *Annona squamosa* and their powders showed insecticidal, antifeedant and repellent activity against number of insects and pests (Vijayalakshmi *et al.*, 2002). Boomiraj and Christopher Lourduraj (2006) reported that application of high level of poultry manure and neem cake along with spraying of herbal leaf extract reduced the population of leaf hopper, whitefly and aphids in bhendi. Rahuman *et al.* (2009) suggested that hot water and petroleum ether extracts of *Ipomea carnea* had the potential to be used as an ideal eco-friendly approach for the control of the major lymphatic filariasis vector. The occurrence of pest incidence as well as weeds was also very low in the combination of *Albizia lebbek* + *Annona squamosa*, which also contributed to higher yield in maize (Nandhakumar, 2010). Tomato plants treated with the extracts of *Artemisia nilagirica* were found to have lengthier shoots and increased shoot weight due to reduction in the nematode populations causing root gall (Sukul *et al.*, 2001). They also indicated that foliar spray with *Acacia auriculiformis* extract significantly increased the number of leaves per plant, leaf and root protein content, decreased the number of root galls, nematode population in roots and rhizosphere soil compared to control.

Vrikshayurveda for Disease Management

Vijaylakshmi *et al.* (2005) reported effective control of chillies leaf spot by 10 per cent, cow urine spray once in 10 days thrice followed by half-litre cow urine along with half-litre sour buttermilk mixed with nine litres of water once in 7 days twice. Hema *et al.* (2006) studied soil application of neem cake 10 kg ha^{-1} + *Trichoderma harzianum* 2.5 kg ha^{-1} reduced fusarium wilt and increased seed germination, fruit yield and decreased seedling mortality and disease intensity. Nargis *et al.* (2006) reported that the extracts of *Adhatoda vasica, Zingiber officinale, Vinca rosea* and *Azadirachta indica* in combination with cow dung, *Calotropis procera* and cow urine posses high ability to inhibit conidial germination of *Bipoloris sorokiniana* which might be used for controlling phytopathogens of crop plants. Kannan *et al.* (2007) reported from their pot and field studies, combined application of buffalo urine and sheep urine as foliar spray (1:1 v/v) at 5 per cent concentration on peanut have completely inhibited the mycelial growth, production and germination of sclerotia of *Sclerotium rolfsii* when compared to control and chemical fungicide, mancozeb (0.05 per cent). Subsequently it has increased the pod yield. Patil (2008) found that among the seven commercially available neem based products tested *viz.*, neem oil, Margotricure, Nimbicidine and Neem Gold at 0.5 per cent and Wanis at 1.0 per cent, sprayed thrice at an interval of 10 days starting from the onset of disease were found promising in reducing the soybean rust severity with significant increase in seed yield and 100 seed weight. Rathod *et al.* (2010) tested leaf extracts of *Azadirachta indica, Ocimum sanctum, Polyalthia longifolia, Tridax procumbens, Catharanthus roseus* and *Vitex negundo* and found that *Azadirachta indica* was more effective than other plant extracts showing inhibitory effect on growth of fungi.

Vrikshayurveda on Soil Fertility and Nutrient Uptake

The leaves of *Albizia lebbek, Delonix regia, Leucaena,* neem and many other species were commonly used as green leaf manure trees (Nair, 1993). Reddy and Reddy (1997) reported that application of tree leaves, as green leaf manure, in conjunction with inorganic fertilizer in tree based system had immense potential in supplementing part of N to the associated crop through efficient mineralization. Prakash and Bhusan (2003) reported that incorporation of *Leucaena* green leaves as 100 per cent substitution for fertilizer N increased soil organic carbon (0.21 to 0.31) content, soil available nitrogen, phosphorous and potassium. In a long term field experiment with different manurial treatments, decline in soil pH and EC was observed in maize - mustard cropping system. Maximum reduction was recorded in the plot receiving 100 per cent recommended nitrogen through FYM in rainy season (maize) and 100 per cent recommended N, P_2O_5 through fertilizer in winter season (mustard) which may be ascribed to the formation of acid during decomposition of organic matter (Kumpawat, 2004). Pandey *et al.* (2006) reported that application of manures, irrespective of sources and rates, recorded significantly higher soil organic carbon, available N, P_2O_5 and K_2O compared to control. Higher content of organic carbon in soil may be due to increased yield of roots and plant residues and external application of organic manures.

Nutrient Uptake of Crops

Karim and Savill (1991) reported that alley cropping and incorporation of *Gliricidia sepium* @ 10t ha^{-1} improved soil fertility and significantly improved N uptake by 40 kg ha^{-1} in rice. Vani and Bheemaiah (2003) revealed that alley cropping + green leaf manuring (*Dalbergia* and *Leucaena*) + 80 kg of N ha^{-1} increased uptake of nitrogen in castor (116.4 kg N ha^{-1}) and P and K were also found significantly more under alley cropping with *Dalbergia* green leaf manure and N application. Patel *et al.* (2003) reported the effect of fertilizer N and *Gliricidia* leaves on yield of cluster bean. It has been recorded that nutrient uptake of cluster bean (96, 30.6, 152) NPK kg ha^{-1} with *Gliricidia* leaves @ 10 t ha^{-1} compared to 100 per cent of RDN + *Gliricidia* leaves (134, 44, 211 kg ha^{-1} NPK uptake). Devipriya and Kumarasamy (2004) observed that application of 37.5 kg of rock phosphate with green manure @ 12.5 t ha^{-1} increased available P uptake of rice (77.2 kg P ha^{-1}) compared to control (33.8 kg P ha^{-1}). Sarawad *et al.* (2005) in rabi sorghum - chickpea yearly crop rotation, observed significant increase in carbon and available nutrients with incorporation of sunhemp as green manure. It also reduced the bulk density and improved the infiltration rate, water stable aggregates, hydraulic conductivity, available water content and water holding capacity of the soil. Combined application of green manure and inorganic fertilizer increased N and K uptake of rice and the highest value was recorded as 21.56 kg ha^{-1} in *Gmelina arborea* and 53.04 kg ha^{-1} in *Trewia nudiflora* treated plot respectively. Significant increase of organic carbon and NPK content of soil was also recorded in combined treatment (Indrani *et al.*, 2008). Incorporation of green leaf manure to castor recorded significantly higher nitrogen uptake (72.35 kg ha^{-1}) over control (Mohan Chavan *et al.*, 2010). The uptake of nitrogen, phosphorus and potassium was high in *Albizia lebbek* as green leaf manure and foliar spraying of *Annona squamosa* leaf extract in maize (Nandhakumar, 2010).

Availability of Nutrients

Application of green manure significantly increased the P content in soil (Atvar Singh and Bahl, 1993). In laboratory condition, application of *Sesbania aculeatea* as green leaf manure significantly increased the available K content in soil (Debnath and Hajra, 1972). According to Tolsanur and Badanur (2003) application of 50 per cent N through *Leucaena* + recommended dose of fertilizer (RDF) 50 per cent, in combination increased organic carbon, phosphorus, potassium and lowered nitrogen. Vani and Bheemaiah (2003) reported that application of green leaf manure (*Leucaena*,

Albizia and *Dalbergia*) @ 5 t ha^{-1}, improved the soil available NPK and also significantly increased soil organic carbon. The incorporation of *Gliricidia sepium* leaves as green manure improved soil quality by a rising both soil nitrogen and organic matter content (Bandara Krmu *et al.*, 2008)

Vrikshayurveda on Soil Physical and Chemical

Physical Properties

Organic farming lowers soil bulk density thereby increases water holding capacity, soil aggregation, drainage and decreased vulnerability of soil to mechanical pressure. It has been well documented that addition of organic manure and green leaf manure imparted good physical soil environment for crop growth. Sultani *et al.* (2007) found that application of green manure to crops, reduced soil bulk density (5 per cent), enhance total porosity (8 per cent) and macropores and large mesopores (28 per cent). Maximum reduction (7 per cent) in soil bulk density and an increase (11 per cent) in total soil porosity and available water content (17 per cent) were observed in plots where *Sesbania* was incorporated as green manure. Incorporation of green manure (*Gliricidia sepium*) improved the soil properties of home gardens significantly when compared to cropping fields, although the impact declined with increasing inclination and the soil depth (Wijesinghe *et al.*, 2009).

Chemical Properties

Sriramachandrasekran *et al.* (1996) compared different organic manures for micronutrient availability and reported higher availability of Zn, Fe and Mn on *Sesbania aculeata* treated soil as compared to paddy straw and composted coir pith.

Organic Carbon

Increased soil organic carbon three weeks after application and incorporation of tree leaves was observed in wheat (Belachew and Abera, 2011).

Vrikshayurveda on Soil Microbes

Cow's urine and milk contain microorganisms, that are useful for digestion. These organisms also produce substances, that have wide antibiotic activity. Some of the organisms may produce substances, which can either kill or inhibit the growth of microbes involved in food poisoning. Urine of cow contains microorganisms, which probably help the growth of plants. Fresh cow's urine exhibits anti-microbial activity. Effective micro-organisms (EMO) are the mixed culture of naturally occurring beneficial microbes (predominantly lactic acid bacteria (*Lactobacillus*), yeast (*Saccharomyces*), actinomycetes (*Streptomyces*) photosynthetic bacteria (*Rhodopseudomonas*) and certain fungi (*Aspergillus*) improved soil quality, growth and yield of sweet corn, which was equal to or higher than chemical fertilizers (Xu and Xu, 2000). Significant differences in total aerobic bacteria, fungal and actinomycetes were observed in soil amended with green manure *viz.*, *Sesbania rostrata* and *Crotalaria juncea* and inorganic fertilizers such as urea super granules and sulphur application (Ramalingam and Kannaiyan, 2006). The soil microbial population was influenced by the application *Albizia lebbek* as green leaf manures and higher concentration of micronutrients in *Annona squamosa* as foliar spray in maize (Nandakumar, 2010).

Other Recipes Used in *Vrikshayurveda*

Panchagavya

Panchagavya is a combination of five products obtained from the cow like, cow dung, cow urine, cow milk, curd and ghee. This is also mentioned in *Vrikshayurveda* texts and it has been commonly used by many of the farmers. It had a significant role in providing resistance to disease, pests and increasing over all yields (Subhashini Sridhar *et al.*, 2001). *Panchagavya* has a pH of 3.7 and EC 0.40 per cent, N 1.28 per cent, P 0.72 per cent, K 2.23 per cent and organic carbon 17.45 per cent (Perumal *et al.*, 2006). When sprayed with panchagavya, the plants produce larger leaves and develop denser canopy. The stem produces more lateral shoots and much more sturdy branches to bear heavy yields. Application of *panchagavya* in cassava as sett treatment and in field significantly increased number of tillers, number of tubers with increased size and weight and three per cent spray of panchagavya on *Jaffna morniga* before the end of fruiting stage produced 100 fruits per harvest (Natarajan, 2002). Panchagavya was an important one that enhanced the biological efficiency of crops and the quality of fruits and vegetables (Swaminathan *et al.*, 2007a). Effective micro-organism present in *panchagavya* might have produced bioactive substances that increased the photosynthetic efficiency and ultimately increased yields (Xu and Xu, 2000); the *panchagavya* application enhanced the biological efficiency of crop plants and quality of fruits and vegetables (Natarajan,1999). It is considered as a growth stimulant and insecticidal principle that increased yield of crops (Natarajan, 2002; Somasundaram *et al.*, 2003a; Selvaraj, 2004). *Panchagavya* spray resulted in significantly higher productive tillers hill^{-1}, panicle length, number of filled grains panicle^{-1} and seed test weight as well as grain yield in rice over control (Christopher Lourduraj *et al.*, 2005). While Ganesh (2007) found that rice cv. White Ponni gave higher grain yield when sprayed with *panchagavya*. Higher grain yield in maize and sunflower were obtained when treated with biogas slunny and *panchagavya* (Somasundaram *et al.*, 2007). A spray of 2 per cent was effective to enhance the growth and yield of rice (Vivekanandan, 1999b) and to advance the paddy harvest by 10 days (Vivekanandan, 1999a). *Panchagavya* spray @ 1.0 per cent reduced the flower drop, increased fruit size, retained freshness and enhanced taste, prevented fruit drops from green worms attack in Hosakere village of Karnataka (http://www. greenconserve.com) The experiment carried out by Cynthia (2003) in *Withania somnifera* (L.) Dunal indicated that application of panchagavya at four per cent produced the highest dry matter and total alkaloid content. Growth characters and physiological parameters were also increased. *Panchagavya* sprayed on chillies produced dark green leaves and new growth within 10 days (Subhashini Sridhar *et al.*, 2001) while advancement of days to first flowering and 50 per cent flowering was observed in annual moringa (Beaulah, 2001). Crude fibre, protein, ascorbic acid, carotene content and shelf life in annual moringa were higher under organic manures applied with *panchagavya* as spray (Beaulah *et al.*, 2002). Vijaykumar (2002) reported that application of *panchagavya* significantly improved the growth, development and yield of onion as intercrop in turmeric field. *Maha Pancha Gavya* -3 (MPG-3) improved soil health and productivity (http://www.cowindia.org). The microorganisms present in the rhizosphere environment around the roots influence the plant growth and crop yield. The beneficial micro organisms from *panchagavya* and their 105 establishment in the soil improved the sustainability of agriculture.

Amirthakaraisal **(in Tamil)**

Amirthakaraisal acts as an organic pesticide. Application through irrigation, reduced the root borne diseases and also increased soil enzymatic activities; acted as a nutrient and growth promoter (Sethuraman, 2002). The materials required for preparing *amirthakaraisal* includes: Fresh cow dung, cow urine, jaggery and water. The extract would be ready for use in a day or two.The product is applied through irrigation water to improve soil fertility (Tony Cisse, 2004). Rajareega (2008) opined that the *amirthakaraisal* mixed with irrigation water acted as a tonic for soil and enrich with nutrients. It was an effective medium for thousands of beneficial micro organisms and bacteria essential for crop growth. It was also suggested that 500 litres of the solution was sufficient for one hectare.

Thaemorkaraisal **(in Tamil)**

Thaemorkaraisal is one of the organic formulations used to induce flowering. The required materials for the preparation of *thaemorkaraisal* for one hectare spray includes: coconut, fermented milk and palm jaggery. The product is ready for spray in 15 days. It is used as spray before three months. There were also evidences to indicate that coconut water spray increased the chlorophyll content of rice (Thangaraj and Siva Subramanian, 1992). Mamaril and Lopez (1997) reported that coconut water acted as growth hormone, increased the biomass yield of sweet pepper and enhanced the photosynthetic activity of soybean (Kalarani and Jeyakumar, 1998). Tender coconut water spray enabled to maintain uniform flowering in rice (Tony Cisse, 2004).

Source Material for *Vrikshayurveda*

The word *Vrikshayurveda* literally means "*Ayurveda for plants*" is used to denote knowledge of plant life in all its varied aspects. In Varahamithira's *Brihat Samhitha* as well as in the *Agni Purana* there are separate sections dealing with *Vrikshayurveda*. The very same term with the word *gulma* preceding it - *Gulmavrikshayurveda* occurs in Kautilya's *Arthasastra* listing the functions of the officer in charge of agriculture and his assistants. The three works in which this term is used and a complete section has been devoted to the subject, are not texts which are devoted to *Vrikshayurveda* completely. The first text *Agni Purana,* is a popular encyclopedia of all kinds of knowledge and practices; the second, the *Brihat Samhitha* is a manual containing directions for the applications of the knowledge of astronomy and astrology in practice; the third, the *Arthasastra,* is a handbook discussing matters relating to royal polity and the art of governance. Thus, all of them are nonbotanical treatises; they are intended not so much to acquaint the students with theories as with practices. In all the three, there is an exclusive section dealing with '*Vrikshayurveda*". Besides these texts there are texts like *Upavana vinoda, Krishisukthi, Krishiparasara, Vrikshayurveda* of *surapala* which deal exclusively with different aspects of *Vrikshayurveda* (Subashini *et al.*, 2000).

Conclusion

Vrikshyurveda farming is a method of farming which primarily uses trees and plants in whole or in part besides the extracts, decoctions of parts of trees and plants, smokes produced from burning of the tree parts and excludes the use of harmful chemicals such as chemical fertilizers, pesticides and herbicides for crop growth, soil health building, control of pests and diseases and to maintain ecological balance and to provide stability in the production level without polluting soil, water and air. With the growing popularity and patronage for Indian Systems of Medicine like *Siddha, Ayurveda* and *Unani* for body health; agriculturists, farm scientists and scholars would also look back to one of the ancient time tested practices so called *Vrikshyurveda* farming to ensure better quality food for future generation.

References

Antonysamy, T. 2004. Organic aggregatum onion better packaging and quality. The Hindu, May 2004, 10p. Atvar Singh and Bahl, G.G. 1993. Phosphate equilibria in soils related to added phosphorus, *Sesbania aculeate* incorporation and cropping - A study of solubility relationships. *J. Indian Soc. Soil. Sci.*, **41**: 233-237.

Bandara Krmu, Henri Eisenbeissb, Martijn Sonnevelt, Chaminda Egodawattad. 2008. Model tropical legume trees for reversing soil degradation as a solution for increasing rural farmers' income by GIS analysis. *The International Archives of the Photogrammetry, Remote Sensing and Spatial Information Sciences.* Vol. 37. Part B2.

Beaulah, A. 2001. Growth and development of moringa (*Moringa oleifera.* Lam.) under organic and inorganic systems of culture. Ph.D., Thesis, Tamil Nadu Agricultural University, Coimbatore.

Beaulah, A., Vadivel,E. and Rajadurai. K.R. 2002. Studies on the effect of organic manures and inorganic fertilizers on the quality parameters of moringa (*Moringa oleifera* Lam) cv. PKM 1. In: Abst.National Seminar on Emerging Trends in Horticulture, Department of Horticulture, Annamalai University, Annamalai Nagar, Tamil Nadu. p.127-128.

Belachew Taye and Yifru Abera. 2011. Effect of application tree leaves and incorporation into soil in combination with nitrogen on soil fertility and yield of bread wheat (*Triticum aestivum*) under double cropping system. *J. Biodiv. and Environ. Sci.*, **1**: 1-11. Boomiraj, K. and Christopher Lourduraj, A. 2006. Effect of organic manures, *Panchagavya* and herbal leaf extract spray on insect incidence in bhendi. *J. Ecobiol.*, **18**: 269-276.

Brewbaker, J.L. and Glover, N. 1988. In: Sustainable Agriculture Green manures in rice farming, IRRI, Philippines. 31-45p. 106

Chelladurai, M. 2006. Studies on performance of (*Allium cepa var.aggregatum*) cv. Co 5 under '*Vrikshyurveda* farming'., M.Sc. (Hort.) Thesis, Department of Horticulture, Agricultural College and Research Institute, Tamil Nadu Agric. Univ., Madurai.

Christopher Lourduraj, A., Boomiraj, K. and Pannerselvam, S. 2005. Effect of organic manures on the production potential of bhendi (*Abelmoshus escullntus* L). OASIS Proc. *Seminar on Organic Agric. Peninsular Indiapromotion.* Tamil Nadu Agric. Univ., Coimbatore, 59- 163p.

Cynthia, A.S.E. 2003. Standardization of organic production packages for *Wilthania somnifera* Dunal. M.Sc (Hort.) Thesis, Tamil Nadu Agricultural University, Coimbatore. Debnath, N.C. and Hajra, J.N. 1972. Transformation of organic matter in soil in relation to mineralization of carbon and nitrogen availability. *J. Indian Soc. Soil Sci.,* **20**: 95-102.

Devarani, N. and Rangasamy, A. 1998. Nutrient and drought management practices for rainfed sorghum under vertisol. *Madras. Agric. J.,* **85**: 391-393.

Devipriya, P. and Kumarasamy, K. 2004. Grain yield of rice (*Oryza sativa* L.) crop as influenced by source of P with and without organic source. *Oryza.,* **41**: 24-27.

Ganesh. 2007. Use of animal products for pest control. Centre for Indian Knowledge Systems, **8**: 24-27.

Hazara, A.H, Shakkela, M., Khan, J., Igbal, M. and Khan, S. 1999. Effect of non-chemical methods and botanical insecticides on onion thrips, (*Thrips tabaci* Lind, Thysanoptera: Thripiae) in onion crop in Balochistan. *Sarhad. J. Agri.,* **15**: 619-624.

Hema, T., Jaganathan, R., Mohan, S. and Sandasakthi, A. 2006. Efficacy of fungal biocontrol agents and oil cakes against fusarium wilt (*Fusarium solani*) of brinjal. National seminar on Recent Advances on Plant Biology., p. 64. http://www.greenconserve.com. http://www.cowindia.org.

Indrani P., Bora, Arundhati Baruah and Jasbir Singh. 2008. Integrated use of legume green manure and inorganic fertilizer on soil health, nutrient uptake and productivity of rice (*Oryza sativa* L.) in shifting cultivation of Assam. *Indian J. Agric. Res.,* **42**: 260-265.

Kalarani, M.K. and Jeyakumar, R. 1998. Effect of nutrient and NAA spray on physiological changes in soybean. *Indian J. PI. physiol.,* **3**: 226-228.

Kannan, C., Rajkumar, V. and Raja, J. 2007. Management of stem rot disease of peanut by use of animal urine. ISOR National Seminar., 29-31p.

Karim, A.B. and Savill, P.S. 1991. Effect of spacing on growth and biomass production of *Gliricidia sepium* (Jacq) Walp in an alley cropping system sierra leone- *Agrofor. Syst.,* **16**: 213-222.

Kavitha, M., Natarajan, S. and Kavitha, P.S. 2005. Effect of organics on growth and yield of amaranthus (OASIS). In: Proc. *Seminar on Organic Agric. Peninsular Indiapromotion. Tamil Nadu Agri. Univ.* Coimbatore. p. 71- 78.

Kumpawat, B.S. 2004. Integrated nutrient management for maize-Indian mustard cropping systems. *Indian J. Agron.,* **49**: 18-21.

Mamaril, J.C. and Lopez., A.M. 1997. The effect of coconut water growth hormones (CWGH) on the growth, development and yield of sweet pepper (*Capsicum annum* L.) *The Phillipines J. Coco. Stud.* **222**: 18-24.

Manikandan, S. 2002. To promote fruiting in bottlegourd. *Indigenous Agriculture News*. **1**: 6.

Meera,T., George Ancy, P. and Udhayakumar, R. 2004. Antifungal activity of plant products against *Colltetrichum capsici* the incitant of fruti rot of chilli. *J. Mycol. Pl. Pathol.*, **34**: 973.

Mohan Chavan, B.T., Pujari, Ashok Surwenshi and Jalageri, B.R. 2010. Nitrogen uptake and available nitrogen content in soil as influenced by green manures and nitrogen levels. *Asian J. Soil Sci.*, **5**: 186-188.

Nair, P.K.R. 1993. An introduction to Agro Forestry, Kluwer Academic publishers, USA 26p.

Nandhakumar, M.R. 2010. Utilisation of leaves and leaf extracts of tropical trees for eco-friendly *Vrkshayur vedic* farming in maize (*Zea mays* L.). M.Sc. (Ag.). Thesis, Tamil Nadu Agricultural University, Department of Agronomy, Agricultural College and Research Institute, Madurai.

Nargis, A., Ferdousi, M., Begum Shahidul Alam and Alam, M.S. 2006. Inhibitory effect of different plant extracts, cow dung and cow urine on conidial germination of *Bipolaris sorokiniana*. *J. Bio-Sci.*, **14**: 87-92.

Natarajan, K. 1999. *Panchagavya. Nam Vazhi Velanmai* (Tamil)., 3-4p. Natarajan, K. 2002. *Panchagavya* - A manual. Other India Press, Mapusa, Goa, India. p.33.

Nene, Y.L. 2004. Ancient classics on Indian Agriculture. Abs. International Conference on Agricultural Heritage of Asia. ASIA-AHF, Secunderabad, India. pp.155-169.

Pal., S.K., Chowdhury, A. and Gunri, S.K. 2005. Effect of integrated nitrogen management and plant density on yield and nitrogen balance of rice under lowland situation. *Oryza.*,**42**: 41-47.

Palanisamy, V. and Punithavathi, N.1998. Effect of seed hardening and seed pelleting on seed quality in ragi. *Madras. Agri. J.*, **85**: 583-585.

Pandey, A.K., Gopinath, K.A., Chattacharya, P., Hooda,K.S. Sushil, S.N., Kundu, S., Selvakumar, G. and Gupta, H.S. 2006. Effect of source and rate of organic manures on yield attributes, pod yield and economics of organic garden pea (*Pisum sativum* subsp. *hortense*) in north west Himalaya. *Indian J. Agric.Sci.*, **76**: 230-234.

Patel, J.J., Patel,B.M.,Patel,B.T. and Patil, R.G. 2003. Studyon use of *Gliricidia sepium* leaves for leaf manuring in cluster bean-pearl millet rotation under Dryland condition. *Agri. Sci. Digest.*, **23**: 10-13.

Patil, P.V., 2008, Evaluation of botanical products againstsoybean rust caused by *Phakopsora pachyrhizi* Syd.*J. Ecofriendly Agric.*, **3**: 62-64.

Pawar, D.B., Warade, S.D., Lawande, K.E. and Patil, S.K. 2003. Effect of different organic products on onion thrips. In: Kirti singh, Lawande, K.E., Pandey, U.B., Lallan singh and Bhonde, S.K. **(Eds.)** *Approaches for Sustainable Development of Onion and Garlic.*

Perumal, K.K. Praveena, V. Stalin and B. Janarthanam. 2006. Assessment of selected organic manure hormone and their impact on the growth attributes of *Allium cepa*. *Biodynamic Association of India.*, pp. 2-19.

Prakash, O.M. and L.S. Bhushan. 2003. Effect of fertilizer substitution through white lead tree (*Leucaena leucocephala*) green biomass on growth, yield and 107 economics of wheat (*Triticum aestirurm*) crop in degraded lands. *Indian J. Agric. Sci.*, **73**: 311-314.

Prasad. 2002. Influence of leaf extracts on cultivated turnip, seed germination, seedling survival and seed weight. *Allelo J.*, **9**: 256-257.

Rahuman, A.A., Bagavan, A., Kamaraj, C., Saravanan, E., Zahir, A.A. and Elango, G. 2009. Efficacy of larvicidal botanical extracts against *Culex quinquefasciatus* Say (Diptera: Culicidae). *Parasitol. Res.*, 104:1365- 1372.

Rajareega. 2008. Farmers' Note Book. *The Hindu*. 131(44): 21. Ramalingam, N. and Kannaiyan, S. 2006. Impact of green manures and sulphur application on soil enzyme activity and grain yield of rice. *Mysore J. Agric. Sci.*, **40**: 68-73.

Ramamurthy, R. and Rajaram, V. 2001. Effect of plant products on the management of leaf hopper of eggplant. *Medicinal and Aromatic Plants, Abst.*, **23**: 758.

Rathod, L.R., Jadhav, M.D., Awate, M.K., Surywanshi, A.M., Deshmukh, P.S. 2010. Utilization of medicinal plants to control seed borne pathogens of selected legumes seeds. *Inter.J. Adv. Biotech.and Res.*, **l1**: 57-59.

Reddy, P.R. and Reddy, G.H.S. 1997. Principles of Agronomy, Kalyani Publications Ltd, New Delhi: pp. 509.

Sarwad, I.M., Guled, M.B. and Gundlur, S.S. 2005. Influence of integrated nutrient supply system for rabi sorghum chickpea crop rotation on crop yields and soil properties. *Kar. J. Agri.Sci.*, **18**: 673-679.

Selvaraj, P. 2004. *Panchakavya* an organic insect repellent. *Num vazhi velanmai* (Tamil)., **13** :6.

Sethuraman, M. 2002. Nothing succeeds like 'do nothing' technique. *Asian J. Exp. Biol. Sci.*, **1**: 387-392.

Sharmarao Jahagirdar, M.R., Ravikumar and Siddaramaiah, A.L. 2003. Traditional methods in the management of plant diseases - A Review. *Agric.Rev.*, **24**: 142-146.

Singh, N. and Attrey, D.P. 2002. Studies on round the year organic production of beet leaf in trenches (underground green house) in cold desert high altitude conditions of Ladakh. **In**: International Seabuckthorn Association meeting, Berlin (Germany). p. 30.

Somasundaram, E., Mohamed Amanullah, M., Vaiyapuri, K., Thirukkumaran, K. and Sathyamoorthi, K. 2007b. Influence of organic sources of nutrients on the yield and economics of crops under maize based cropping system *J. App. Sci. Res.*, **3**: 1774-1777.

Somasundaram, E., Sankaran, N. and Meena, S. 2003. Efficiency of *panchagavya* (organic nutrition) spray on green gram. In: Abst. of UGC sponsored State level seminar on Indigenization of India farming: Problems and Prospects Gandhigram Rural Institute, Deemed University, Gandhigram, Tamil Nadu. p.41.

Sriramachandrasekaran, M.V., Ramanathan, G. and Ravichandran, M. 1996. Effect of organic manures on the availability of nutrients in rice rhizosphere soil. *Oryza.*, **33**: 126-131.

Subhashini Sridhar, S., Arumugasamy, S., Saraswathy, H. and Vijayalakshmi, K. 2000. Organic vegetable gardening. Centre of Indian Knowledge Systems, Chennai.

Sritharan, N., Rajavel, M. and Chandrasekhar, C.N. 2010. Impact of Bioregulators on Phytochemicals and quality of Black Night shade solanum nigrum.*Madras Agric. J.*, **97**: 93-96.

Stoll, G. 2000. National protection in the tropics. *Cur. Sci. Germany.*, **9**. pp. 5-8.

Subhashini Sridhar, Arumugasamy, A., Vijayalakshmi, K. and Balasubramanian, A.V. 2001. *Vrikshayurveda*ayurveda for plants. Centre for Indian Knowledge Systems. Chennai, Tamil Nadu. 47p.

Sukul, N.C., Sinha Babu, S.P., Datta, S.C., Nandi, B. and Sukul, A. 2001. Nematotoxic effect of *Acacia auriculiformis* and *Artemisia nilagirica* against rootknot nematode. *Allelopathy J.*, **8**: 65-72.

Sultani, M.I., Gill, M.A., Anwar, M.M. and Athar, M. 2007. Evaluation of soil physical properties as influenced by various application tree leaves and incorporation into soil legumes and phosphorus fertilization under rain fed conditions. Int. *J. Environ. Sci. Tech.*, **4**:109-118.

Swaminathan, C. 2000. Efficacy of tree origin botanicals on food grain yield in cowpea *Vigna unigulculata* (1) walp. Paper Presented in First World Congress of Agro Forestry, University of Florida, USA, Abst.p.

Swaminathan, C. and Gururajan, B. 2005. Revisiting and researching Indian systems of agriculture through *Vrikshayurveda*. 93rd Indian science congress" integrated rural development and science and technology.Acharya N.G. Ranga Agri.Univ. Rajendra nagar. Andhra Pradhesh. 35-39p.

Swaminathan, C., Swaminathan, V. and Chelladurai, M. 2008. *Vrikshayurveda*. In National Seminar on Organic Farming and Sustainable Agriculture for Food Security. Gandhigram Rural University, DIndigul, India. Abst.p.

Swaminathan, C.,Swaminathan, V. and Vijayalakshmi, K. 2007a. *Panchagavya* Boon to Organic farming. IBD Pub. Network. 28p.

Swaminathan, C.,Swaminathan, V. and Vivekanandan, P. 2007b. *Vrikshayurveda* for crop health and heavy foods. NAAS 8th Agric. Science Congress, Tamil Nadu Agricultural University. p. 198

Thambidurai, G. and Jayaraj, S. 2000. Efficacy of certain botanical pesticides in the management of major pests of brinjal, bhendi and chillis. Agri. College and Res. Instt., Madurai.

Thangaraj, M. and Sivasubramanian, K. 1992. Effect of growth regulators on productivity in Samba and Thaladi rice. *Madras Agric.J.*, **78**: 71-76.

Tolsanur, S.L. and Badanur, V.P. 2003. Effect of integrated use of organic manure, green manure and fertilizer nitrogen on sustaining productivity of rabi sorghumchickpea system and fertility of a vertisol. *J. Indian. Soc. Soil. Sci.*, **51**: 41-44.

Tony cisse. 2004. Techniques for organic paddy cultivation. *Indigenous Agriculture News*. **4**: 1-4.

Tripathi, S., Tripathi, A., Kari, D.C. and Paroha, S. 2000. Effect of *Dalbergia sissoo* extracts rhizobhium and nitrogen on germination, growth and yield of *Vigna radiata*. *Allelo J.*, **7**: 255-264.

Upendra Shenoy, U.K., Purushothama Rao, Aruna Kumara and Anand, A.S. 2000. *Krishi Prayoga Pariwara*: A Group of Experimenting Farmers. 9p.

Vaidya, K., Shyamsunder, M. and Vijayalakhsmi, K. 1994. Enhancement of growth and flowering in *Impatiens sultanii* (Balsam plant). *Vrikshayurveda*. Ayruveda for plants. Centre for Indian Knowledge Systems. Chennai. 51-52p.108

Vani, K.P. and Bheemaiah, G. 2003. Efficient use of fertilizerthrough Alley cropping and green leaf manures for increased productivity of rainfed castor. *Ann. Agric.Res. Sci.*, **124**: 598-604.

Vijayakumar. 2002. Iyarkai Vivasayathil *panchagavya* (Panchagavya in organic farming) p.10.

Vijayalakshmi, K. 1993a. *Vrikshayurveda*. Ayurvdea for plants. Centre for Indian Knowledge System. Chennai. pp.19-20.

Vijayalakshmi, K. 1993b. *Vrikshayurveda*. An introduction to Indian Plant Science, lok *Swasthya Parampara Samuardhan Samithi*, Centre for Indian Knowledge Systems. Chennai. 101p.

Vijayalakshmi, K. and Shyamsundar,K.M. 1994. Pest control and diseases management in *Vrikshayurveda lok swasthya parama para samuardhan samithi*. Centre for Indian Knowledge System. Chennai. 86p.

Vijayalakshmi, K., Subshashini, B. and Koul, S. 2002. Plants in pest control. Centre for Indian Knowledge System. Chennai, India.

Vijaylakshmi, K., Arumugasmy, S. and Subhashini, S. 2005. Received: February 2, 2011; Accepted: May 5, 2011organic methods of pest and disease management, Centre for Indian Knowledge Systems.8p.

Vivekanandan, P. 1999a. *Panchagavya* advance paddy harvest by 10 days. *Agri. News.*, **2**: 11. Vivekanandan, P. 1999b. '*Panchagavya*'. *Nam Vazhi Velanmai* (Tamil). pp. 4.

Wijesinghe, D.B., Egodawatta, W.C.P., Sangakkara, U.R. Pushpakumara, D.K.N.G. and Stamp, P. 2009. Productivity of maize (*Zea mays* L.) and mung bean (*Vigna radiata* L. Wikazek) in home gardens and cropping fields under subsistence conditions. *Tro. Agric. Res.*, **21**: 168-176.

Xu, Hui Lian and Xu, H.L. 2000. Effects of microbial inoculants and organic fertilizers in the growth, photosynthesis and yield of sweet corn. *J. Crop Prod.*, **3**: 183-214.

Zodape, S.T., Soumita Mukhopadhyay, K. Eswaran, M.P. Reddy and Chikara, J. 2010. Enhanced yield and nutritional quality in green gram (*Phaseolus radiate* L.) treated with seaweed (*Kappahycus alvarezii*) extract. *J. Sci. Indu. Res.*, **69**: 468-471.

Chapter 5
Physiological Parameters

The influence of Vrikshayurvedic farming practices in various physiological parameters are discussed.

Dry Matter Production

The Dry matter production (DMP) of a plant is worked out as given below with a help of the formula. The whole plant samples were collected at randomly in net plot area. These sample were air-dried initially and then oven dried at 65 ± 5°C to attain a constant weight. The DMP can be recorded during various growth stages of crop and also at harvest stage.

In barnyard millet, the incorporation of green leaf manures had significant influence at 90 DAS (harvest stage). Among the green leaf manure sources, *Peltophorum ferrugineum* recorded the highest DMP of 6318 kg ha^{-1}. It was followed by *Delonix regia* (6050 kg ha^{-1}) and all other treatments were on par with each other. Significant variation in DMP was observed due to foliar application of different leaf extracts. Among the four leaf extracts the maximum DMP was registered in *Annona squamosa* with the DMP of 6262 kg ha^{-1}. It was followed by *Moringa oleifera* with 6018 kg ha^{-1} of DMP and minimum 5558 kg ha^{-1} DMP was recorded in *Morinda tinctoria*. At harvest stage the DMP was significantly influenced due to interaction effect. Among the 20 treatment combinations maximum DMP of 6717 kg ha^{-1} was recorded by the combination *Peltophorum ferrugineum* + *Annona squamosa* and it was followed by *Delonix regia* + *Annona squamosa* (6441 kg ha^{-1}) which was on par with *Peltophorum ferrugineum* +*Moringa oleifera* (6328 kg ha^{-1}). The minimum DMP of 5170 kg ha^{-1} was recorded by *Glyricidia sepium*+ *Morinda tinctoria*. The absolute control zero input recorded 4991 kg ha^{-1} of DMP at harvest stage (90 DAS). When compared to Control, *Peltophorum ferrugineum* + *Annona squamosa* showed 25.6 per cent increased DMP at harvest stage. The recommended dose of inorganic fertilizers recorded 7204 kg ha^{-1}. When compared to this, *Peltophorum ferrugineum* + *Annona squamosa* showed 6.7 per cent decrease in DMP at harvest stage.

In finger millet (ragi), the green leaf manure application had significant effect on dry matter production at harvest stage of crop. Among the green leaf manure incorporation, *Delonix regia* recorded highest DMP of 4665 kg ha^{-1}. This was followed by *Leucaena leucocephala*, which was on par with *Albizia lebbek. Peltophorum ferrugineum* registered lowest DMP of 2618 kg ha^{-1}. Foliar spraying of leaf extracts had significant effect on dry matter production. Among the treatments, *Aegle marmellos* recorded highest DMP (3817 kg ha^{-1}) and all other treatments were on par with each other. *Annona squamosa* recorded lowest DMP of 3210 kg ha^{-1}. The interaction effect between green leaf manure application and foliar spraying of leaf extracts showed that *Delonix regia + Moringa oleifera* recorded highest DMP (5030 kg ha^{-1}). This was on par with *Delonix regia + Aegle marmellos*, which was followed by *Delonix regia + Morinda tinctoria. Peltophorum ferrugineum + Morinda tinctoria* recorded lowest DMP of 2079 kg ha^{-1}, it was comparable with *Peltophorum ferrugineum + Moringa oleifera* at harvest stage. The absolute control zero input recorded 1985 kg ha^{-1} of DMP. When compared to zero input, *Delonix regia + Moringa oleifera* showed 60.54 percent increased DMP. The recommended dose of inorganic fertilizers recorded 4599 kg ha^{-1}. When compared to this, *Delonix regia + Moringa oleifera* showed 8.57 per cent increase in DMP at harvest stage.

In maize crop, the incorporation of green leaf manures had significant effect on dry matter production (DMP) at 45 DAS. *Albizia lebbek* registered higher DMP of 2537 kg ha^{-1} and which was followed by *Peltophorum ferrugineum* which recorded a DMP of 2286 kg ha^{-1}. The minimum value of DMP was recorded in *Leucaena leucocephala* with 1953 kg ha^{-1}. The same trend continued at 60 DAS and also at harvest. The foliar spraying of leaf extract had a significant effect on DMP of maize. Among the treatments *Annona squamosa* recorded the highest DMP of 2339 kg ha^{-1}. It was followed by *Morinda tinctoria* recorded a DMP of 2310 kg ha^{-1} and the lowest DMP of 2073 kg ha^{-1} was recorded in *Aegle marmellos* at 45 DAS. The same trend continued at 60 DAS and also at harvest. The lowest DMP of (5924 kg ha^{-1}) was recorded in *Aegle marmellos*. The interaction effect was significant. Among the treatment, highest DMP (2836 kg ha^{-1}) was recorded in *Albizia lebbek + Annona squamosa* at 45 DAS. It was followed by *Albizia lebbek + Morinda tinctoria* recorded the DMP of 2686 kg ha^{-1} and the lowest DMP of 1702 kg ha^{-1} was recorded in *Leucaena leucocephala + Alangium salvifolium*. The interaction effect was significant for DMP at 60 DAS and also at harvest stage. Among the treatment highest DMP of 5673 and 8104 kg ha^{-1} was recorded in *Albizia lebbek + Annona squamosa* at 60 DAS and also at harvest stage. It was followed by *Albizia lebbek + Morinda tinctoria* recorded the DMP of 5373 and 7676 kg ha^{-1}. The lowest DMP of 3405 and 4865 kg ha^{-1} was recorded in *Leucaena leucocephala + Alangium salvifolium*. The absolute control zero input recorded 344 kg ha^{-1} of DMP at 45 DAS when compared to Control, *Albizia lebbek + Annona squamosa* showed decreased DMP (2492 kg ha^{-1}). The recommended dose of inorganic fertilizers CPG recorded 3475 kg ha^{-1} when compared, *Albizia lebbek + Annona squamosa* showed increased level of DMP (638 kg ha^{-1}). The absolute control zero input recorded 678 and 835 kg ha^{-1} of DMP at 60 DAS and at harvest stage. When compared to control, *Albizia lebbek + Annona squamosa* showed increased DMP (4994 and 7268 kg ha^{-1}). The recommended dose of inorganic fertilizers CPG recorded 8573 and 11532 kg ha^{-1} at 60 DAS and harvest stage, respectively. When compared to this, *Albizia lebbek* + *Annona squamosa* showed decrease in DMP with values of 2900 and 3427 kg ha^{-1} at 60 DAS and at harvest stage, respectively.

In black gram, the incorporation of green leaf manures had significant effect on dry matter production at 50 DAS. *Pongamia pinnata* registered higher DMP of 693 kg ha^{-1} and which was followed by *Albizia lebbek* with a value of 559 kg ha^{-1}. The minimum value of DMP was recorded in *Delonix regia* (384 kg ha^{-1}). The same trend continued at harvest stage also. At harvest stage, the higher DMP (1176 kg ha^{-1}) was recorded in *Pongamia pinnata*, DMP of 1089 kg ha^{-1} which was followed by *Albizia lebbek* and the lowest DMP (713 kg ha^{-1}) was recorded in *Gliricidia sepium* at harvest stage. The foliar spraying of leaf extract had a significant effect on DMP of black gram. Among the treatments, *Moringa oleifera* recorded the highest DMP of 568 kg ha^{-1}. It was followed by *Annona squamosa*, with a value of 489 kg ha^{-1} and the lowest DMP (462 kg ha^{-1}) was recorded in *Morinda tinctoria* at 50 DAS. The same trend continued at harvest stage also. At harvest stage of higher DMP (975 kg ha^{-1}) was recorded in *Moringa oleifera*, which was followed by *Annona squamosa* (966 kg ha^{-1}) and the lowest DMP of (874 kg ha^{-1}) was recorded in *Aegle marmellos*. The interaction effect was significant. Among the treatments highest DMP 764 kg ha^{-1} was recorded in *Pongamia pinnata* + *Moringa oleifera* at 50 DAS. It was followed by *Pongamia pinnata* + *Annona squamosa* which recorded the DMP of 738 and the lowest DMP (272 kg ha^{-1}) was recorded in *Gliricidia sepium* + *Morinda tinctoria*. The interaction effect was significant harvest stage. Among the treatments highest DMP of 1300 kg ha^{-1} was recorded in *Pongamia pinnata* + *Moringa oleifera*. It was followed by *Albizia lebbek* + *Annona squamosa* recorded the DMP of 1222 kg ha^{-1}. The lowest DMP of 702 kg ha^{-1} was recorded in *Gliricidia sepium* + *Moringa oleifera* at harvest stage. The absolute control zero input recorded 165 kg ha^{-1} of DMP. When compared Control *Pongamia pinnata* + *Moringa oleifera* produced 599 kg ha^{-1} more. The recommended dose of inorganic fertilizers CPG recorded 865 kg ha^{-1}. When compared to this, *Pongamia pinnata* + *Moringa oleifera* showed a decreased (101 kg ha^{-1}) DMP at 50 DAS. The absolute control zero input recorded 437 kg ha^{-1} of DMP at harvest stage. When compared to Control, *Pongamia pinnata* + *Moringa oleifera* produced more DMP (862 kg ha^{-1}) at harvest stage. The recommended dose of inorganic fertilizers recorded 1764 kg ha^{-1} at harvest stage. When compared to this, *Pongamia pinnata* + *Moringa oleifera* showed a decreased DMP of 463 kg ha^{-1}.

In bhendi crop, during 90 DAS, the highest dry matter production (1641 kg ha^{-1}) was recorded in *Albizia lebbek* and lowest (1600 kg ha^{-1}) was recorded in *Leucaena leucocephala*. At 120 DAS, the highest dry matter production (2403 kg ha^{-1}) was recorded in *Albizia lebbek* and lowest (2278 kg ha^{-1}) was recorded in *Leucaena leucocephala*. While, foliar spraying of *Annona squamosa leaf extracts* recorded 1662 kg ha^{-1} of dry matter production and the lowest (1585 kg ha^{-1}) was recorded in *Ocimum sanctum* leaf extract spray at 90 DAS. The highest dry matter production (2411 kg ha^{-1}) was recorded in *Annona squamosa* and the lowest (2307 kg ha^{-1}) was recorded in *Morinda tinctoria* at 120 DAS. The interaction effect was significant, the treatment combinations, *Albizia lebbek* + *Annona squamosa* recorded 1803 kg ha^{-1} the highest dry matter production and the lowest (1559 kg ha^{-1}) was recorded in *Gliricidia sepium* +

Alangium salvifolium at 90 DAS. At 120 DAS the highest dry matter production of 2713 kg ha^{-1} was recorded in *Albizia lebbek* + *Annona squamosa* and the lowest (2225 kg ha^{-1}) was recorded in *Gliricidia sepium* + *Alangium salvifolium*. The absolute control recorded 810 and 1162 kg ha^{-1} of dry matter production at 90 and 120 DAS, respectively. While comparing with *Albizia lebbek* + *Annona squamosa*, it showed increased dry matter production of 993 and 1551 kg ha^{-1} respectively. The recommended dose of inorganic fertilizers recorded 1975 and 2950 kg ha^{-1} at 90 and 120 DAS respectively. When compared to *Albizia lebbek* + *Annona squamos*, it showed decreased level of 171 and 237 kg ha^{-1}.

In onion crop the incorporation of green leaf manure had significant effect on dry matter production of aggregatum onion at 60 and 90 DAS. While at 60 DAS the highest dry matter production of 1189 kg ha^{-1} was recorded in *Gliricidia sepium* and the lowest dry matter production of 1100 kg ha^{-1} was recorded in *Senna siamea*. The application of leaf extract through seed treatment and foliar sprays had a significant effect on dry matter production of aggregatum onion. At 60 DAS, the highest dry matter production of 1177 kg ha^{-1} was recorded in *Aegle marmelos* and the lowest dry matter production of 1098 kg ha^{-1} was recorded in *Alangium salvifolium*. The interaction effect was significant. Among the treatment highest dry matter production at 60 DAS (1300 kg ha^{-1}) was recorded in *Gliricidia sepium* + *Aegle marmelos* and the lowest dry matter production of 1030 kg ha^{-1} was recorded in *Leucaena leucocephala* + *Annona squamosa* at 60 DAS. The absolute control recorded 356 and 988 kg ha^{-1} of dry matter production at 30 and 60 DAS. When compared to control, *Gliricidia sepium* + *Aegle marmelos* showed an increased dry matter production (212 and 315 kg ha^{-1}). The recommended dose of inorganic fertilizers recorded 1440 kg ha^{-1} at 60 DAS, when compared to this, *Gliricidia sepium* + *Aegle marmelos* showed a decrease of 100 kg ha^{-1}. At 90 DAS the highest dry matter production (1910 kg ha^{-1}) was recorded in *Gliricidia sepium* and lowest dry matter production (1719 kg ha^{-1}) was recorded in *Delonix regia*. Among the treatments, *Aegle marmelos* recorded 2752 kg ha^{-1} of dry matter production and the lowest dry matter production (2461 kg ha^{-1}) was recorded in *Alangium salvifolium*. The interaction effect was significant, The treatment *Gliricidia sepium* + *Aegle marmelos* recorded (3173 kg ha^{-1}) the highest dry matter production and the lowest dry matter production (2163 kg ha^{-1}) was recorded in *Leucaena leucocephala* + *Annona squamosa*. The absolute control recorded 1389 kg ha^{-1} of dry matter production when compared to, *Gliricidia sepium* + *Aegle marmelos* showed a increased dry matter production of 1784 kg ha-$^{-1}$. The recommended dose of inorganic fertilizers was found to produce 3223 kg ha^{-1} of dry matter which was higher than that of *Gliricidia sepium* + *Aegle marmelos* by 149.67 kg ha^{-1}.

Crop Growth Rate

The crop growth rate during the growth period may be calculated during growth stages of crops by using the formula given by Buttery (1970) and expressed in g m^{-2} day^{-1}.

$$CGR = \frac{W_2 - W_1}{P(t_2 - t_1)}$$

where,

W_1 and W_2 = Initial and final dry weight of the plant at time t_1 and t_2 per unit ground area (g) respectively

P = Land area occupied by the plant in m^2

t_1 and t_2 = Initial and final day of a period of observation, respectively

In bhendi crop, the incorporation of green leaf manure had significant effect on crop growth rate (30-60, 60-90 and 90-120 DAS). The highest crop growth rate (15.53 g cm^{-2} day^{-1}) was recorded in *Albizia lebbek* and the lowest crop growth rate (14.74 g cm^{-2} day^{-1}) was recorded in *Delonix regia* at first phase, while at second-phase the highest crop growth rate (16.03 g cm^{-2} day^{-1}) was recorded in *Delonix regia* and the lowest crop rate (15.39 g cm^{-2} day^{-1}) was recorded in *Leucaena leucocephala.* The highest crop growth rate (18.81 g cm^{-2} day^{-1}) was recorded in *Albizia lebbek* and the lowest crop rate (16.75 g cm^{-2} day^{-1}) was recorded in *Leucaena leucocephala* during third phase. The application of leaf extract through seed treatment and foliar sprays had significant effect on crop growth rate of Bhendi. Among the treatments, *Aegles marmellos* recorded the highest crop growth rate (15.64 g cm^{-2} day^{-1}) and the lowest crop growth rate (14.77 g cm^{-2} day^{-1}) was recorded in *Alangium salvifolium* during first phase, while at second-phase (60-90 DAS) the highest crop growth rate (16.28g $cm^{-2}$$day^{-1}$) was recorded in *Annona squamosa* and the lowest crop growth rate (15.21 g cm^{-2} day^{-1}) *Ocimum sanctum.* The highest crop growth rate (18.51 g cm^{-2} day^{-1}) was recorded in *Annona squamosa* and the lowest crop growth rate (17.27 g cm^{-2} day^{-1}) was recorded in *Alangium salvifolium* at third stage. The interaction effect was significant. Among the treatment combinations, *Albizia lebbek + Annona squamosa* recorded the highest crop growth rate (17.56 g cm^{-2} day^{-1}), which was on par with *Gliricidia sepium + Aegles marmellos* recorded 17.15 g cm^{-2} day^{-1} and lowest crop growth rate (14.16 g cm^{-2} day^{-1}) was recorded in *Delonix regia+ Alangium salvifolium* during first phase (30-60 DAS), while at second phase (60-90 DAS), the highest crop growth rate (17.53 g cm^{-2} day^{-1}) was recorded in *Albizia lebbek + Annona squamosa,* which was on par with *Gliricidia sepium + Aegless marmllos* and *Delonix regia+ Alangium salvifolium* was recorded in 17.49 g cm^{-2} day^{-1} and lowest crop growth rate (14.95 g cm^{-2} day^{-1}) was recorded in *Gliricidia sepium + Alangium salvifolium.* The highest crop growth rate (22.46 g cm^{-2} day^{-1}) was recorded in *Albizia lebbek + Annona squamosa,* which was on par with *Delonix regia +Morinda tinctoria* recorded 18.86 g cm^{-2} day^{-1}. The lowest crop growth rate (16.34 g cm^{-2} day^{-1}) recorded *Leucaena leucocephala+ Morinda tinctoria* during third stage. The absolute control recorded 6.98 g cm^{-2} $day^{-1}$of the crop growth rate at first phase (30-60 DAS), second phase (60-90) it recorded 7.28 g cm^{-2} day^{-1} and third phase (90-120) it recorded 8.39 g cm^{-2} day^{-1}. When control was compared with *Albizia lebbek + Annona squamosa* in all three phases the increased level of 10.58 g cm^{-2} day^{-1}, 10.25 g cm^{-2} $day^{-1}$and 14.07 g cm^{-2} day^{-1} over the control. While the recommended dose of inorganic fertilizers recorded 17.83 g cm^{-2} day^{-1} at first phase, second phase recorded 21.23 g cm^{-2} day^{-1} and third phase recorded 24.07 g cm^{-2} day^{-1} of crop growth rate. When itwas compared with *Albizia lebbek + Annona squamosa* in all three phases there was decreased crop growth rate of 0.27 g cm^{-2} day^{-1}, 3.70 g cm^{-2} day^{-1} and 1.61 g cm^{-2} day^{-1}, respectively.

In onion, the incorporation of green leaf manure had significant effect on crop growth rate of aggregaum onion at both the phases (30-60 DAS and 60-90 DAS). The highest crop growth rate (69.8 g m^{-2} day^{-1}) was recorded in *Gliricidia sepium* and the lowest crop growth rate (60.4 g m^{-2} day^{-1}) was recorded in *Senna siamea* at first phase, while at second phase the highest crop growth rate (156.9 g m^{-2} day^{-1}) was recorded in *Gliricidia sepium* and the lowest crop rate (128.443 g m^{-2} day^{-1-1}) was recorded in *Delonix regia*. The application of leaf extract through seed treatment and foliar sprays had a significant effect on crop growth rate of aggregatum onion. Among the treatments, *Aegle marmelos* recorded the highest crop growth rate (69. g m^{-2} day^{-1}) and the lowest crop growth rate (61.79 g m^{-2} day^{-1}) was recorded in *Annona squamosa* during first phase (30-60 DAS), while at second phase (60-90 DAS) the highest crop growth rate (146.612 g m^{-2} day^{-1}) was recorded in *Morinda tinctoria* and *Alangium salvifolium* recorded the lowest crop growth rate (133.6 g m^{-2} day^{-1}). The interaction effect was significant. The treatment, *Gliricidia sepium* + *Aegle marmelos* recorded the highest crop growth rate (77.57 g m^{-2} day^{-1}), which was on par with *Leucaena leucocephala* + *Morinda tinctoria* recorded (76.250 g m^{-2} day^{-1}) during first phase (30-60 DAS), while at second phase (60-90 DAS), the highest crop growth rate (187.610 g m^{-2} day^{-1}) was recorded in *Gliricidia sepium* + *Aegles marmllos*, which was on par with *Albizia lebbek* + *Annona squamosa* recorded 169.53 g m^{-2} day^{-1}. The absolute control recorded 45.629 g m^{-2} $day^{-1}$of the crop growth rate at first phase (30-60 DAS) and at second phase (60-90) it recorded 99.836 g m^{-2} day^{-1}. When it was compared to Control, *Gliricidia sepium* + *Aegles marmllos* in both phases had increased 31.941 and 87.7 g m^{-2} day^{-1}. While the recommended dose of inorganic fertilizers recorded (79.93 g m^{-2} day^{-1}) at first phase while second phase recorded (175.83 g m^{-2} day^{-1}) of crop growth rate. When compared to this, *Gliricidia sepium* + *Aegles marmllos* in both phase showed decreased 2.35 g m^{-2} day^{-1} and 11.78 g m^{-2} day^{-1}.

Relative Growth Rate

The relative crop growth rate during the growth period may be calculated during growth stages of the crop by using the formula given by Enyi (1982) and expressed in g g^{-1} day^{-1}

$$\text{Relative growth rate} = \frac{\text{Log e } W_2 - \text{Log e } W_1}{t_2 - t_1}$$

W_1 and W_2 = Dry weight of the plant material at time t_1 and t_2 per unit ground area (g), respectively

t_1 and t_2 = Initial and final day of a period of observation respectively.

In bhendi, the incorporation of green leaf manure had significant effect on relative growth rate at the three phases (30-60 DAS, 60-90 DAS and 90-120 DAS) studied. The highest relative growth rate (0.0332 mg g^{-1} day^{-1}) was recorded in *Albizia lebbek* and the lowest (0.0319 mg $g^{-1}day^{-1}$) was recorded in *Delonix regia* during first phase, while at second phase, the highest relative growth rate (0.0171 mg g^{-1} day^{-1}) was recorded in *Delonix regia* and the lowest relative growth rate (0.0164 mg g^{-1} day^{-1}) was recorded in *Leucaena leucocephala*. The highest relative rate (0.0127 mg g^{-1} day^{-1}) was recorded in *Albizia lebbek* and the lowest (0.0118 mg g^{-1} day^{-1}) was recorded in *Leucaena leucocephala* during third stage. The application of leaf extract through seed treatment and foliar sprays had a significant effect on relative growth rate of Bhendi. Among the treatments, *Annona squamosa* recorded the highest relative growth rate (0.0332 mg g^{-1} day^{-1}) and the lowest (0.0322 mg g^{-1} day^{-1}) was recorded in *Ocimum sanctum* at first phase (30-60 DAS), while at second phase (60-90 DAS) the highest relative growth rate (0.0171 mg g^{-1} day^{-1}) was recorded in *Alangium salvifolium* and the lowest relative growth rate (0.0163 mg g^{-1} day^{-1}) was recorded in *Ocimum sanctum*. In the third phase (90-120 DAS) the highest relative rate (0.0126 mg g^{-1} day^{-1}) was recorded in *Ocimum sanctum* and the lowest (0.0120 mg g^{-1} day^{-1}) was recorded in *Alangium salvifolium*. The interaction effect was significant. The treatment combinations, *Albizia lebbek+ Annona squamosa* recorded the highest relative growth rate (0.0350 mg g^{-1} day^{-1}) and whereas, *Delonix regia +Alangium salvifolium* and *Gliricidia sepium + Alangium salvifolium* recorded the lowest (0.0313mgg^{-1}day^{-1}) during first phase (30-60 DAS), while at second-phase (60-90 DAS) the highest relative growth rate (0.0187 mg g^{-1} day^{-1}) was recorded in *Albizia lebbek+ Annona squamosa* and where as the lowest and similar (0.0160 mg g^{-1} day^{-1}) relative growth rate was recorded *Albizia lebbek + Ocimum sanctum, Delonix regia + Ocimum sanctum, Leucaena leucocephala + Aegles marmellos* and *Leucaena leucocephala + Morinda tinctoria* and these were found to be on par. The highest relative growth rate (0.0136 mg g^{-1} day^{-1}) were recorded in *Albizia lebbek+ Annona squamosa* and *Gliricidia sepium + Ocimum sanctum* and the lowest, *Leucaena leucocephala + Aegles marmellos* was recorded (0.0115 mg g^{-1} day^{-1}) during third phase. The absolute control recorded relative growth rate of 0.0123, 0.0151 and 0.0116 mg g^{-1}day^{-1} at first phase (30-60 DAS), second phase (60-90 DAS) and third phase (90-120 DAS) respectively. When it was compared with *Albizia lebbek+ Annona squamosa* showed an increased value of 0.0227, 0.0036 and 0.0020 mg g^{-1}day^{-1}. The recommended dose of inorganic fertilizers recorded relative growth rate of 0.0368, 0.0191 and 0.0138 mg g^{-1} day^{-1} When it was compared with *Albizia lebbek+ Annona squamosa* showed a decreased relative growth rate of 0.0018, 0.0004 and 0.0002 mg g^{-1} day^{-1}.

In onion, the incorporation of green leaf manure had significant effect on relative growth rate of aggregaum onion at both the phase (30-60 DAS and 60-90 DAS). The highest relative growth rate (0.0333 mg g^{-1} day^{-1}) was recorded in *Gliricidia sepium* and the lowest relative growth rate (0.0287 mg g^{-1}day^{-1}) was recorded in *Senna siamea* during first phase, while at second phase the highest relative growth rate (0.0287 mg g^{-1} day^{-1}) was recorded in *Gliricidia sepium* and the lowest relative rate growth (0.0263 mg g^{-1} day^{-1}) was recorded in *Leucaena leucocephala*. The application of leaf extract through seed treatment and foliar sprays had a significant effect on relative growth rate of aggregatum onion. Among the treatments, *Aegle marmelos* recorded the highest relative growth rate (0.0333 mg g^{-1} day^{-1}) and the lowest relative growth rate (0.0291mg g^{-1} day^{-1}) was recorded in *Annona squamosa* at first phase (30-60 DAS), while at second phase (60-90 DAS) the highest relative growth rate (0.0289 mg g^{-1} day^{-1}) was recorded in *Aegle marmelos* and *Alangium salvifolium* recorded the lowest relative growth rate (0.0265 mg g^{-1} day^{-1}). The interaction effect was significant. The treatment *Gliricidia sepium* + *Aegle marmelos* recorded the highest relative growth rate (0.0384 mg g^{-1} day^{-1}), which was on par with *Leucaena leucocephala* + *Morinda tinctoria* recorded (0.0357mg g^{-1} day^{-1}), during first phase (30-60 DAS), while at second-phase (60-90 DAS) the highest relative growth rate (0.0314 mg g^{-1} day^{-1}) was recorded in *Gliricidia sepium* + *Aegle marmelos*, which was on par with *Senna siamea* + *Aegle marmelos* which recorded (0.0310 mg g^{-1} day^{-1}). The absolute control recorded relative growth rate of 0.0230 and 0.0123 mg $g^{-1}day^{-1}$ at first phase (30-60 DAS) and second phase (60-90 DAS). When compared to Control, *Gliricidia sepium* + *Aegle marmelos* showed increased values for crop growth rate, 0.015 and 0.0191 mg $g^{-1}day^{-1}$, respectively. The recommended dose of inorganic fertilizers recorded relative growth rates of 0.0398 and 0.03105 mg g^{-1} day^{-1} during the two respective phases, when compared to this, *Gliricidia sepium* + *Aegle marmelos* showed decreased values of 0.0014 and 0.0003 mg g^{-1} day^{-1} in aggregatum onion.

Leaf Chlorophyll Content

The chlorophyll contents *viz.*, Chlorophyll *a*, *b* and total chlorophyll content in the fresh leaves were determined at each treatment using the method advocated by Arnon (1949) using UV – VIS Spectrophotometer.

In barnyard millet, the incorporation of green leaf manures did not have significant effect on leaf chlorophyll content at 60 DAS (flowering stage). However, *Peltophorum ferrugineum* recorded higher leaf chlorophyll content of 25.72. The same trend was followed at harvest stage also. The higher leaf chlorophyll content (23.31) was recorded by *Peltophorum ferrugineum* at 90 DAS. The foliar spraying of leaf extract did not have any significant effect on leaf chlorophyll content of barnyard millet at 60 and 90 DAS. However, *Annona squamosa* recorded higher leaf chlorophyll content of 25.83 at 60 DAS and 23.20 at 90 DAS. The interaction effect was not significant at 60 and 90 DAS. However, the higher leaf chlorophyll content in the combination *Peltophorum ferrugineum* + *Annona squamosa* (26.10 and 23.74) was recorded at 60 and 90 DAS, respectively. The absolute control, zero input recorded the leaf chlorophyll content of 22.5 and 18.31 at 60 DAS and 90 DAS respectively. The recommended dose of inorganic fertilizers recorded leaf chlorophyll content of 26.36 and 25.6 at 60 DAS and 90 DAS, respectively.

In finger millet (ragi), the incorporation of green leaf manures did not have significant effect on leaf chlorophyll content at 60 DAS. However, *Moringa oleifera* recorded numerically higher leaf chlorophyll content of 27.81, while higher leaf chlorophyll content (24.23) was recorded by *Albizia lebbek* at 90 DAS. The foliar spraying of leaf extract did not have any significant effect on leaf chlorophyll content of finger millet at 60 and 90 DAS. However, *Aegle marmellos* recorded numerically higher leaf chlorophyll content of 27.86 at 60 DAS and *Moringa oleifera* registered maximum chlorophyll content with the value of 24.16 at 90 DAS. The absolute control, zero input recorded the leaf chlorophyll content of 26.13 and 22.99 at 60 DAS and 90 DAS respectively. The recommended dose of inorganic fertilizers recorded leaf chlorophyll content of 29.15 and 25.89 at 60 DAS and 90 DAS, respectively.

In maize, the incorporation of green leaf manures had significant effect on leaf chlorophyll content at 80 DAS. *Albizia lebbek* recorded higher leaf chlorophyll content of 30.40. It was followed by *Peltophorum ferrugineum* which recorded the leaf chlorophyll content of 27.74 and minimum leaf chlorophyll content was recorded in *Leucaena leucocephala* (23.77). The foliar spraying of leaf extract did not have significant effect on leaf chlorophyll content of maize. *Annona squamosa* recorded higher leaf chlorophyll content of 28.79 and it was followed by *Alangium salvifolium*, which recorded the leaf chlorophyll content of 28.09 and minimum leaf chlorophyll content was recorded in *Aegle marmellos* (24.75). The interaction effect was found to be significant. The combination *Albizia lebbek* + *Annona squamosa* recorded the highest leaf chlorophyll content of (32.10). It was followed by *Albizia lebbek* + *Morinda tinctoria* which recorded leaf chlorophyll content of 31.42 and the lowest leaf chlorophyll content of 13.83 was recorded in *Leucaena leucocephala* + *Aegle marmellos* at 80 DAS of the crop. The absolute control zero input recorded the leaf chlorophyll content of 20.20 and the recommended dose of inorganic fertilizers recorded the leaf chlorophyll content of 45.70 in maize at 80 DAS of the crop. When compared to control *Albizia lebbek* + *Annona squamosa* showed increased leaf chlorophyll content (11.90).

In black gram, the incorporation of green leaf manures did not have significant effect on leaf chlorophyll content. However *Pongamia pinnata* recorded higher leaf chlorophyll content of 39.63. It was followed by *Albizia lebbek*, recorded the leaf chlorophyll content of 39.38 and minimum leaf chlorophyll content was recorded by *Delonix regia* (38.13). The foliar spraying of leaf extract did not have significant effect on leaf chlorophyll content of black gram. *Moringa oleifera* recorded higher leaf chlorophyll content of 39.24. It was followed by *Annona squamosa* which recorded the leaf chlorophyll content of 39.00 and minimum leaf chlorophyll content was recorded in *Morinda tinctoria* (38.58). The interaction effect was found to be significant. The combination *Albizia lebbek* + *Moringa oleifera* recorded the highest leaf chlorophyll content of (41.10). It was followed by *Pongamia pinnata* + *Moringa oleifera* which recorded the leaf chlorophyll content of 40.70 and the lowest leaf chlorophyll content of 36.50 was recorded in *Gliricidia sepium* + *Moringa oleifera*. The absolute control zero input recorded the leaf chlorophyll content of 36.50. The recommended dose of inorganic fertilizers recorded the leaf chlorophyll content of 42.30 in black gram. When compared to control *Albizia lebbek* + *Moringa oleifera* showed increased leaf chlorophyll content (1.86).

Leaf Area Index

The Leaf Area Index (LAI) for any crop can be calculated using the formula given by Palanisamy *et al.* (1978).

$$\text{LAI} = \frac{\text{L x W x N x K}}{\text{Spacing adopted (cm)}}$$

where,

L = Length of the leaf in cm

B = Width of the leaf in cm

N = Number of leaves $plant^{-1}$

K = Constant factor

In barnyard millet, the incorporation of green leaf manure had significant effect on leaf area index (LAI) at 60 DAS. The maximum LAI of 4.69 was recorded in *Peltophorum ferrugineum* and lower LAI of 3.72 was recorded in *Glyricidia sepium*. All the other treatments are on par with each other. Foliar spraying of leaf extract had significant effect on LAI. The maximum LAI of 4.77 recorded by *Annona squamosa* at 60 DAS, it was followed by *Moringa oleifera* with LAI of 4.38 and the lower LAI of 3.52 was recorded *Morinda tinctoria*. The interaction of treatments significantly influenced LAI at 60 DAS. The highest LAI of 5.30 was recorded by *Peltophorum ferrugineum* + *Annona squamosa* and it was on par with *Delonix regia* + *Annona squamosa* with 5.25 LAI and the lowest LAI 3.03 was recorded by *Gliricidia sepium* + *Morinda tinctoria* at 60 DAS. The absolute control recorded LAI of 2.86 at 60 DAS. When compared to control, *Peltophorum ferrugineum* + *Annona squamosa* showed an increased LAI of 2.44. While the recommended dose of inorganic fertilizers recorded LAI of 5.53 at 60 DAS. On comparison, *Peltophorum ferrugineum* + *Annona squamosa* showed decreased LAI of 0.23 at 60 DAS, over recommended dose of inorganic fertilizers

In finger millet (ragi), the incorporation of green leaf manures had significant effect on leaf area index (LAI) at 60 and 90 DAS. *Delonix regia* registered higher LAI of 3.18, which was on par with *Leucaena leucocephala* at 60 days. At 90 days, *Delonix regia* again registered highest LAI (4.91) and it was followed by *Leucaena leucocephala* and *Albizia lebbek* and both were on par with each other. The minimum value of LAI was recorded in *Peltophorum ferrugineum* with 1.99, 2.95 at 60, 90 DAS. The foliar spraying of leaf extract had a significant effect on LAI in finger millet. Among the treatments *Aegle marmellos* recorded higher LAI of 2.72. It was on par with *Moringa oleifera* and *Morinda tinctoria* on 60 DAS. At 90 DAS higher LAI of 4.24 was recorded in *Aegle marmellos*, which was on par with *Moringa oleifera* with a value of 3.91. The lower LAI of 2.22 and 3.53 at 60 and 90 DAS, respectively were observed with *Annona squamosa*. The interaction effect was significant and the treatment combination *Delonix regia* + *Moringa oleifera* recorded highest LAI of 4.01 and 5.68 at 60 and 90 DAS, respectively. The lower value of LAI was recorded in *Peltophorum ferrugineum* + *Morinda tinctoria* with LAI of 1.54 at 60 DAS. At 90 DAS, the lower value of LAI was recorded in *Gliricidia sepium* + *Annona squamosa* with LAI of 2.47, which was on par with *Peltophorum ferrugineum* + *Moringa oleifera*. The absolute control zero input recorded the LAI was 2.06 and 3.54 at 60 and 90 DAS, respectively. When compared to zero input, *Delonix regia* + *Moringa oleifera* showed an increased LAI of 1.95 and 2.14. While the recommended dose of inorganic fertilizers recorded the LAI of 4.00 and 4.80 at 60 and 90 DAS, resepctively.

In maize, the incorporation of green leaf manures had significant effect on leaf area index (LAI) of maize at 45, 60 and 90 DAS. *Albizia lebbek* registered higher LAI of 4.78, 3.51 and 5.19 which was followed by *Peltophorum ferrugineum* which recorded the LAI of 4.55, 3.24 and 4.90 at 45, 60 and 90 DAS, respectively. The minimum value

Table 5.1: Influence of Vrikshyurveda Farming on Harvest Index of Barnyard Millet

Treatments	*Moringa*	*Aegle*	*Morinda*	*Annona*	*Mean*
Albizia	0.25	0.24	0.22	0.27	**0.24**
Gliricidia	0.24	0.22	0.22	0.26	**0.24**
Pongamia	0.26	0.23	0.21	0.28	**0.24**
Delonix	0.29	0.26	0.25	0.31	**0.28**
Peltophorum	0.31	0.32	0.31	0.33	**0.32**
Mean	**0.27**	**0.25**	**0.24**	**0.29**	
	M	***S***	***M at S***	***S at M***	
SEd	**0.03**	**0.05**	**0.07**	**0.08**	
CD (P≤0.05)	NS	NS	NS	NS	
Zero input	0.24				
Fertiliser application	0.35				

Table 5.2: Influence of Vrikshyurveda Farming on Dry Matter Production (DMP) at Harvest in Finger Millet

Treatments	*Leaf Extract Spray*				
GLM Incorporation	*Moringa*	*Morinda*	*Annona*	*Aegle*	*Mean*
Albizia	3242	3145	3369	3502	**3314**
Gliricidia	2455	3563	2492	3114	**2906**
Leucaena	3870	3482	3171	4248	**3693**
Delonix	5030	4546	4361	4723	**4665**
Peltophorum	2237	2079	2656	3499	**2618**
Mean	**3367**	**3363**	**3210**	**3817**	
	SEd		***CD (P≤0.05)***		
M	175.15		486.31		
S	73.89		157.50		
M at S	226.17		569.57		
S at M	165.23		352.18		
Fertiliser application		**4599**			
Zero input		**1985**			

of LAI was recorded in *Leucaena leucocephala* with 2.78, 4.21 at 45, 90 DAS, respectively and the least value of LAI was recorded at 60 DAS by *Gliricidia sepium* 2.58.

The foliar spraying of leaf extract had a significant effect on LAI in maize. Among the treatments *Annona squamosa* recorded higher LAI of 4.76. It was followed by *Morinda tinctoria* with a value of 3.83 on 45 DAS. At 60 DAS, higher LAI of 3.12 was recorded in *Annona squamosa*, which was followed by *Alangium salvifolium* with a

Table 5.3: Influence of Vrikshyurveda Farming on Leaf Area Index of Finger Millet

Treatments	*60 Days*					*90 Days*				
	Leaf Extract Spray					*Leaf Extract Spray*				
GLM Incorporation	*Moringa*	*Morinda*	*Annona*	*Aegle*	*Mean*	*Moringa*	*Morinda*	*Annona*	*Aegle*	*Mean*
Albizia	2.11	2.47	2.23	2.29	**2.27**	4.18	3.99	4.79	3.33	**4.07**
Gliricidia	2.37	2.31	1.68	2.61	**2.24**	3.19	3.61	2.47	3.85	**3.28**
Leucaena	3.00	3.19	2.50	3.08	**2.94**	4.05	4.03	3.52	5.11	**4.17**
Delonix	4.01	3.10	2.39	3.23	**3.18**	5.68	4.61	3.71	5.65	**4.91**
Peltophorum	1.72	1.54	2.30	2.40	**1.99**	2.48	2.93	3.16	3.25	**2.95**
Mean	**2.64**	**2.51**	**2.22**	**2.72**		**3.91**	**3.83**	**3.53**	**4.24**	
	SEd		***CD (P≤0.05)***			***SEd***		***CD (P≤0.05)***		
M	0.17		0.47			0.22		0.61		
S	0.18		0.37			0.23		0.49		
M at S	0.38		0.86			0.49		1.12		
S at M	0.39		0.84			0.51		1.10		
Fertiliser application		**4.00**					**4.80**			
Zero input		**2.06**					**3.54**			

Table 5.4: Influence of Vrikshyurveda Farming on Leaf Chlorophyll Content (SPAD 502 Readings) of Finger Millet

Treatments	*Leaf Chlorophyll Content*									
	60 DAS					*90 DAS*				
	Leaf Extract Spray					*Leaf Extract Spray*				
GLM Incorporation	*Moringa*	*Morinda*	*Annona*	*Aegle*	*Mean*	*Moringa*	*Morinda*	*Annona*	*Aegle*	*Mean*
Albizia	27.84	27.40	27.42	27.61	**27.57**	24.75	23.72	23.84	24.62	**24.23**
Gliricidia	27.32	27.35	27.30	27.84	**27.45**	24.70	22.37	23.28	24.70	**23.76**
Leucaena	27.63	27.56	27.76	27.88	**27.71**	24.15	22.36	24.77	22.30	**23.40**
Delonix	27.95	27.89	27.50	27.89	**27.81**	24.83	22.21	23.19	25.49	**23.93**
Peltophorum	27.47	27.55	27.34	28.10	**27.61**	22.35	22.07	24.54	23.26	**23.06**
Mean	**27.74**	**27.55**	**27.46**	**27.86**		**24.16**	**22.55**	**23.92**	**24.07**	
	SEd		***CD (0.05)***			***SEd***		***CD (0.05)***		
M	0.56		NS			0.75		NS		
S	0.40		NS			0.65		NS		
M at S	0.96		NS			1.47		NS		
S at M	0.90		NS			1.45		NS		
Fertiliser application		**29.15**					**25.89**			
Zero input		**26.13**					**22.99**			

Table 5.5: Influence of Vrikshyurveda Farming on Harvest Index of Finger Millet

Treatments	*Harvest Index*				
	Leaf Extract Spray				
GLM Incorporation	*Moringa*	*Morinda*	*Annona*	*Aegle*	*Mean*
Albizia	0.127	0.092	0.098	0.115	**0.108**
Gliricidia	0.114	0.058	0.074	0.067	**0.078**
Leucaena	0.089	0.082	0.097	0.085	**0.088**
Delonix	0.165	0.065	0.111	0.129	**0.118**
Peltophorum	0.132	0.090	0.078	0.063	**0.091**
Mean	**0.125**	**0.077**	**0.092**	**0.092**	

	SEd	*CD (0.05)*
M	0.009	0.025
S	0.010	0.021
M at S	0.021	0.048
S at M	0.022	0.048
Fertiliser application	**0.62**	
Zero input	**0.04**	

Table 5.6: Influence of Vrikshyurveda Farming on Leaf Chlorophyll Content (SPAD 502 Readings) of Maize

GLM Application	*Leaf Chlorophyll Content*				
	Alangium	*Aegle*	*Morinda*	*Annona*	*Mean*
Albizia	30.46	27.64	31.42	32.10	**30.40**
Gliricidia	26.07	26.73	24.40	27.26	**26.11**
Leucaena	27.19	13.83	25.26	28.83	**23.77**
Delonix	29.68	27.49	21.44	29.13	**26.93**
Peltophorum	28.95	28.09	29.15	24.80	**27.74**
MEAN	**28.47**	**24.75**	**26.33**	**28.41**	
	M	***S***	***M at S***	***S at M***	
SE.d	1.60	1.56	3.43	3.50	
CD (0.05)	3.25	NS	6.83	6.93	
Fertiliser application					45.7
Zero input					20.2

value of 3.08. At 90 DAS, *Annona squamosa* recorded higher LAI of 5.15 and it was followed by *Morinda tinctoria* with 4.77. The lower LAI of 3.52, 2.63 and 4.30 at 45, 60 and 90 DAS, respectively were observed with *Aegle marmellos*.

Table 5.7: Influence of Vrikshyurveda Farming on Leaf Area Index (LAI) of Maize

GLM Application	*60 DAS*					*90 DAS*				
	Alangium	*Aegle*	*Morinda*	*Annona*	*Mean*	*Alangium*	*Aegle*	*Morinda*	*Annona*	*Mean*
Albizia	3.66	3.14	3.35	3.88	**3.51**	5.87	4.23	4.15	6.50	**5.19**
Gliricidia	2.66	2.50	2.54	2.61	**2.58**	3.56	5.13	4.74	4.82	**4.56**
Leucaena	2.60	2.21	2.78	2.80	**2.59**	3.62	4.22	4.64	4.34	**4.21**
Delonix	2.26	2.90	2.90	2.99	**2.76**	4.44	3.91	5.51	4.88	**4.69**
Peltophorum	4.19	2.43	3.02	3.33	**3.24**	5.59	4.00	4.81	5.21	**4.90**
Mean	**3.08**	**2.63**	**2.92**	**3.12**		**4.62**	**4.30**	**4.77**	**5.15**	
	M	***S***	***M at S***	***S at M***		***M***	***S***	***M at S***	***S at M***	
SEd	0.18	0.14	0.34	0.33		0.34	0.26	0.61	0.58	
CD(P≤0.05)	0.37	0.29	0.68	0.66		0.70	0.51	1.21	1.15	
Fertiliser application					3.78					5.51
Zero input					2.60					4.56

Table 5.8: Influence of Vrikshyurveda Farming on Dry Matter Production (Kg ha^{-1}) of Maize

GLM Application	*60 DAS*					*At Harvest Stage*				
	Alangium	*Aegle*	*Morinda*	*Annona*	*Mean*	*Alangium*	*Aegle*	*Morinda*	*Annona*	*Mean*
Albizia	4688	4567	5373	5673	**5074**	6697	6525	7676	8104	**7251**
Gliricidia	3742	4743	4719	4231	**4359**	5347	6775	6741	6045	**6227**
Leucaena	3405	4000	4184	4039	**3907**	4865	5714	5977	5771	**5582**
Delonix	5027	3445	4193	5106	**4443**	7182	4922	5990	7295	**6347**
Peltophorum	4742	3978	4630	4941	**4573**	6774	5683	6614	7059	**6533**
Mean	**4321**	**4147**	**4679**	**4738**		**6173**	**5924**	**6600**	**6855**	
	M	***S***	***M at S***	***S at M***		***M***	***S***	***M at S***	***S at M***	
SEd	89	50	132	112		127	72	188	161	
CD(P≤0.05)	179	107	320			255	153	458	343	
Fertiliser application	8573					11532				
Zero input	678					835				

Table 5.9: Influence of Vrikshyurveda Farming on Harvest Index of Maize

GLM Application	*Harvest Index*				
	Alangium	*Aegle*	*Morinda*	*Annona*	*Mean*
Albizia	0.220	0.247	0.234	0.258	**0.240**
Gliricidia	0.197	0.267	0.232	0.210	**0.227**
Leucaena	0.185	0.185	0.151	0.176	**0.174**
Delonix	0.220	0.220	0.200	0.238	0.219
Peltophorum	0.244	0.233	0.234	0.268	**0.245**
Mean	**0.213**	**0.230**	**0.210**	**0.230**	
	M	***S***	***M at S***	***S at M***	
SEd	0.015	0.013	0.030	0.030	
CD (P≤0.05)	0.043	NS	NS	NS	
Fertiliser application					0.302
Zero input (1)					4.168

Table 5.10: Influence of Vrikshyurveda Farming on Leaf Chlorophyll (SPAD 502 Readings) of Black Gram

Treatments	*Foliar Applications*				
	Moringa	*Aegle*	*Morinda*	*Annona*	*Mean*
Basal applications (GLM)					
Albizia	41.10	38.10	37.90	40.40	**39.38**
Gliricidia	36.50	40.40	39.30	40.00	**39.05**
Pongamia	40.70	39.30	39.00	39.50	**39.63**
Delonix	40.60	38.20	37.90	35.80	**38.13**
Peltophorum	37.30	38.00	38.80	39.30	**38.35**
Mean	**39.24**	**38.80**	**38.58**	**39.00**	
	M	***S***	***M at S***	***S at M***	
SEd	0.40	0.50	1.05	1.12	
CD(P≤0.05)	NS	NS	2.33	2.38	
Fertiliser application					42.3
Zero input					36.5

The interaction effect was significant and the treatment combination *Albizia lebbek* + *Annona squamosa* was most effective in increasing LAI by 6.66. It was followed by *Peltophorum ferrugineum* + *Annona squamosa* which recorded the LAI of 5.64 and the lower value of LAI was recorded in *Leucaena leucocephala* + *Alangium salvifolium* with LAI of 2.17 at 45 DAS. At 60 DAS, the interaction effect was significant and the

Table 5.11: Influence of Vrikshyurveda Farming on Leaf Area Index of Black Gram

Treatments	*Leaf Area Index*									
	Foliar Applications									
	Vegetative Stage					*Flowering Stage*				
	Moringa	*Aegle*	*Morinda*	*Annona*	*Mean*	*Moringa*	*Aegle*	*Morinda*	*Annona*	*Mean*
Basal applications (GLM)										
Albizia	0.336	0.505	0.535	0.673	**0.512**	1.483	2.185	1.253	2.196	**1.779**
Gliricidia	0.596	0.448	0.352	0.488	**0.471**	2.330	1.418	1.224	0.729	**1.425**
Pongamia	0.964	0.660	0.582	0.539	**0.686**	2.934	1.333	2.174	2.598	**2.260**
Delonix	0.414	0.543	0.318	0.475	**0.437**	1.566	1.198	0.883	1.352	**1.250**
Peltophorum	0.358	0.518	0.405	0.488	**0.442**	1.366	2.243	1.530	1.096	**1.559**
MEAN	**0.533**	**0.535**	**0.438**	**0.533**		**1.936**	**1.675**	**1.413**	**1.594**	
	M	***S***	***M at S***	***S at M***		***M***	***S***	***M at S***	***S at M***	
SEd	0.04	0.03	0.07	0.07		0.20	0.15	0.36	0.35	
CD(P≤0.05)	0.10	0.06	0.16	0.14		0.56	0.34	0.85	0.75	
Fertiliser application					0.84					2.99
Zero input					0.37					1.28

Table 5.12: Influence of Vrikshyurveda Farming on Dry Matter Production (kg ha^{-1}) of Black Gram

Treatments	*Dry Matter Production*									
	Foliar Applications									
	Flowering Stage					*At Harvest Stage*				
	Moringa	*Aegle*	*Morinda*	*Annona*	*Mean*	*Moringa*	*Aegle*	*Morinda*	*Annona*	*Mean*
Basal applications (GLM)										
Albizia	628	321	606	679	**559**	1028	1022	1084	1222	**1089**
Gliricidia	489	404	272	374	**385**	702	726	704	719	**713**
Pongamia	764	639	630	738	**693**	1300	1086	1149	1170	**1176**
Delonix	457	485	296	299	**384**	922	719	728	674	**761**
Peltophorum	501	460	503	355	**455**	924	817	831	1046	**905**
Mean	**568**	**462**	**462**	**489**		**975.70**	**874**	**899**	**966**	
	M	***S***	***M at S***	***S at M***		***M***	***S***	***M at S***	***S at M***	
SEd	0.3	0.5	1.1	1.2		0.32	0.55	1.10	1.22	
CD(P≤0.05)	0.9	1.1	2.4	2.6		0.89	1.17	2.42	2.61	
Fertiliser application					865					1764
Zero input					165					494

Table 5.13: Influence of *Vrikshayurveda* Treatments on Harvest Index of Black Gram

Treatments	*Harvest Index*				
	Foliar Applications				
	Moringa	*Aegle*	*Morinda*	*Annona*	*Mean*
Basal applications (GLM)					
Albizia	0.430	0.398	0.411	0.419	**0.414**
Gliricidia	0.411	0.402	0.408	0.413	**0.408**
Pongamia	0.450	0.410	0.403	0.440	**0.426**
Delonix	0.397	0.398	0.398	0.402	**0.399**
Peltophorum	0.416	0.421	0.397	0.411	**0.411**
Mean	**0.421**	**0.406**	**0.403**	**0.417**	
	M	***S***	***M at S***	***S at M***	
SEd	0.004	0.008	0.016	0.018	
CD (0.05)	0.012	NS	NS	NS	
Fertiliser application	0.483				
Zero input	0.398				

treatment combination *Peltophorum ferrugineum* + *Alangium salvifolium* was most effective in increasing LAI by 4.19. It was followed by *Albizia lebbek* + *Annona squamosa* which recorded the LAI of 3.88. The lower value of LAI was recorded in *Leucaena leucocephala* + *Aegle marmellos* with LAI of 2.21 at 60 DAS.

At 90 DAS, also the interaction effect was significant and the treatment combination *Albizia lebbek* + *Annona squamosa* was most effective in increasing LAI by 6.50. It was followed by *Albizia lebbek* + *Alangium salvifolium* which recorded the LAI of 5.87 at 90 DAS. The absolute control zero input recorded the LAI was 2.31, 2.60 and 4.56 at 45, 60 and 90 DAS, respectively. When compared to zero input, *Albizia lebbek* + *Annona squamosa* showed an increased LAI of 2.47, 0.91 and 0.63. While the recommended dose of inorganic fertilizers recorded the LAI of 3.81, 3.78 and 5.51 at 45, 60 and 90 DAS. On comparison *Albizia lebbek* + *Annona squamosa* showed an increased LAI of 0.97 at 45 DAS, decreased LAI of 0.27 and 0.32 at 60 and 90 DAS, over recommended dose of fertilizers.

In black gram, the incorporation of green leaf manures had significant effect on leaf area index (LAI) at 35 and 50 DAS. However *Pongamia pinnata* registered higher LAI of 0.69 and 2.26 which was followed by *Albizia lebbek* which recorded the LAI of 0.51 and 1.78 at 35 and 50 DAS, respectively. The minimum value of LAI was recorded in *Delonix regia* with 0.43 and 1.25 at 35 and 50 DAS, respectively. The foliar spraying of leaf extract had a significant effect on LAI in black gram. Among the treatments *Aegle marmellos* recorded higher LAI of 0.54. It was followed by *Moringa oleifera* with a value of 0.53 on 35 DAS. At 50 DAS, higher LAI of 1.93 was recorded in *Moringa oleifera*, which was followed by *Annona squamosa* with a value of 1.59. The lower LAI

Table 5.14: Influence of Vrikshyurveda Farming on Dry Matter Production (kg ha^{-1}) in Aggregatum Onion cv. CO 5

GLM Incorporation	At 60 DAS					At 90 DAS				
	Alangium	*Annona*	*Aegles*	*Morinda*	*Mean*	*Alangium*	*Annona*	*Aegles*	*Morinda*	*Mean*
Albizia	1050	1160	1130	1156	1124	2270	2940	2550	2763	1769
Senna	1090	1136	1126	1050	1100	2473	2836	2610	2270	1799
Gliricidia	1113	1140	1300	1203	1189	2573	2590	3173	2816	1910
Leucaena	1200	1030	1203	1276	1177	2863	2163	2816	2976	1853
Delonix	1036	1113	1126	1140	1104	2126	2506	2610	2590	1719
Mean	**1098**	**1116**	**1177**	**1165**	**1139**	**2461**	**2607**	**2752**	**2683**	**1810**
	M	***S***	***M X S***			***M***	***S***	***M X S***		
SEd	25.61	22.53	50.59			58.55	65.28	139.33		
CD (P≤0.05)	59.07	46.01	106.75			135.01	133.33	291.04		
Zero input					980					1389
Fertiliser application					1440					3223

Table 5.15: Influence of Vrikshyurveda Farming on Relative Growth Rate (mg^{-1} g^{-1} day^{-1}) in Aggregatum Onion cv. CO 5

GLM Incorporation	*I–Stage*					*II–Stage*				
	Alangium	*Annona*	*Aegles*	*Morinda*	*Mean*	*Alangium*	*Annona*	*Aegles*	*Morinda*	*Mean*
Albizia	0.0330	0.0355	0.0309	0.0301	0.0324	0.0257	0.0309	0.0291	0.0287	0.0286
Senna	0.0283	0.0256	0.0277	0.0331	0.0287	0.0273	0.0305	0.0310	0.0257	0.0287
Gliricidia	0.0351	0.0289	0.0384	0.0309	0.0333	0.0303	0.0274	0.0314	0.0258	0.0287
Leucaena	0.0321	0.0271	0.0345	0.0357	0.0324	0.0271	0.0247	0.0250	0.0282	0.0263
Delonix	0.0273	0.0283	0.0349	0.0303	0.0320	0.0240	0.0271	0.0280	0.0271	0.0265
Mean	**0.0312**	**0.0291**	**0.0333**	**0.0320**	**0.0314**	**0.0265**	**0.0281**	**0.0289**	**0.0271**	**0.0277**
	M	***S***	***M X S***			***M***	***S***	***M X S***		
SEd	0.00073	0.00072	0.00157			0.00049	0.00062	0.00129		
CD (p≤0.05)	0.00168	0.00147	0.00330			0.00113	0.00126	0.00269		
Zero input					0.0230					0.0123
Fertiliser application					0.0398					0.0317

Table 5.16: Influence of Vrikshyurveda Farming on Crop Growth Rate (g^{-1} m^{-1} day^{-1}) in Aggregatum Onion cv. CO 5

GLM Incorporation	*I–Stage*					*II–Stage*				
	Alangium	*Annona*	*Aegles*	*Morinda*	*Mean*	*Alangium*	*Annona*	*Aegles*	*Morinda*	*Mean*
Albizia	62.920	72.450	64.090	75.340	68.700	116.190	169.530	135.230	150.330	142.820
Senna	59.490	58.120	67.640	56.540	60.448	131.790	161.900	141.260	116.190	137.785
Gliricidia	69.080	63.020	77.570	69.680	69.838	157.493	138.090	187.610	144.440	156.908
Leucaena	69.690	54. 630	75.430	76.250	69.000	158.880	107.960	127.700	161.900	139.110
Delonix	55.270	60.750	64.140	70.550	62.678	103.800	132.690	141.260	136.020	128.443
Mean	**63.290**	**61.794**	**69.774**	**69.672**	**66.133**	**133.631**	**142.034**	**146.612**	**141.776**	**141.013**
	M	***S***	***M X S***			***M***	***S***	***M X S***		
SEd	1.477	1.610	3.450			3.809	4.227	9.030		
CD (p≤0.05)	3.407	3.288	7.213			8.785	8.634	18.866		
Zero input					45.629					99.856
Fertiliser application					79.925					175.826

Table 5.17: Influence of Vrikshyurveda Farming on Shoot to Bulb Ratio and Harvest Index in Aggregatum Onion cv. CO 5

GLM Incorporation	*Shoot to Bulb Ratio*					*Harvest Index*				
	Alangium	*Annona*	*Aegles*	*Morinda*	*Mean*	*Alangium*	*Annona*	*Aegles*	*Morinda*	*Mean*
Albizia	0.78	0.67	0.79	0.83	0.76	0.56	0.60	0.54	0.56	0.56
Senna	0.64	0.64	0.73	0.57	0.72	0.61	0.61	0.53	0.58	0.58
Gliricidia	0.55									
		0.57								
			0.89							
				0.69	0.67	0.65	0.64	0.66	0.64	0.65
Leucaena	0.77	0.77	0.73	0.69	0.74	0.59	0.52	0.58	0.56	0.56
Delonix	0.48	0.53	0.55	0.56	0.53	0.59	0.62	0.65	0.65	0.64
Mean	**0.64**	**0.63**	**0.73**	**0.66**	**0.66**	**0.60**	**0.61**	**0.59**	**0.56**	**0.60**
	M	***S***	***M X S***			***M***	***S***	***M X S***		
SEd	0.02	0.01	0.04			0.02	0.010	0.028		
CD (p≤0.05)	0.05	0.03	0.08			0.046	0.021	0.062		
Zero input					0.15					0.36

Table 5.18: Influence of *Vrikshayurveda* Farming on Dry Matter Production (kg ha^{-1}) in Bhendi cv. Arka Anamika

GLM Incorporation	*90*						*120*					
	S_1	S_2	S_3	S_4	S_5	*Mean*	S_1	S_2	S_3	S_4	S_5	*Mean*
Albizia	1623	1803	1615	1598	1566	1641	2323	2713	2411	2320	2247	2403
Delonix	1646	1625	1625	1590	1590	1615	2405	2326	2305	2360	2325	2344
Gliricidia	1559	1606	1783	1617	1606	1634	2225	2303	2621	2313	2414	2375
Leucaena	1619	1613	1613	1576	1577	1600	2293	2305	2280	2238	2275	2278
Mean	**1612**	**1662**	**1659**	**1595**	**1585**	**1622**	**2311**	**2411**	**2404**	**2307**	**2315**	**2350**
Zero input						810						1162
Fertiliser application						1975						2950
	M	***S***	***M xS***				***M***	***S***	***M xS***			
SEd	3.8	10.1	18.5				12.3	11.9	24.6			
CD(p≤0.05)	9.3	20.6	38.1				30.1	24.2	52.6			

of 0.44 and 1.41 at 35 and 50 DAS, respectively were observed with *Morinda tinctoria*. At 35 DAS, the interaction effect was significant and the treatment combination *Pongamia pinnata* + *Moringa oleifera* was most effective in increasing LAI by 0.96. It was followed by *Pongamia pinnata* + *Aegle marmellos* which recorded the LAI of 0.66 and the lower value of LAI was recorded in *Delonix regia* + *Morinda tinctoria* with LAI of 0.32. At 50 DAS the interaction effect was significant and the treatment combination *Pongamia pinnata* + *Moringa oleifera* was most effective in increasing LAI by 2.93. It was followed by *Pongamia pinnata* + *Annona squamosa* which recorded LAI of 2.60. The lower value of LAI was recorded in *Delonix regia* + *Morinda tinctoria* with LAI of 0.88 at 50 DAS. The absolute control (zero input) recorded LAI of 0.37 and 1.28 at 35 and 50 DAS. When compared to zero input, *Pongamia pinnata* + *Moringa oleifera* showed an increased LAI of 0.96 and 2.93, while the recommended dose of inorganic fertilizers recorded LAI of 0.84 and 2.99 at 35 and 50 DAS, respectively. On comparison, *Pongamia pinnata* + *Moringa oleifera* showed an increased LAI of 0.12 at 35 DAS over control and decreased LAI of 0.06 at 50 DAS, over recommended dose of fertilizers.

Discussion on Various Physiological Parameters

In barnyard millet, the maximum number of tillers was observed in the treatment combination *Pheltophorum ferrugineum* + *Annona squamosa*. The better nutrient availability might have resulted in better tillering. The number of tillers silmultaneously increased the LAI (5.3) and leaf chlorophyll content (23.74) at flowering stage. The leaf chlorophyll content was found to be highest in *Pheltophorum ferrugineum* + *Annona squamosa* at flowering stage. The relative increase in chlorophyll content due to supply of higher nutrients especially nitrogen and micronutrients to the growing tissues which lead to the synthesis of leaf chlorophyll. There were also evidences to indicate that coconut water spray increased the chlorophyll content of crops. The least values of chlorophyll content under the treatment by *Gliricidia sepium* + *Morinda tinctoria* was noticed due to less availability of nutrients and the low LAI as compared to other treatments. The growth regulators present in the *Morinda tinctoria* might have reduced the leaf chlorophyll content. The maximum dry matter production was observed under *Pheltophorum ferrugineum* + *Annona squamosa* at harvest stage. It was followed by *Delonix regia* + *Annona squamosa*. This might be due to faster decomposition of the green leaf manures, subsequent availability and better uptake of nutrients there by increased plant height, LAI as observed in this study, which could have utilized the photosynthesis process and increased dry matter production in barnyard millet. Harvest index was highest in the treatment combination *Pheltophorum ferrugenieum* and *Annona squamosa*. This might be due to the existence of better source-sink relationship and better photosynthates accumulation in the economic part of the plant.

In finger millet the LAI which determines the total assimilating area available for translocation to the sink showed positive response to green leaf manure application. The higher LAI was observed at 60 DAS in treatment containing combination of *Delonix regia* and *Moringa oleifera*. This might be due to availability of nutrients to the crop during different growth phases and could have improved the

LAI. Similar results were quoted by Patel *et al.* (2003) and also Shameem banu (2006) in cluster bean.

Moringa leaf extract is rich in zeatin, ascorbates, carotenoids, phenols, calcium and potassium. It might have plant growth capabilities, which delayed leaf area by decreasing the chlorophyll degeneration. *Moringa oleifera* leaf extract sprayed plants showed the maximum leaf area compared to control. This might be due to the presence of cytokinins in the *Moringa* leaf extract, which improved the leaf area under normal and mild stress conditions (Foilde *et al.*, 2001).

The transition of photosynthates from the leaves which is the source, to the other parts of the plant could be explained by the physiological changes in the crop. This physiological change is reflected through accumulation of dry matter production and harvest index.

The present study revealed that dry matter production was highest in the treatment with *Delonix regia* as green leaf manures and foliar spraying of *Moringa oleifera*. This might be due to faster decomposition of the green leaf manures and subsequent availability and better uptake of nutrients by crops increased plant height and number of leaves which could have utilized the photosynthesis process and thus increased dry matter production in the crop supported by Bheemaiah and Subramaniyam, 2003. Similar observation was recorded by Patel *et al.* (2003) in cluster bean. The crop growth rate is a function of dry matter production as reported by Watson (1958). This might be due to the production of higher number of leaves as the plants age and tallest plants which would have accumulated from the leaf.

In maize, the leaf chlorophyll content was found to be highest in *Albizia lebbek* + *Annona squamosa*. The relative increase in chlorophyll content due to supply of higher nutrients especially nitrogen and micronutrients to the growing tissues which lead to the synthesis of leaf chlorophyll. The least values of chlorophyll content under the treatment *Leucaena leucocephala* + *Aegle marmellos* was noticed, since low availability of nutrients and less LAI as compared to other treatments. Harvest index were highest in the treatment with *Peltophorum ferrugineum* as green leaf manure and foliar spray of *Annona squamosa*. This might be due to the existence of better source-sink relationship and better photosynthates accumulation in the economic part of the plant. The recommended dose of inorganic fertilizers recorded highest values of all physiological parameters. This might be due to nutrient availability and better uptake of nutrients and increased LAI and plant height with better photosynthesis activity and accumulation. As usual, the zero input recorded the lowest in all physiological parameters. This might be due to the non-availability of the nutrients from the soil to the plant.

In black gram, the maximum production of dry matter was observed under *Pongamia pinnata* + *Moringa oleifera* at flowering and harvest stage. It was followed by *Albizia lebbek* + *Annona squamosa*. This might be due to faster decomposition of the green leaf manures and subsequent availability and better uptake of nutrients, increased plant height, LAI which could have utilized the photosynthesis process and increased dry matter production in black gram.

The leaf chlorophyll content was found to be highest in *Albizia lebbek* + *Moringa oleifera* at flowering stage. The relative increase in chlorophyll content due to supply of higher nutrients especially nitrogen and micronutrients to the growing tissues which lead to the synthesis of leaf chlorophyll. The least values of chlorophyll content under the treatment *Delonix regia* + *Annona squamosa* was noticed, since low availability of nutrients and less LAI as compared to other treatments. The recommended dose of inorganic fertilizers recorded highest values for all physiological parameters.

Application of recommended dose of fertilizers for bhendi resulted in the highest values of physiological parameters than Vrikshyurveda farming. The increase in fruit size and weight specially with the application of fertilizers might be owing to the balanced and better mineral utilization by plants resulting in enhancement of photosynthesis, other metabolic activity and greater diversion of food material to the fruits that ultimately lead to increase in cell elongation and cell division which is also responsible for increase in dry matter production. Further better absorption of potassium would have helped in the translocation of photosynthates to the fruits.

Biomass yield of a crop is the result of availability of the nutrients during the crop growth period which in turn is governed by the nutrient contents and uptake by the crop. The transition of photosynthates from the leaves which is the source to the other parts of the plant could be explained all by the physiological changes in the crops. Growth of bhendi in terms of dry matter production, crop growth rate and relative growth rate was significantly influenced by the imposed treatments. The study with bhendi crop revealed that dry matter production and crop growth rate, were high in the treatment with *Albizia lebbek* as green leaf manures and seed soaking and foliar sprays of *Annona squamosa*. This might be due to the gradual release of nutrients from the decomposing green leaf manures and subsequent availability of nutrients and better uptake of nutrients which increased plant height, number of leaves ultimately resulting in increased dry matter production. In addition, the presence of secondary metabolites in the leaf extracts might have enhanced the vegetative growth like number of leaves and plant height, which would have also increased the dry matter production. The crop growth rate is a function of dry matter production and it showed an increasing trend with advancement of the crop. This might be attributed to the production of more number of leaves in the tallest plants which would have accumulated the photosynthates in the leaf.

The study on onion revealed that dry matter production and crop growth rate, were highest in the treatment with *Gliricidia sepium* as green leaf manures and bulb soaking and foliar sprays of *Aegles marmellos*. This might be due to faster decomposition of the green leaf manures and subsequent availability and better uptake of nutrients, increased plant height, number of leaves which could have utilized the photosynthesis process and increased dry matter production in the crop. The presence of secondary metabolites in the leaf extracts might have enhanced the vegetative growth *i.e.* number of leaves and plant height, which would have increased the dry matter production. The crop growth rate is a function of dry matter production and it showed on increasing trend with advancement of the crop.The production of higher number of leaves and tallest plants which would have accumulated from the leaf would have been responsible. The recommended dose of inorganic fertilizer recorded highest in all physiological parameters. This might be due to nutrient availability and better uptake of nutrients and increased number leaves and plant height with better photosynthesis activity and accumulation. On the contrary the absolute control recorded the lowest in all physiological parameters. This might be due to the non - availability of the nutrients from the soil to the plant.

Chapter 6

Allelopathy and Weed Management

Introduction

Green plants produce numerous secondary metabolites, many of which are capable of initiating chemical warfare among the neighboring plants growing in a community. These chemicals have been designated as allelo-chemicals and the process as allelopathy. Both, crop and weedy plants may possess such allelopathic compounds, but, unfortunately, the weed species often possess much higher levels of these than our present day crop varieties which have lost such metabolites during their breeding by man for high yield quality. This is now considered a very important mechanism available with most weeds to stunt the growth of crops. The allelopathic compounds may be released from plants into the soil as either root exudates or as decomposition products of their dead and worn-out tissues. Some weedy plants have also been found to release volatile allelopathic compounds from their foliage which prove unhealthy to the nearby crop plants. The allelopathic compounds studied so far are found to largely belong to two chemical groups *viz.*, the phenolic compounds and the Terpenoids, which comprise molecules like benzoic acids, cinnamic acids, phenolic acids, cumarins hydroquinones, benzoquinones, thiopenes, juglone and cineoles etc. Some authors prefer to call the harmful plant interactions of such chemicals as teletoxy, but the border term allelopathy is still widely accepted. The allelopathic effects of several weeds on specific crop plants have been demonstrated by way of preparing their hydroleachates and irrigating the test plants with these. A few, recent laboratory demonstrations of allelopathy includes: inhibition of germination of certain small grains with *Cyperus rotundus* tuber *leachates*; inhibition of growth of many agronomic crops with *Parthenium* sp. extracts reduction in the germination of cabbage and egg plants with *Amaranthus retroflexus* extracts and reduction in the growth of wheat plants with exudates of *Avena fatua*, *Phalaris minor* and *Chenopodium album*. Further, sometimes a daughter plant hydroleachates may exhibit teletoxy to its parent plant itself. This process is called autotoxy. This phenomenon is exhibited by *Parthenium hysterophorus*.

While one is concerned about the adverse allelopathic influence of weeds on the crop plants, there are several instances when such a phenomenon is observed amongst the weeds themselves, giving scientists opportunity to use it in Allelopathic control of certain weeds using specific botanicals. However, in this chapter our concern is with the deleterious effects of weed exudates on crop plants. But, despite numerous examples of laboratory demonstrations of such allelopathy, direct proff of the extent of existence of the phenomenon in field conditions is more difficult to obtain. It is so because under field conditions the secondary metabolites from weeds may be subjected in soil to biodegradation, adsorption, leaching and like phenomena; leaving little amount, if at all, for damaging meaningfully the crop growth. Moreover, such weed exudates are usually crop specific.

Allelopathy, in true sense, does not form any aspect of weed-crop competition, although it causes weed-crop interference. Therefore, the weed scientists now prefer to talk of weed-crop interference rather than weed-crop competition. The term weed-crop interference includes competition as well as the possible allelopathy.

More recently the term 'allelopathy' has been redefined by the International Allelopathy Society, making it more accommodative than before. It now includes the secondary metabolites produced by both living and non-living tissues of green plants, micro organism, virus and fungi that influence the growth and development of biological system. Also, now the promotive effects of such metabolites on plants have been suggested for inclusion in the subject of allelopathy (Kathiresan Rm, 2002).

Nature of Allelopathy

Commonly cited effects of allelopathy include reduced seed germination and seedling growth. Like synthetic herbicides, there is no common mode of action or physiological target site for all allelochemicals. However, known sites of action for some allelochemicals include cell division, pollen germination, nutrient uptake, photosynthesis and specific enzyme function. For example, one study that examined the effect of an allelochemical known in velvet bean, 3-(3′,4′-dihydroxyphenyl)-l-alanine (l-DOPA), indicated that the inhibition by this compound is due to adverse effects on amino acid metabolism and iron concentration equilibrium. Allelopathic inhibition is complex and can involve the interaction of different classes of chemicals, such as phenolic compounds, flavonoids, terpenoids, alkaloids, steroids, carbohydrates and amino acids, with mixtures of different compounds sometimes having a greater allelopathic effect than individual compounds alone. Furthermore, physiological and environmental stresses, pests and diseases, solar radiation, herbicides and less than optimal nutrient, moisture and temperature levels can also affect allelopathic weed suppression. Different plant parts, including flowers, leaves, leaf litter and leaf mulch, stems, bark, roots, soil and soil leachates and their derived compounds, can have allelopathic activity that varies over a growing season.

Allelochemicals can also persist in soil, affecting both neighboring plants as well as those planted in succession. Although derived from plants, allelochemicals may be more biodegradable than traditional herbicides, but allelochemicals may also have undesirable effects on non-target species, necessitating ecological studies before widespread use. Selective activity of tree allelochemicals on crops and other plants has also been reported. For example, *Leucaena leucocephala*, the miracle tree promoted for revegetation, soil and water conservation and livestock nutrition in India, contains a toxic, non-protein amino acid in its leaves that inhibits the growth of other trees but not its own seedlings. *Leucaena* species have also been shown to reduce the yield of wheat but increase the yield of rice. Many invasive plants may have allelopathy as a feature for their ecological success. Time, environmental conditions and plant tissue all factor into variations in allelochemical concentrations in the producer plant. Foliar and leaf litter leachates of Eucalyptus species, for example, are more toxic than bark leachates to some food crops. The allelopathic potential of Ipomoea is significantly greater at higher environmental temperatures.

Research Strategies and Potential

The basic approach used in allelopathic research for agricultural crops has been to screen both crop plants and natural vegetation for their capacity to suppress weeds. To demonstrate allelopathy, plant origin, production and identification of allelochemicals must be established as well as persistence in the environment over time in concentrations sufficient to affect plant species. In the laboratory, plant extracts and leachates are commonly screened for their effects on seed germination, with further isolation and identification of allelochemicals from greenhouse tests and field soil confirming laboratory results. Interactions among allelopathic plants, host crops and other non-target organisms must also be considered. Furthermore, allelochemistry may provide basic structures or templates for developing new synthetic herbicides. Studies have elucidated specific allelochemicals involved in weed suppression, including benzoxanoids in rye; diterpenoid momilactones in rice; tabanone in cogon grass; alkaloids and flavonoids in fescue; anthratectone and naphthotectone in teak (*Tectona grandis*); abscisic acid beta-d-glucopyranosyl ester in red pine; cyanamide in hairy vetch; and a cyclopropene fatty acid in hazel sterculia (*Sterculia foetida*). Incorporation of allelopathic traits from wild or cultivated plants into crop plants through traditional breeding or genetic engineering methods could also enhance the biosynthesis and release of allelochemicals. Genetic basis of allelopathy has now been demonstrated in winter wheat and rice. An allelopathic crop can potentially be used to control weeds by planting a variety with allelopathic qualities, either as a smother crop, in a rotational sequence, or when left as a residue or mulch, especially in low-till systems, to control subsequent weed growth. In a multiseason field study, when applied as a soil amendment, mustard seed meal derived from white mustard was effective for weed suppression in organic sweet onion, but crop injury was also significant. Alternatively, application of allelopathic compounds before, along with, or after synthetic herbicides could increase the overall effect of both materials, thereby reducing application rates of synthetic herbicides. Some attempts have been reported on application of aqueous extracts of allelopathic plants on crops for weed suppression. In one study, an extract of brassica (*Brassica napus*), sorghum and sunflower was used on rain-fed wheat to successfully reduce weed pressure. When an allelopathic plant water extract was tank-mixed with atrazine, a significant degree of weed control was achieved in wheat with a reduced dose of herbicide. Sunflower residues with a pre-plant herbicide (Treflan®) enhanced weed suppression in broad bean.

By utilizing the principles of allelopathy the observations on weed growth and spread in plots supplied with various green leaf manures where made in crops and discussed below.

Weed Management

Trees with Allelopathic Potential for Weed Control

Through allelopathy, trees can regulate the germination, growth and development of weeds. Ethanol extracts of seeds of *Annona squamosa, Carica papaya, Coffea Arabica,* and *Tamarindus indica* were found to inhibit germination of *Amaranthus spinosus* by 13, 58, 100 and 36 per cent, respectively. They isolated and identified an allelochemical, 1, 3, 7 – trimethylxanthine (1, 3, 7-T) from the seeds of C. *arabica* and tested for herbicidal potential, it completely inhibited the seed germination of *A. spinosus* at 1200 ppm. The compound also suppressed the germination of *Avena fatua* (40.1 per cent), *Echinochloa colonum* (100 per cent) and *Echinochloa crusgalli* (91.2 per cent) at a concentration of 2000 ppm.

Eucalyptus spp., because of high allelopathic activity, is expected to control weeds *Eucalyptus* leaf leachate and oil showed differential effects on the growth of two weeds. The leaf leachate of 20 per cent concentration, suppressed the biomass production of *Cynodon dactylon* by about 50 per cent, whereas 1.0 per cent oil caused 68 per cent reduction. Application of 1.0 per cent oil significantly inhibited shoot and root length, leaf chlorophyll, and total biomass production of *Cyperus rotundus.* Aqueous extracts of its bark, leaves and oil inhibited *Parthenium hysterophorus.* Allelochemicals like mimosine, 3, 4-dihydroxy pyridine, and phenolics present in *Leucaena leucocephala* have been found to totally exclude weeds like *Ageratum conyzoides* and *Mimosa pudica.* Ailanthone, another chemical (isolated from *Ailanthus ultissisma*), has been reported to possess a post-emergence herbicidal property similar to glyphosate and paraquat. Dried mango (*Mangifera indica*) leaf powder @ 25 g mixed with one kilogram of soil in pot culture completely inhibited the germination of *Cyperus rotundus* upto 21 days, and its application to seedling considerably inhibited the tillering by 60 per cent. Seed germination and seedling growth of *Amaranthus retroflexux, Avena sativa, Chenopodium album* and *Cynodon dactylon* were found to be inhibited by aqueous extracts, decaying materials and volatile compounds of senescent and non-senescent leaves of *Citrus aurantium.*

Weed Management in Maize

To get information on the spread and weed population of weeds in the experimented plots, weed counts were taken at 75 DAS and expressed in number/ sq.m.

Broad Leaf Weed

The incorporation of green leaf manures had significant effect on Broad Leaf Weed occurrence at 80 DAS. *Albizia lebbek* recorded the lowest mean population of BLW 3.7/sq.m. It was followed by *Delonix regia* recorded the BLW population of 4.5/ sq.m. and highest BLW population was recorded in *Gliricidia sepium* (4.8/sq.m.). The foliar spraying of leaf extract had significant effect on the occurrence of BLW at 80

DAS. *Annona squamosa* recorded the lowest mean population of BLW 3.9/sq.m. It was followed by *Alangium salvifolium* recorded the BLW population of 4.1/sq.m and highest BLW population was recorded in *Alangium salvifolium* (4.9/sq.m.). The interaction effect was significant on BLW occurrence, *Albizia lebbek* + *Annona squamosa*, recorded the lowest mean population (3.5/sq.m.). It was followed by *Albizia lebbek* + *Alangium salvifolium* which recorded the BLW population of 3.6/sq.m. The highest BLW population was recorded in *Gliricidia sepium* + *Morinda tinctoria* 7.5/sq.m. at 80 DAS. The absolute control zero input recorded BLW population of 13.0/sq.m. When compared to control, *Albizia lebbek* + *Annona squamosa* showed a decreased BLW population of 16.5/sq.m. The recommended dose of inorganic fertilizers recorded (10.0/sq.m.) BLW population. When compared to this, *Albizia lebbek* + *Annona squamosa* showed a decreased BLW population of 6.5/sq.m. at 80 DAS.

Table 6.1: Weed Population (Broad Leaved Weeds) in Maize Field

GLM Incorporation	*Population of Broad Leaf Weeds (number of weeds/sq.m)*				
	Alangium	*Aegle*	*Morinda*	*Annona*	*Mean*
Albizia	3.60 (0.728)	4.20 (0.771)	3.85 (0.747)	3.50 (0.721)	**3.78**
Gliricidia	4.20 (0.771)	5.20 (0.834)	5.95 (0.876)	4.10 (0.764)	**4.86**
Leucaena	3.90 (0.750)	5.90 (0.874)	4.90 (0.817)	4.00 (0.757)	**4.67**
Delonix	5.10 (0.829)	4.00 (0.757)	4.70 (0.804)	4.20 (0.771)	**4.50**
Peltophorum	3.90 (0.750)	5.29 (0.840)	5.10 (0.829)	4.00 (0.757)	**4.57**
Mean	**4.14**	**4.91**	**4.90**	**3.96**	
	M	***S***	***M at S***	***S at M***	
SEd	**0.00096**	**0.00067**	**0.00161**	**0.00149**	
CD (p≤0.05)	0.0026	0.0014	0.00380	0.00317	
Fertiliser application	10.00				
Zero input	13.00				

Sedges Population

The incorporation of green leaf manures had significant effect on sedge weed occurrence at 80 DAS. *Albizia lebbek* recorded the lowest mean population of sedge 6.4/sq.m. It was followed by *Leucaena leucocephal,a* recorded sedge population of 6.5/sq.m. and highest sedge population was recorded in *Gliricidia sepium* (7.4/sq.m.). The foliar spraying of leaf extract had significant effect on the occurrence of sedge at 80 DAS. *Annona squamosa* recorded the lowest mean population of sedges (6.2/sq.m). It was followed by *Aegle marmellos* with a sedges population of 6.6/sq.m. and highest sedges population was recorded in *Morinda tinctoria* (7.4/sq.m.). The interaction effect was significant on sedge weed occurrence, *Albizia lebbek* + *Annona squamosa* recorded the lowest mean population (5.2/sq.m.). It was followed by *Leucaena leucocephala* + *Annona squamosa* which recorded the sedge population of 5.5/sq.m. and the highest sedge population was recorded in *Gliricidia sepium* + *Morinda tinctoria* and *Peltophorum ferrugineum* + *Alangium salvifolium* 8.2/sq.m. at 80 DAS. The absolute control zero input recorded sedge population of 20.0/sq.m. When compared to this, *Albizia lebbek* + *Annona squamosa* showed a decreased sedge population of 14.5/sq.m. The recommended dose of inorganic fertilizers recorded (15.0/sq.m.) sedge population. When compared to this *Albizia lebbek* + *Annona squamosa* showed a decreased sedge population of 9.5/sq.m. at 80 DAS.

Table 6.2: Weed Population (Sedges) in Maize Field (number of weeds / sq.m.)

GLM Incorporation	*Weed Population (Sedges)*				
	Alangium	*Aegle*	*Morinda*	*Annona*	*Mean*
Albizia	6.50 (0.905)	6.75 (0.917)	7.20 (0.938)	5.20 (0.834)	**6.41**
Gliricidia	7.40 (0.947)	6.85 (0.922)	8.20 (0.982)	7.32 (0.944)	**7.44**
Leucaena	6.80 (0.905)	6.30 (0.895)	7.40 (0.947)	5.50 (0.852)	**6.50**
Delonix	6.50 (0.905)	6.80 (0.919)	7.00 (0.929)	7.40 (0.947)	**6.92**
Peltophorum	8.20 (0.982)	6.40 (0.900)	7.30 (0.943)	5.80 (0.868)	**6.92**
Mean	**7.08**	**6.62**	**7.42**	**6.24**	
	M	***S***	***M at S***	***S at M***	
SE d	**0.00069**	**0.00062**	**0.00139**	**0.00139**	
CD (P≤0.05)	0.00191	0.00133	0.00318	0.00296	
Fertiliser application	15				
Zero input	20				

Grass Population

The incorporation of green leaf manures had significant effect on grass weed occurrence at 80 DAS. *Albizia lebbek* recorded the lowest mean population of grass 4.4/sq.m. It was followed by *Leucaena leucocephala which* recorded the grass population of 5.6/sq.m. and highest grass population was recorded in *Gliricidia sepium* (6.0/ sq.m.). The foliar spraying of leaf extract had significant effect on the occurrence of grass at 80 DAS. *Annona squamosa* recorded the lowest mean population of sedges 4.9/sq.m. It was followed by *Alangium salvifolium* recorded the grass population of 4.9/sq.m. and highest grass population was recorded in *Morinda tinctoria* (6.2/sq.m.). The interaction effect was significant, *Albizia lebbek* + *Annona squamosa* recorded the lowest mean population (3.8/sq.m.). It was followed by *Albizia lebbek* + *Morinda tinctoria,* recorded the grass population of 4.1/sq.m. and the highest population was recorded in *Gliricidia sepium* + *Morinda tinctoria* 3.5/sq.m. at 80 DAS. The absolute control zero input recorded grafs population of 19/sq.m. When compared to this, *Albizia lebbek* + *Annona squamosa* showed a decreased grass population of 15.2/sq.m. The recommended dose of inorganic fertilizers recorded (14.0/sq.m.) grass population. When compared to this *Albizia lebbek* + *Annona squamosa* showed a decreased grass population of 10.2/sq.m. at 80 DAS.

Table 6.3: Weed Population (Grass) in Maize Field (Number of weeds/sq.m. in maize)

GLM Incorporation	*Weed Population (Grass)*				
	Alangium	*Aegle*	*Morinda*	*Annona*	*Mean*
Albizia	4.30 (0.778)	5.40 (0.846)	4.10 (0.764)	3.80 (0.743)	**4.40**
Gliricidia	5.50 (0.852)	6.60 (0.910)	7.10 (0.934)	5.04 (0.825)	**6.06**
Leucaena	4.30 (0.778)	7.00 (0.929)	6.10 (0.884)	5.20 (0.834)	**5.65**
Delonix	6.12 (0.885)	5.30 (0.840)	5.90 (0.874)	5.40 (0.846)	**5.68**
Peltophorum	4.50 (0.791)	7.00 (0.929)	6.10 (0.884)	5.20 (0.834)	**5.70**
Mean	**4.94**	**6.26**	**5.86**	**4.92**	
	M	***S***	***M at S***	***S at M***	
SE d	**0.00128**	**0.00073**	**0.00191**	**0.00164**	
CD (0.05)	0.00355	0.00156	0.00462	0.00349	
Fertiliser application	14				
Zero input	19				

Further the occurrence of weeds evidenced from this observation is also very low in the combination *Albizia lebbek* + *Annona squamosa,* which also contributed higher yield. Leaf and seed extract of *Annona squamosa* in water and also their powders were reported to show insecticidal, antifeedant and repellent activity against number of insects and pests.

References

Fraenkel, G. S. 1959. "The Raison d'Etre of Secondary Plant Substances." *Science* 129: 1466–1470.

Stamp, N. 2003. "Out of the Quagmire of Plant Defense Hypotheses." *The Quarterly Review of Biology* 78: 23–55.

Willis, R. J. 2010. *The History of Allelopathy*. Dordrecht, the Netherlands: Springer.

Chapter 7
Quality Analysis

One of the most important reasons for analyzing foods from both the consumers and the manufacturer's standpoint is to ensure that they are safe. It would be economically disastrous, as well as being rather unpleasant to consumers, if a food manufacturer sold a product that was harmful or toxic. A food may be considered to be unsafe because it contains harmful microorganisms (*e.g.*, Listeria, Salmonella), toxic chemicals (*e.g.*, pesticides, herbicides) or extraneous matter (*e.g.*, glass, wood, metal, insect matter). It is therefore important that food manufacturers do everything they can to ensure that these harmful substances are not present, or that they are effectively eliminated before the food is consumed. This can be achieved by following good manufacturing practice regulations specified by the government for specific food products and by having analytical techniques that are capable of detecting harmful substances. In many situations it is important to use analytical techniques that have a high sensitivity, *i.e.*, that can reliably detect low levels of harmful material. Food manufacturers and government laboratories routinely analyze food products to ensure that they do not contain harmful substances and that the food production facility is operating correctly.

The food industry is highly competitive and food manufacturers are continually trying to increase their market-share and profits. To do this they must ensure that their products are of higher quality, less expensive and more desirable than their competitors, whilst ensuring that they are safe and nutritious. To meet these rigorous standards food manufacturers need analytical techniques to analyze food materials before, during and after the manufacturing process to ensure that the final product meets the desired standards. In a food factory one starts with a number of different raw materials, processes them in a certain manner (*e.g.* heat, cool, mix, dry), packages them for consumption and then stores them. The food is then transported to a warehouse or retailer where it is sold for consumption.

If there are a number of alternative methods available for measuring a certain property of a food, the choice of a particular method will depend on which of the above criteria is most important. For example, *accuracy* and use of an *official method* may be the most important criteria in a government laboratory which checks the validity of compositional or nutritional claims on food products, whereas *speed* and the ability to make *nondestructive* measurements may be more important for routine quality control in a factory where a large number of samples have to be analyzed rapidly.

Millet

Protein Content

The Protein content of the barnyard millet seeds was estimated using the Lowry's method and expressed in per cent (per cent) suggested by Lowry (1951).

Crude Fibre

The crude fibre content of the barnyard millet grains was estimated by using acid digestion method. The value was expressed in terms of per cent (per cent).

$$\text{Crude Fibre \%} = \frac{\text{Loss in weight on ignition (g)}}{\text{Wt. of original sample (g)}} \times 100$$

Iron and Zinc Content

The grain samples were collected in the individual treatment randomly and iron and zinc content was estimated by using Atomic Absorption Spectrophotometer (AAS). The value was expressed in terms of mg 100 g^{-1}.

Crude Protein Content

Nitrogen content was estimated by Kjeldhal method by Ma and Zuazaga (1942). The nitrogen value multiplied by factor 6.25 gives the crude protein content of the sample. The crude protein content value was expressed at g 100 g^{-1}.

Onion

Total Soluble Solids

The total soluble solids of onion bulb in each treatment was found out through a hand refractometer and expressed as degree of Brix.

Ascorbic Acid

Ascorbic acid content of the bulb was estimated by using the procedure described in AOAC (1975) and was expressed as mg g^{-1} of fresh sample.

Total Sugars

The amount of total sugars present in the bulb was estimated by anthrone and phenol sulphuric acid method (Dubois *et al.*, 1951) and expressed in mg g^{-1}.

Pungency

The pyruvate analysis was done according to Randle and Bussard (1993). The content of pyruvic acid released after blending flesh scale with water was used as a measure of pungency. One half to one third of the bulb was weighed and placed in manson jars with equal weight of distilled water and blended for 49 seconds. After 12 jars have been blended, samples were let to stand for 10 minutes to allow the formation of pyruvic acid. The open juice was then filtered into a test tube through No. 4 filter paper.

Using a pipette, 2 ml of filtrate was transferred into a volumetric flask, diluted to 100 ml with distilled water and then shaken. Two ml of the dilute solution was then placed in a test tube to which 1 ml of DNPH (0.0125 per cent 2, 4 dinitrophenyl hydrazine in 2 N HCL) was added. A rack of 12 tubes was placed into a 37°C water bath. After exactly 10 minutes the reaction was stopped by adding 5 ml of NaOH to each tube. Samples were then allowed to stand for five minutes for colour development.

Absorbance of each sample was measured with a spectrophotometer set at 420 nm using a reagent blank set at 0 absorbance (2 ml water + 1 ml DNPH + 5 ml NaOH). A standard curve for calibration was obtained by using sodium pyruvate as reagent with gradient concentrations of pyruvic acid. Abosrbance was read to get a standard curve of pyruvic acid for transforming the spectrometer reading into pyruvic acid concentration and expressed in μ mol g^{-1}.

Sulphur

The colorimetric procedure of Chopra and Kanwar (1991) was used to determine the sulphur content in bulb and expressed in mg g^{-1}.

Carbohydrate

The carbohydrate content present in the seed samples was estimated by Anthrone method (Hedge and Hofreiter, 1962). Amount of carbohydrate was expressed in percentage.

Crude Fibre

The crude fibre content was determined by Fibraplus method described by Manyard (1970). The value was expressed in terms of per cent (g $100g^{-1}$).

Crude Fat

The fat content of the sample was estimated by the method described by Cohen (1917). The lipid in the sample was extracted with petroleum ether (60-80°C) in Soxhlet apparatus for two hours. Then the solvent was evaporated and the remaining residue was weighed. The fat content was expressed as g $100g^{-1}$.

Calcium Content

The samples were collected in the individual treatment randomly and it was estimated by using Versenate method. The value was expressed in terms of mg 100 g^{-1}.

Crude Fibre

The crude fibre content of bhandi fruit was analyzed as per the method described by Chobra and Kanwar (1976) and expressed in percentage.

Vitamin C

The vitamin C content in the fresh palatable fruits of bhendi was analyzed as per the colorimetric method described by Sadasivam and Manickam (1992) and expressed in mg 100 g^{-1} sample.

The results of quality analysis done in various crop produces were discussed below for better understanding.

Quality Analysis in Barnyard Millet

Protein Content

The incorporation of green leaf manure did not have any significant effect on protein content of barnyard millet. The highest value was recorded by *Peltophorum ferrugineum* with 7.53 per cent. The foliar spraying of different leaf extracts significantly influenced the iron content of barnyard millet grain. *Annona squamosa* recorded higher protein content of 7.50 per cent and it was followed by *Moringa oleifera* with 7.47 per cent and the other treatments *Morinda tinctoria* and *Aegle marmellos* were on par with each other. There was no interaction effect on protein content on barnyard millet grain. However the highest value (7.82 per cent) was obtained by *Peltophorum ferrugineum* + *Annona squamosa*). The absolute control zero input recorded protein content of 6.38 per cent and recommended dose of inorganic fertilizers recorded protein content of 8.25 per cent at harvest of the crop.

Table 7.1: Effect of *Vrikshayurveda* Farming on Protein Content of Barnyard Millet Grains Protein (g 100g $^{-1}$)

GLM Incorporation	*Moringa*	*Aegle*	*Morinda*	*Annona*	*Mean*
Albizia	7.41	7.38	7.40	7.67	**7.46**
Glricidia	7.45	7.39	7.34	7.66	**7.46**
Pongamia	7.44	7.42	7.37	7.65	**7.46**
Delonix	7.53	7.37	7.38	7.70	**7.50**
Peltophorum	7.48	7.42	7.38	7.82	**7.53**
Mean	**7.47**	**7.39**	**7.38**	**7.50**	
SEd	0.05	0.01	0.07	0.06	
CD (p≤0.05)	NS	0.02	NS	NS	
Zero input					6.38
Fertiliser application					8.25

Crude Fibre

Significant variation was found in crude fibre of the barnyard millet due to incorporation of green leaf manures. The higher level of crude fibre of 9.14 per cent was recorded by *Peltophorum ferrugineum*. It was followed by *Delonix regia* and *Pongamia pinnata* (9.09 per cent). All other treatments were on par with each other. Foliar spraying of leaf extract significantly influenced the crude fibre of the barnyard millet grain. The higher level of crude fibre of 9.17 per cent was registered by *Annona squamosa* and it was on par with *Moringa oleifera* with 9.14 per cent and the lower level of crude fibre 8.97 per cent was recorded by *Morinda tinctoria* and it was on par with *Aegle marmellos* (9.01 per cent). The interaction effect was significantly influence the crude fibre of the barnyard millet grain. The higher level of crude fibre was registered by *Pheltophorum ferrugineum* + *Annona squamosa* and *Delonix regia* + *Annona squamosa* (9.20 per cent) and the lower level of crude fibre was obtained by *Gliricidia sepium* + *Morinda tinctoria* with 8.84 per cent. The control zero input recorded the crude fibre 8.96 per cent and the recommended dose of inorganic fertilizers recorded the crude fibre of 10.74 per cent at harvest of the crop.

Table 7.2: Effect of *Vrikshayurveda* Farming on Crude Fiber (g 100 g $^{-1}$) of Barnyard Millet Grains

GLM Incorporation	*Moringa*	*Aegle*	*Morinda*	*Annona*	*MEAN*
Albizia	9.05	8.86	9.08	9.18	9.04
Gliricidia	9.18	8.94	8.84	9.10	9.02
Pongamia	9.19	9.02	8.95	9.18	9.09
Delonix	9.12	9.04	8.99	9.20	9.09
Peltophorum	9.18	9.17	9.02	9.20	9.14
Mean	**9.14**	**9.01**	**8.97**	**9.17**	
	M	***S***	***M at S***	***S at M***	
SEd	0.01	0.02	0.03	0.05	
CD (p≤0.05)	0.04	0.04	0.06	0.10	
Zero input					8.96
Fertiliser application					10.74

Iron Content (mg 100 g^{-1})

The incorporation of green leaf manure did not have any significant effect on iron content of barnyard millet. However, the numerically highest value was recorded by *Peltophorum ferrugineum* with 13.59 mg. The foliar spraying of different leaf extracts significantly influenced the iron content of barnyard millet grain. *Annona squamosa* recorded higher iron content of 14.07 and it was followed by *Moringa oleifera* with 13.37 mg and the lower iron content was obtained by *Morinda tinctoria* with 12.03 mg. There was no interaction effect on iron content on barnyard millet grain. However, the highest value (14.90 mg) was obtained by *Peltophorum ferrugineum* + *Annona squamosa*. The control zero input recorded the iron content 12.27 mg and recommended dose of inorganic fertilizers recorded the iron content of 15.37 mg.

Table 7.3: Effect of *Vrikshayurveda* Farming on Iron Content (mg100 g^{-1}) of Barnyard Millet Grains

GLM Incorporation	*Moringa*	*Aegle*	*Morinda*	*Annona*	*MEAN*
Albizia	13.15	13.19	12.87	13.89	**13.27**
Gllricidia	12.94	12.20	11.45	13.46	**12.51**
Pongamia	13.48	12.86	12.09	14.04	**13.12**
Delonix	13.70	13.27	12.63	14.09	**13.17**
Peltophorum	13.61	13.26	12.60	14.90	**13.59**
MEAN	**13.37**	**12.96**	**12.03**	**14.07**	
	M	**S**	**M at S**	**S at M**	
SEd	0.25	0.15	0.38	0.34	
CD (p≤0.05)	NS	0.322	NS	NS	
Zero input					12.27
Fertiliser application					15.37

Zinc Content (mg 100 g^{-1})

The incorporation of green leaf manure significantly influenced the zinc content of barnyard millet. The maximum zinc content (5.78 mg) was recorded by *Peltophorum ferrugineum*, it was followed by *Delonix regia* with zinc content of 5.48 and all other treatments were on par with each other. Foliar application of barnyard millet significantly influences the zinc content. The higher value of 5.67 mg was recorded in *Annona squamosa* and it was followed by *Moringa oleifera* with 5.52 mg zinc content and the lower level of zinc content (5.11 mg) was recorded by *Morinda tinctoria*. The interaction effect was not found in zinc content of barnyard millet. However, the

Table 7.4: Effect of *Vrikshayurveda* Farming on Zinc Content (mg 100 g^{-1}) of Barnyard Millet Grains

GLM Incorporation	*Moringa*	*Aegle*	*Morinda*	*Annona*	*MEAN*
Albizia	5.53	5.25	4.85	5.45	**5.27**
Gllricidia	5.35	5.25	4.90	5.50	**5.25**
Pongamia	5.31	5.28	5.08	5.6	**5.32**
Delonix	5.60	5.25	5.25	5.80	**5.48**
Peltophorum	5.80	5.83	5.50	6.00	**5.78**
MEAN	**5.52**	**5.37**	**5.11**	**5.67**	
	M	***S***	***M at S***	***S at M***	
SEd	0.05	0.06	0.13	0.13	
CD (p≤0.05)	0.15	0.13	NS	NS	
Zero input					4.20
Fertiliser application					6.35

numerically the highest value of 6.00 mg was recorded in *Peltophorum ferrugineum* + *Annona squamosa*. The absolute control zero input recorded the zinc content 4.2 mg and the recommended dose of inorganic fertilizers recorded the zinc content of 6.35 mg at harvest of the crop.

Discussion Points

The iron and zinc content was high in the treatment combination *Pheltophorum ferrugineum* and *Annona squamosa*. The higher microbial decomposition of green leaf manure with simultaneous release of organic acids might have increased the availability of micro nutrients in soil and uptake by plants. The increased availability of zinc with the green leaf manure application in the formation of zinc stable chelates, which reduces absorption, fixation and precipitation of zinc in the soil. The unavailable form of ferric iron was reduced by green leaf manure incorporation. This may be ascribed to formation of stable chelates, which decreased the susceptibility of Fe to adsorption, fixation or precipitation reaction in soil resulting in greater availability of Fe. Further iron content was high in barnyard millet which is treated with *Pheltophorum ferrugineum* + *Annona squamosa*. The leaf extracts of *Annona squamosa* itself having high micro nutrients, so it increases the iron content in barnyard millet. The protein content was high in the treatment combination *Pheltophorum ferrugineum* and *Annona squamosa*. This might be due to more nitrogen uptake by plants. The nitrogen is the essential constituent which influenced protein content and the protein content was high in fertilized plot due to high nitrogen available soil nitrogen and uptake by plants.

Quality Parameters in Finger Millet

Carbohydrate Content

Green leaf manure incorporation and foliar application of leaf extracts significantly influenced carbohydrate content of finger millet grains.

Maximum content of carbohydrate (69.75 per cent) was recorded in the treatment of *Delonix regia*, which was followed by *Leucaena leucocephala* and minimum content of carbohydrate (63.75 per cent) was registered in *Peltophorum ferrugineum*. Among the foliar spray sof leaf extract, the highest content of 67.80 per cent was recorded at *Moringa oleifera*, on contrary *Annona squamosa* showed lowest content of carbohydrate (64.60 per cent) in finger millet grains. *Annona squamosa*, *Morinda tinctoria* and *Aegle marmellos* were on par with each other.

Carbohydrate content was significantly influenced by the interaction between green leaf manure incorporation and foliar spraying of different leaf extracts. *Delonix regia* + *Moringa oleifera* showed maximum content of carbohydrate (72.00 per cent), which was on par with *Delonix regia* + *Aegle marmellos* and *Leucaena leucocephala* + *Aegle marmellos*. The lowest content of carbohydrate (61.00 per cent) was recorded *Albizia lebbek* + *Aegle marmellos* in finger millet grains. The absolute control zero input recorded the carbohydrate content of 52.00 per cent and the recommended dose of fertiliser application registered the carbohydrate content 65.00 per cent.

Table 7.5: Effect of Vrikshyurveda Farming on Carbohydrate Content (per cent) of Finger Millet Grains

Treatments	*Leaf Extract Spray*				
GLM Incorporation	*Moringa*	*Morinda*	*Annona*	*Aegle*	*Mean*
Albizia	68.00	62.00	65.00	61.00	**64.00**
Gliricidia	69.00	64.00	62.00	65.00	**65.00**
Leucaena	66.00	68.00	63.00	70.00	**66.75**
Delonix	72.00	69.00	67.00	71.00	**69.75**
Peltophorum	64.00	63.00	66.00	62.00	**63.75**
Mean	67.80	65.2	64.60	65.80	
	SEd		***CD(P≤0.05)***		
M	0.92		2.56		
S	0.69		1.47		
M at S	1.62		3.79		
S at M	1.54		3.28		
Fertiliser application		**65.00**			
Zero input		**52.00**			

Crude Protein Content

Crude protein content of finger millet was significantly influenced by incorporation of green leaf manures. The maximum content of crude protein was recorded by *Delonix regia* with value of 10.88 g 100 g^{-1}. All other treatments were on par with each other. The foliar spraying of different leaf extracts significantly influenced the crude protein content of finger millet grain. *Moringa oleifera* recorded maximum crude protein content of 10.45 g 100 g^{-1}. This treatment was on par with *Aegle marmellos* and *Morinda tinctoria*, while *Annona squamosa* registered low content of crude protein (9.73 g 100 g^{-1}).

Interaction between green leaf manure incorporation and different leaf extract spray significantly influenced crude protein content of finger millet grains. However, the maximum value (11.78 g 100 g^{-1}) was obtained by *Delonix regia* + *Moringa oleifera*, which was on par with *Delonix regia* + *Aegle marmellos* and *Leucaena leucocephala* + *Aegle marmellos*. Low content of crude protein with a value of 8.8 g 100 g^{-1} was recorded in the treatment of *Albizia lebbek* + *Moringa oleifera*. The absolute control zero input recorded the crude protein content of 9.22 g 100 g^{-1} and the recommended dose of fertilizer application registered the crude protein content 7.64 g 100 g^{-1} in finger millet grains.

Crude Fibre

Significant variation was found in crude fibre content of the finger millet grains due to incorporation of green leaf manures. The higher content of crude fibre of 3.27 g 100 g^{-1} was recorded by *Delonix regia*. It was followed by *Gliricidia sepium*,

Table 7.6: Effect of Vrikshayurveda Farming on Crude Protein Content of Finger Millet Grains

Treatments	*Leaf Extract Spray*				
GLM Incorporation	*Moringa*	*Morinda*	*Annona*	*Aegle*	*Mean*
Albizia	8.80	10.8	10.56	9.62	**9.94**
Gliricidia	10.75	11.00	9.26	9.24	**10.06**
Leucaena	10.97	9.12	8.90	11.44	**10.11**
Delonix	11.78	9.60	10.56	11.58	**10.88**
Peltophorum	9.98	10.74	9.36	9.72	**9.95**
Mean	**10.45**	**10.25**	**9.73**	**10.32**	
	SEd		***CD(P≤0.05)***		
M	0.19		0.52		
S	0.25		0.52		
M at S	0.51		1.14		
S at M	0.55		1.17		
Fertiliser application		**9.22**			
Zero input		**7.64**			

Table 7.7: Effect of Vrikshyurveda Farming on Crude Fibre (g $100g^{-1}$) of Finger Millet Grains

Treatments	*Leaf Extract Spray*				
GLM Incorporation	*Moringa*	*Morinda*	*Annona*	*Aegle*	*Mean*
Albizia	3.06	3.14	3.08	3.00	**3.07**
Gliricidia	3.18	3.14	3.18	3.14	**3.14**
Leucaena	3.08	3.07	3.22	3.19	**3.14**
Delonix	3.49	3.08	3.24	3.27	**3.27**
Peltophorum	3.10	3.11	3.11	3.04	**3.09**
Mean	**3.18**	**3.11**	**3.17**	**3.11**	
	SEd		***CD(P≤0.05)***		
M	0.023		0.064		
S	0.022		0.047		
M at S	0.049		0.111		
S at M	0.050		0.106		
Fertiliser application		**3.08**			
Zero input		**2.96**			

Leucaena leucocephala and *Peltophorum ferrugineum*. The three treatments were on par with each other. *Albizia lebbek* registered minimum content of crude fibre (3.07 g 100 g^{-1}) in finger millet grains. Foliar spraying of leaf extract significantly influenced the crude fibre of the finger millet grain. The higher level of crude fibre of 3.18 g 100 g^{-1} was registered by *Moringa oleifera* and it was on par with *Annona squamosa* and the lower level of crude fibre with value of 3.11 g 100 g^{-1} was recorded by *Morinda tinctoria*, which was on par with *Aegle marmellos*. The interaction effect was significant for crude fibre of the finger millet grains. The maximum content of crude fibre (3.49 g 100 g^{-1}) was registered by *Delonix regia* + *Moringa oleifera* and the lower level of crude fibre (3.00 g 100 g^{-1}) was obtained by *Albizia lebbek* + *Aegle marmellos*. The control zero input recorded the crude fibre (3.08 g 100 g^{-1}) and recommended dose of inorganic fertilizers recorded a crude fibre content of 2.96 g 100 g^{-1}.

Crude Fat

A positive and significant variation was found in crude fat content of finger millet grain due to incorporation of green leaf manures and foliar spraying of different leaf extract. The maximum crude fat content (1.23 g 100 g^{-1}) was recorded by *Delonix regia*, which was on par with *Leucaena leucocephala* and *Gliricidia sepium*. The minimum (1.06 g 100 g^{-1}) was recorded under *Peltophorum ferrugineum* in finger millet grains. The maximum crude fat (1.18 g 100 g^{-1}) content was registered by *Aegle marmellos*, which was on par with two other treatments such as *Moringa oleifera* and *Morinda tinctoria*. *Annona squamosa* registered less crude fat (1.14 g 100 g^{-1}) content. The interaction effect significantly influenced crude fat content of finger millet grains due to green leaf manure incorporation and different leaf extract spray. However, the maximum value (1.29 g 100 g^{-1}) was obtained by *Delonix regia* + *Moringa oleifera*, which was on par with *Delonix regia* + *Aegle marmellos* and *Leucaena leucocephala* + *Moringa oleifera*. Less amount of crude fat with a value of 1.07 g 100 g^{-1} was recorded

Table 7.8: Effect of Vrikshyurveda Farming on Crude Fat Content (g 100g^{-1}) of Finger Millet Grains

Treatments	*Leaf Extract Spray*				
GLM Incorporation	*Moringa*	*Morinda*	*Annona*	*Aegle*	*Mean*
Albizia	1.07	1.13	1.16	1.15	**1.13**
Gliricidia	1.14	1.21	1.14	1.20	**1.17**
Leucaena	1.27	1.25	1.12	1.26	**1.23**
Delonix	1.29	1.18	1.17	1.28	**1.23**
Peltophorum	1.07	1.16	1.14	1.18	**1.06**
Mean	**1.17**	**1.16**	**1.14**	**1.18**	
	SEd		***CD(P≤0.05)***		
M	0.023		0.064		
S	0.013		0.028		
M at S	0.034		0.084		
S at M	0.030		0.063		
Fertiliser application		**1.10**			
Zero input		**0.92**			

under the treatment of *Peltophorum ferrugineum* + *Moringa oleifera* and *Albizzia lebbek* + *Moringa oleifera*. The control zero input recorded crude fat content of 0.92 g 100 g^{-1} and recommended dose of inorganic fertilizers recorded the crude fat of 1.10 g 100 g^{-1}.

Calcium Content (mg 100 g^{-1})

The incorporation of green leaf manure did significantly influence the calcium content of finger millet. The maximum calcium content (296 mg) was recorded by *Delonix regia* and the minimum with *Peltophorum ferrugineum*. Foliar application on finger millet significantly influenced the calcium content. The higher value of 287 mg was recorded in *Moringa oleifera* and the lower amount of calcium (232 mg) was recorded by *Morinda tinctoria*. The highest value of 342 mg was recorded in *Delonix regia* + *Moringa oleifera*. The lowest value of 205 mg was registered in combination of treatments such as *Peltophorum ferrugineum* + *Morinda tinctoria*, *Leucaena leucocephala* + *Morinda tinctoria* and *Albizia lebbek* + *Morinda tinctoria*. The control zero input recorded the calcium content 205 mg and the recommended dose of inorganic fertilizers recorded the calcium content of 273 mg in finger millet grains.

Iron Content (mg 100 g^{-1})

The incorporation of green leaf manure did not have significant effect on iron content of finger millet. However, the numerically highest value was recorded by *Delonix regia* with 318 mg. The foliar spraying of different leaf extracts significantly influenced the iron content of finger millet grain. *Aegle marmellos* recorded higher iron content of 327 mg and the lower iron content was obtained by *Morinda tinctoria* with 312 mg. The interaction effect significantly influenced crude fat content of finger millet grains due to green leaf manure incorporation and different leaf extract sprays. Among the treatments, *Delonix regia* + *Moringa oleifera* showed maximum iron content (336 mg) in grains which was on par with *Delonix regia* + *Aegle marmellos*. The minimum iron content (301 mg) was recorded in *Delonix regia* + *Annona squamosa*. The control zero input recorded the iron content of 297 mg and recommended dose of inorganic fertilizers recorded the iron content of 307 mg. The increasing consciousness about health hazards due to usage of agrochemicals has brought a major shift in consumer preferences. The consumers are increasingly looking forward to chemical free food grains that are considered to be safe and hazard free.

The carbohydrate content was high in ragi which was treated with *Delonix regia* with *Moringa oleifera* leaf extract. The protein content was high in the treatment combination *Delonix regia* with *Moringa oleifera*. This might be due to more nitrogen uptake by plants. Being nitrogen is the essential constituent for protein, which influenced the protein content in maize and further the protein content was high in fertilized plots due to high available soil nitrogen and uptake by plants. And then Crude fat and fibre content was high in the treatment combination of *Delonix regia* with *Moringa oleifera*. Calcium and Iron content was also high in *Delonix regia* with *Moringa oleifera*. Foliar application of leaf extracts may also help in yield enhancement and increased concentration of nutrition in crops due to the presence of micro elements and growth regulators.

Table 7.9: Effect of Vrikshyurveda Farming on Calcium and Iron Content (mg $100g^{-1}$) of Finger Millet Grains

Treatments	*Calcium (mg)*					*Iron (mg)*				
GLM Incorporation	*Moringa*	*Morinda*	*Annona*	*Aegle*	*Mean*	*Moringa*	*Morinda*	*Annona*	*Aegle*	*Mean*
Albizia	273.00	205.00	296.00	251.50	**256.37**	320.00	308.00	328.50	311.50	**317.00**
Gliricidia	228.00	273.00	273.00	250.00	**256.00**	309.00	324.00	313.50	316.50	**315.75**
Leucaena	296.00	205.00	228.00	319.00	**262.00**	310.00	315.50	313.00	331.00	**317.37**
Delanix	342.00	273.00	250.00	319.00	**296.00**	336.00	304.00	301.50	331.50	**318.25**
Peltophorum	296.00	205.00	252.00	276.00	**245.37**	304.00	311.00	314.00	323.00	**313.00**
Mean	**287.00**	**232.30**	**259.80**	**273.50**		**315.80**	**312.50**	**314.10**	**327.70**	
		SEd		***CD (0.05)***			***SEd***		***CD (0.05)***	
M		1.08		3.01			NS		NS	
S		1.09		2.33			2.06		4.39	
M at S		2.38		5.39			4.97		11.72	
S at M		2.44		5.21			4.61		9.82	
Fertiliser application			**273**					**307**		
Zero input			**205**					**297**		

Quality Analysis in Black Gram

Crude Protein Content

The incorporation of green leaf manures had significant effect on protein content of black gram. The higher per cent of protein (25.79) had been recorded in *Pongamia pinnata*. It was followed by *Albizia lebbek* recorded the protein content of (24.47) and the lowest of per cent of protein (17.35) was recorded in *Peltophorum ferrugineum*. The foliar spraying of leaf extract had a significant effect on crude protein content of black gram. The higher per cent of protein (24.18) had been recorded in *Moringa oleifera*. It was followed by *Annona squamosa* recorded protein content of (22.44) and the lowest per cent of protein (19.78) was recorded in *Morinda tinctoria*. The interaction effect was significant. The combination, *Pongamia pinnata* + *Moringa oleifera* recorded the higher per cent of protein content (30.14). It was followed by *Albizia lebbek* + *Annona squamosa* with 29.74 per cent of protein and the lowest per cent of protein (14.74) was recorded in *Peltophorum ferrugineum* + *Annona squamosa*. The absolute control zero input recorded 17.96 per cent of crude protein and recommended dose of inorganic fertilizers recorded 24.73 per cent of protein in black gram. When compared to Control, *Pongamia pinnata* + *Moringa oleifera* recorded increased per cent of protein (12.18) and when compared to fertilizer application, it showed increased per cent of protein in black gram (5.41).

Table 7.10: Effect of Vrikshyurveda Farming on Protein Content (per cent) of Black Gram Grains

Treatments	*Foliar Application*				
GLM Incorporation	*Moringa*	*Morinda*	*Annona*	*Aegle*	*Mean*
Albizia	25.79	22.53	19.82	29.74	**24.47**
Gliricidia	24.97	18.89	20.49	18.91	**20.81**
Pongamia	30.14	24.63	22.17	26.22	**25.79**
Delonix	20.79	19.79	19.69	22.62	**20.72**
Peltophorum	19.19	18.73	16.75	14.74	**17.35**
MEAN	**24.18**	**20.91**	**19.78**	**22.44**	
	M	***S***	***M at S***	***S at M***	
SEd	0.003	0.445	0.862	0.996	
CD (p≤0.05)	0.008	0.949	1.839	2.123	
Zero input					24.73
Fertiliser application					17.96

Quality Analysis in Onion

Total Soluble Solids (Degree of Brix)

The incorporation of green leaf manures had significant effect on TSS in aggregatum onion at harvest of the crop. The highest TSS (18.5° brix) was recorded in

Senna siamea and *Gliricidia sepium* and the lowest TSS (16.75°brix) was recorded in *Leucaena leucocephala* at harvest of the crop. The application of leaf extract through seed treatment and foliar sprays had a significant effect on TSS of aggregatum onion at harvest of the crop. Among the treatments, *Alangium salvifolium* recorded highest TSS (18.20° brix) and the lowest TSS (17.40° brix) was recorded in *Aegle marmelos* in aggregatum onion at harvest of the crop. The interaction effect was significant. Among treatments *Gliricidia sepium* + *Aegle marmelos* recorded the highest TSS (21° brix). This was on par with *Delonix regia* + *Annona squamosa* which recorded total soluble solids of 20° brix in aggregatum onion. The absolute control zero input recorded TSS of 10° brix. When it was compared with control *Gliricidia sepium* + *Aegle marmelos* showed a increased total soluble solids of 11° brix and recommended dose of inorganic fertilizers recorded TSS of 16° brix. When compared to this, *Gliricidia sepium* + *Aegle marmelos* showed a increased total soluble solids of 5° brix in aggregatum onion at harvest of the crop.

Table 7.11: Effect of Vrikshyurveda Farming on Total Soluble Solids (Brix °) Quality in Aggregatum Onion cv. CO 5

GLM Incorporation	*Alangium*	*Annona*	*Aegle*	*Morinda*	*Mean*
Albizia	19	17	16	17	17.25
Senna	19	18	18	19	18.50
Gliricidia	17	17	21	17	18.00
Leucaena	17	17	16	17	16.75
Delonix	19	20	16	19	18.50
Mean	**18.20**	**17.80**	**17.40**	**17.80**	**17.00**
	M	***S***	***M X S***		
SEd	0.12	0.09	0.22		
CD(P≤0.05)	0.29	0.20	0.48		
Zero input					10
Fertiliser application					16

Total Sugars

The incorporation of green leaf manures had significant effect on total sugar in aggregatum onion. The highest total sugar (5.05 mg 100 g^{-1}) was recorded in *Senna siamea*. This was on par with *Leucaena leucocephala* which recorded total soluble solids of 4.76 mg 100 g^{-1} and the lowest total sugar (4.67 mg 100 g^{-1}) was recorded in *Delonix regia*. The application of leaf extract through seed treatment and foliar sprays had a significant effect on total sugar of aggregatum onion at harvest of the crop. Among the treatments, *Aegle marmelos* recorded highest total sugar (5.08 mg 100 g^{-1}), which was on par with *Morinda tinctoria* (4.87) and the lowest total sugar (4.54 mg 100 g^{-1}) was recorded in *Alangium salvifolium* in aggregatum onion. The interaction effect was significant. Among treatments *Leucaena leucocephala* + *Aegle marmelos* recorded the highest total sugar (5.23 mg 100 g^{-1}), which was on par with *Gliricidia sepium* which recorded (5.19 mg 100 g^{-1}) in aggregatum onion. The absolute control zero input recorded total sugar of 3.99 mg/100g. When it was compared to Control, *Leucaena leucocephala* + *Aegle marmelos* showed a increased total sugar 1.24 mg 100 g^{-1}. The recommended dose of inorganic fertilizers recorded total sugar of (4.95 mg 100 g^{-1}). When compared to this, *Leucaena leucocephala* + *Aegle marmelos* showed a decreased total sugar of 0.28 mg 100 g^{-1} in aggregatum onion at harvest of the crop.

Table 7.12: Effect of Vrikshyurveda Farming on Total Sugar (mg/100 g^{-1}) Quality in Aggregatum Onion cv. CO 5

GLM Incorporation	*Alangium*	*Annona*	*Aegle*	*Morinda*	*Mean*
Albizia	4.46	4.26	5.00	5.03	4.68
Senna	5.11	5.11	5.02	4.93	5.05
Gliricidia	4.34	4.26	5.19	5.02	4.70
Leucaena	4.31	5.04	5.23	4.47	4.76
Delonix	4.48	4.30	4.97	4.93	4.67
Mean	**4.54**	**4.60**	**5.08**	**4.87**	**4.77**
	M	***S***	***M X S***		
SEd	0.10	0.09	0.21		
CD(P≤0.05)	0.25	0.20	0.46		
Zero input					3.99
Fertiliser application					4.95

Pyruvic Acid (Pungency)

The incorporation of green leaf manures had significant effect on pyruvic acid content in aggregatum onion at harvest of the crop. The highest pyruvic acid content 1.96 μ mol g^{-1} was recorded in *Gliricidia sepium* and *Leucaena leucocephala*, while the lowest pyruvic acid content 1.87 μ mol g^{-1} was recorded in *Albizia lebbek*. The application of leaf extract through seed treatment and foliar sprays had a significant effect on pyruvic acid content of aggregatum onion at harvest of the crop. Among the treatments, *Annona squamosa*, recorded highest pyruvic acid content 2.00 μ mol g^{-1} and *Alangium salvifolium* had the lowest pyruvic acid content 1.85 μ mol g^{-1} in aggregatum onion. The interaction effect was significant. Among treatments *Gliricidia sepium* + *Aegle marmelos* recorded the highest pyruvic acid content 2.10 μ mol g^{-1}. This was on par with *Leucaena leucocephala* + *Annona squamosa* which recorded 2.09 μ mol g^{-1} and *Senna siamea* + *Annona squamosa* recorded of pyruvic acid content of 2.08 μ mol g^{-1} in aggregatum onion. The absolute control zero input recorded pyruvic acid content of 1.03 μ mol g^{-1}. When compared to Control *Gliricidia sepium* + *Aegle marmelos* showed a increased pyruvic acid content of 1.07 μ mol g^{-1}. The recommended dose of inorganic fertilizers recorded pyruvic acid content of 1.98 μ mol g^{-1}. When compared to this *Gliricidia sepium* + *Aegle marmelos* showed a decreased pyruvic acid content of content of 0.12 μ mol g^{-1} in aggregatum onion at harvest of the crop.

Table 7.13: Effect of Vrikshyurveda Farming on Pyruvic Acid ((μ mol g^{-1}) Quality in Aggregatum Onion cv. CO 5

GLM Incorporation	*Alangium*	*Annona*	*Aegle*	*Morinda*	*Mean*
Albizia	1.67	1.91	2.01	1.99	1.87
Senna	2.00	2.08	1.99	1.68	1.93
Gliricidia	1.79	1.99	2.10	1.99	1.96
Leucaena	1.91	2.09	1.98	1.97	1.96
Delonix	1.91	2.01	1.80	2.01	1.93
Mean	**1.85**	**2.00**	**1.97**	**1.91**	**1.93**
	M	***S***	***M X S***		
SEd	0.042	0.039	0.087		
CD(P≤0.05)	0.097	0.080	1.84		
Zero input					1.03
Fertiliser application					1.98

Sulphur Content

The incorporation of green leaf manures had significant effect on sulphur content in aggregatum onion. The highest sulphur content (0.55 mg g^{-1}) was recorded in *Leucaena leucocephala* and *Albizia lebbek*. The lowest sulphur content (0.47mg g^{-1}) was recorded in *Delonix regia*. The application of leaf extract through seed treatment and foliar sprays had a significant effect on sulphur content of aggregatum onion at harvest of the crop. Among the treatments, *Aegle marmelos* recorded highest sulphur content (0.54 mg g^{-1}) and the lowest sulphur content (0.49 mg g^{-1}) was recorded in *Annona squamosa* of aggregatum onion. The interaction effect was significant. Among treatments, *Gliricidia sepium* + *Aegle marmelos* recorded the highest sulphur content (0.61 mg g^{-1}), which was on par with *Senna siamea* + *Alangium salvifolium* (0.59 mg g^{-1}). The absolute control zero input recorded sulphur content of 0.36 mg g^{-1}. When compared to control *Gliricidia sepium* + *Aegle marmelos* showed increased sulphur content of 0.25 mg g^{-1}. The recommended dose of inorganic fertilizers recorded sulphur content of 0.49 mg g^{-1}. When compared to this *Gliricidia sepium* + *Aegles marm*ellos showed a decreased sulphur content of 0.12 mg g^{-1} in aggregatum onion at harvest of the crop.

Ascorbic Acid

The incorporation of green leaf manures had significant effect on ascorbic acid content in aggregatum onion at harvest of the crop. The highest ascorbic acid content 7.73 mg g^{-1} was recorded in *Senna siamea*, This was on par with *Albizia lebbek* which recorded 7.69 mg g^{-1}. The lowest ascorbic acid content 6.98 mg g^{-1}) was recorded in *Leucaena leucocephala*. The application of leaf extract through seed treatment and foliar sprays had a significant effect on ascorbic acid content of aggregatum onion at harvest of the crop. Among the treatments, *Annona squamosa* recorded highest ascorbic acid content 7.69 mg g^{-1}, which was on par with (7.58 mg g^{-1}) and lowest value was

Table 7.14: Effect of Vrikshyurveda Farming on Sulphur Content (mg g^{-1}) Quality in Aggregatum Onion cv. CO 5

GLM Incorporation	*Alangium*	*Annona*	*Aegle*	*Morinda*	*Mean*
Albizia	0.56	0.53	0.57	0.55	0.55
Senna	0.59	0.45	0.49	0.47	0.50
Gliricidia	0.34	0.49	0.61	0.55	0.53
Leucaena	0.56	0.53	0.55	0.56	0.55
Delonix	0.45	0.46	0.48	0.51	0.47
Mean	**0.53**	**0.49**	**0.54**	**0.52**	**0.52**
	M	***S***	***M X S***		
SEd	0.010	0.016	0.023		
CD(P≤0.05)	0.024	0.021	0.048		
Zero input					0.36
Fertiliser application					0.49

Table 7.15: Effect of Vrikshyurveda Farming on Ascorbic Acid (mg g^{-1}) Quality in Aggregatum Onion cv. CO 5

GLM Incorporation	*Alangium*	*Annona*	*Aegle*	*Morinda*	*Mean*
Albizia	7.95	7.89	7.65	7.33	7.69
Senna	7.59	7.89	7.56	7.91	7.73
Gliricidia	6.88	7.56	8.02	7.64	7.50
Leucaena	6.99	7.23	7.55	6.45	6.98
Delonix	6.87	7.92	7.05	7.92	7.44
Mean	**7.25**	**7.69**	**7.58**	**7.45**	**7.47**
	M	***S***	***M X S***		
SEd	0.138	0.131	0.290		
CD(P≤0.05)	0.31	0.26	0.41		
Zero input					5.25
Fertiliser application					6.99

recorded in *Alangium salvifolium* with 7.25 mg g^{-1}. The interaction effect was significant. Among treatments *Gliricidia sepium* + *Aegle marmelos* recorded the highest ascorbic acid content 8.02 mg g^{-1}. The lowest ascorbic acid content 6.45 mg g^{-1} was recorded in *Leucaena leucocephala* + *Morinda tinctoria* in *aggregatum* onion. The absolute control zero input recorded ascorbic acid content of (5.25 mg g^{-1}). When compared to Control *Gliricidia sepium* + *Aegle marmelos* showed a increased ascorbic acid of 2.77 mg g^{-1}.

The recommended dose of inorganic fertilizers recorded ascorbic acid content of 6.99 mg g^{-1}. When compared to this *Gliricidia sepium* + *Aegle marmelos* showed a decreased ascorbic acid content of 1.03 mg g^{-1} in aggregatum onion.

Quality Analysis in Bhendi

Crude Fibre

The incorporation of green leaf manures had significant effect on crude fibre in Bhendi at harvest. The highest crude fibre (7.81 per cent) was recorded in *Leucaena leucocephala* and the lowest crude fibre (7.46 per cent) was recorded in *Albizia lebbek* at harvest of the crop. The application of leaf extract through seed treatment and foliar sprays had a significant effect on crude fibre in Bhendi. Among the treatments, *Ocimum sanctum* recorded highest crude fibre (7.86 per cent) and the lowest crude fibre (7.24 per cent) was recorded in *Annona squamosa* in Bhendi at harvest stage. The interaction effect was significant. Among the treatment combinations, *Delonix regia+ Ocimum sanctum* recorded highest crude fibre (8.48 per cent) and lowest crude fibre (6.17 per cent) was recorded in *Albizia lebbek +Annona squamosa*. The absolute control recorded 9.32 per cent crude fibre. When it was compared with *Albizia lebbek +Annona squamosa* it showed a decrease in crude fibre (3.15 per cent) and recommended dose of inorganic fertilizers recorded crude fibre of 10.72 per cent.

Table 7.16: Effect of Vrikshyurveda Farming on Crude Fibre (per cent) Content in Bhendi cv. Arka Anamika

GLM Incorporation	*Alangium*	*Annona*	*Aegles*	*Morinda*	*Ocimum*	*Mean*
Albizia	7.33	6.17	8.24	8.31	7.24	7.46
Delonix	7.33	7.55	8.40	7.06	8.48	7.77
Gliricidia	8.39	7.78	6.65	7.20	7.87	7.58
Leucaena	7.97	7.46	7.76	8.05	7.84	7.81
Mean	**7.76**	**7.24**	**7.76**	**7.66**	**7.86**	**7.65**
	M	***S***	***M x S***			
SEd	0.089	0.147	0.277			
CD(P≤0.05)	0.219	0.299	0.577			
Zero input						9.32
Fertiliser application						10.72

Vitamin C

The incorporation of green leaf manures had significant effect on Vitamin C in Bhendi. The highest Vitamin C (12.48 mg 100 g^{-1}) was recorded in *Albizia lebbek* and the lowest (10.89 mg 100 g^{-1}) was recorded in *Delonix regia*. The application of leaf extract through seed treatment and foliar sprays had a significant effect on Vitamin C in Bhendi at harvest of the crop. Among the treatments, *Annona squamosa* recorded highest Vitamin C (11.89 mg 100 g^{-1}), which was on par with *Aegles marmellos* (11.85) and the lowest Vitamin C (11.09 mg 100 g^{-1}) was recorded in *Alangium salvifolium*. The interaction effect was significant. Among the treatment combinations, *Albizia lebbek +Annona squamosa* recorded the highest Vitamin C (14.10 mg 100 g^{-1}) and the lowest Vitamin C (10.18 mg 100 g^{-1}) was recorded in *Gliricidia sepium + Alangium salvifolium*. The absolute control recorded 8.15 mg 100 g^{-1} of Vitamin C. When it was compared to *Albizia lebbek +Annona squamosa* this treatment showed an increased Vitamin C of 5.95 mg 100 g^{-1}. The recommended dose of inorganic fertilizers recorded (13.85 mg 100 g^{-1}) of Vitamin C. When it was compared with *Albizia lebbek +Annona squamosa,* latter showed an increase in Vitamin C of 0.25 mg 100 g^{-1} in Bhendi at harvest of the crop.

Crude Protein

The incorporation of green leaf manures had significant effect on crude protein in Bhendi at harvest of the crop. The highest crude protein (12.86 per cent) was recorded in *Gliricidia sepium,* while the lowest crude protein (10.73 per cent) was recorded in *Leucaena leucocephala.* The application of leaf extract through seed treatment and foliar sprays had a significant effect on crude protein in Bhendi at harvest. Among the treatments, *Annona squamosa,* recorded highest crude protein (12.83 per cent) and *Ocimum sanctum* had the lowest crude protein (11.51 per cent) in Bhendi. The interaction effect was significant. Among treatments *Albizia lebbek +Annona squamosa* recorded the highest crude protein (14.27 per cent) and *Leucaena leucocephala + Ocimum sanctum* recorded the lowest crude protein (10.07 per cent) in Bhendi. The absolute control recorded the crude protein of 9.61 per cent and recommended dose of inorganic fertilizers recorded the crude protein of 11.00 per cent. When it was compared to *Albizia lebbek + Annona squamosa* there was an increase in level of protein by 4.66 per cent and 3.27 per cent over control and fertilizer application respectively in bhendi at harvest stage.

Table 7.17: Effect of Vrikshyurveda Farming on Crude Protein (per cent) Content in Bhendi cv. Arka Anamika

GLM Incorporation	*Alangium*	*Annona*	*Aegles*	*Morinda*	*Ocimum*	*Mean*
Albizia	12.50	14.27	12.85	11.78	11.94	12.67
Delonix	12.32	12.56	11.09	13.24	11.95	12.23
Gliricidia	13.22	13.36	13.57	12.07	12.08	12.86
Leucaena	10.60	11.13	10.78	11.09	10.07	10.73
Mean	**12.16**	**12.83**	**12.07**	**12.05**	**11.51**	**12.12**
	M	***S***	***M x S***			
SEd	0.423	0.249	0.615			
CD(P≤0.05)	1.036	0.509	1.373			
Zero input						9.61
Fertiliser application						11.00

The biochemical constituents such as crude protein, crude fibre and vitamin C are considered as some important quality parameters both for consumption and processing. The quality decides the market price and high crude protein and vitamin C coupled with low crude fibre would be as ideal vegetable. The studies indicated that the high quality products with high crude protein and Vitamin C and low crude fibre were found in the treatment with *Albizia lebbek* as green leaf manure and seed soaking along with foliar sprays of *Annona squamosa*. The quality and suitability of a vegetable for consumption, is judged based on the crude fibre content of the fruits and the organic manuring considerably decreased the crude fibre content. The highest level of ascorbic acid content may be due to the action of macro and micronutrients and also the growth hormones released from the added manures. Apart from this, the presence of phenols, protein and flavonoids in the leaf extracts would have promoted the nutrient uptake and ultimately increased the quality. The observed average quality of fruits interms of crude protein, vitamin C and crude fibre in the recommended dose of inorganic fertilizers may be due to the supply of macro nutrients alone and devoid of micronutrients and plant hormones. Micronutrients improve the chemical composition of fruits and vegetables and are known to act as catalyst in promoting organic reactions taking place in plants. The absolute control recorded the lowest quality of bhendi due to non- availability of nutrients and soil reserve alone acts as the only source of nutrients.

Chapter 8

Secondary Metabolites of Plant Origin

Premise

Secondary metabolites are a wide range of compounds from different metabolite families that can be highly inducible in response to stresses. These compounds are not essential for cell structure and maintenance of life but are often involved in plant protection against biotic or abiotic stresses. Some secondary metabolite families such as carotenoids and flavonoids are also involved in cell pigmentation in flower and seed, which attract pollinators and seed dispersers. Therefore, they are also involved in plant reproduction. Moreover, plant secondary metabolites have chemical and pharmaceutical properties interesting for human health.Compounds belonging to the terpenoids, alkaloids and flavonoids are currently used as drugs or as dietary supplements to cure or prevent various diseases and in particular some of these compounds seems to be efficient in preventing and inhibiting various types of cancer. Secondary metabolites, also known as **natural products**, are those products (chemical compounds) of metabolism that are not essential for normal growth, development or reproduction of an organism. In this sense they are "secondary". The function or importance of these compounds to the organism's development is usually of ecological nature as they are used as defense against predators (herbivores, pathogens etc.), for interspecies competition and to facilitate the reproductive processes. Contrary to primary metabolites these compounds are not ubiquitous in the living organisms that neither produce them nor are they necessarily expressed continuously.

Secondary metabolites are the intermediates and products of metabolism. The term *metabolite* is usually restricted to small molecules. A primary metabolite is directly involved in normal growth, development and reproduction. A secondary metabolite is not directly involved in those processes, but usually has an important ecological function. Examples include antibiotics and pigments in plants. Further, Secondary metabolites are molecules that are not essential for growth and development of an organism. Secondary metabolism refers to compounds present in specialized cell's that are not necessary for the cells survival but are thought to be required for the plants survival in the environment.

No direct function in growth and development of the plant often limited distribution among plants. A particular metabolite may be restricted to one species or one genus. Usually synthesized from primary products. Variety of functions recognized today but originally thought to be waste products.

What are the Functions of Secondary metabolites?

- ✰ The most common roles for secondary compounds in plants are ecological roles that govern interactions between plants and other organisms.
- ✰ Nicotine and other toxic compounds may protect the plant from herbivores and microbes.
- ✰ Other secondary compounds like rubber and tetrahydrocannabinil (THC) from cannabis plants have no known function in plants.
- ✰ Brightly colored pigments like anthocyanin those attract pollinators and fruit and seed dispersers.
- ✰ Secondary metabolites display antimicrobial activity and Play important role in plant defense with Protection against herbivore and pathogens

How they are Classified?

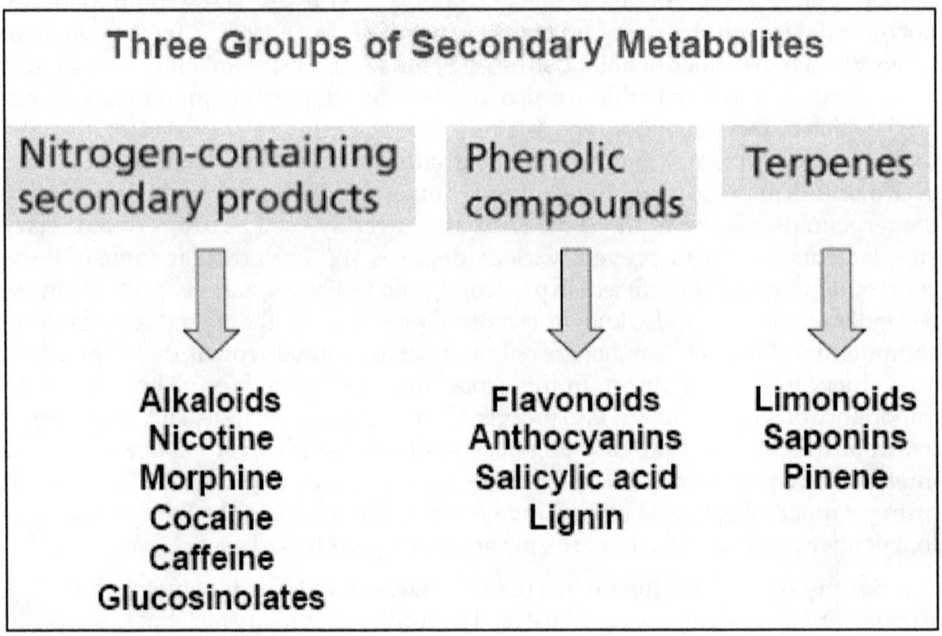

Class	*Example Compounds*	*Example Sources*	*Some Effects and Uses*
		NITROGEN CONTAINING	
Alkaloids	Nicotine cocaine theobromine	Tobacco coca plant chocolate (cocoa)	Interfere with neurotransmission, block enzyme action
		NITROGEN and SULFUR CONTAINING	
Glucosinolates	Sinigrin	Cabbage, relatives	
		TERPENOIDS	
Monoterpenes	Menthol, linalool	Mint and relatives, many plants	Interfere with neurotransmission, block ion transport, anesthetic
Sesquiterpenes	Parthenolid	Parthenium and relatives (*Asteraceae*)	Contact dermatitis
Diterpenes	Gossypol	Cotton	Block phosphorylation; toxic
Tetraterpenoids	Carotene	Many plants	Antioxidant; orange coloring
Sterols	Spinasterol	Spinach	Interfere with animal hormone action
		PHENOLICS	
Phenolic acids	Caffeic, chlorogenic	All plants	Cause oxidative damage, browning in fruits and wine
Coumarins	Umbelliferone	Carrots, parsnip	Cross-link DNA, block cell division
Flavonoids	Anthocyanin, catechin	Almost all plants	Flower, leaf color; inhibit enzymes, anti- and pro-oxidants, estrogenic
Tannins	Gallotannin, condensed tannin	*Acacia* spp, myrobalan, legumes	Bind to proteins, enzymes, block digestion, antioxidants
Lignin	Lignin	Many plants	Structure, toughness, fiber

Role of Plant Derived Natural Products in Industrial Applications

Industry	*Plant Product*	*Plant Species*	*Industrial Use*
Pharmaceuticals	Codeine Diosgenin Quinine	*Papaver somniferum* *Dioscorea deltoidea* *Cinchona ledgeriana*	Anti tussive Anti fertility agen tAnti malarial
Agrochemicals	Pyrethrin	*Crysanthemum cinerariaefolium*	Insecticide
Food and drink	Quinine	*Cinc thaumatococcus*	Bittering agent
Cosmetics	Jasmine	*Jasminum* sp.	Perfume

Pharmacological Activity of Plant Derived Natural Products

Medicinal Agent	*Activity*	*Plant Source*
Atropine	Anti cholinergic	*Atropa belladonna* L.
Codeine	Anti tussive	*Papaver somniferum* L.
Colchicines	*Antigout*	*Colchicum autumnale* L.

Contd...

Medicinal Agent	Activity	Plant Source
Digitoxin	Cardiovascular	*Digitalis purpurea* L.
Digoxin	Cardio tonic	*Digitalis lanata* L.
Hyoscyamine	Anti cholinergic	*Hyoscyamus niger* L.
Morphine	Analgesic	*Papaver somniferum* L.
Pilocarpine	Cholinergic	*Pilocarpus* sp.
Quinine	Anti malarial	*Cinchona ledgeriana* Moens ex Trimen
Reserpine	Antihypertensive	*Rauwolfia serpentina* L.
Scopolamine	Anti cholinergic	*Datura metel* L.
Steroids from Diosgenin	Anti fertility agent	*Dioscorea deltoidea* Wall.
Vincristine/Vinblastine	Anticancer	*Catharanthus roseus* G. Don

Alkaloid	Source
Ajmalicine	*Catharanthus roseus*
Atropine	*Atropa belladonna*
Berberine	*Berberis* spp., *Coptis japonica*

Examples for Pure Chemicals Extracted from Plants

Species	Plant Part Used	Compunds
Aloe barbadensis	Succulent leaves	Aloin
Artemisia annua	Whole plants	Artemisinin, santonin
Atropa belladonn	Roots, leaves	Atropin, hyoscyamine
Cassia angustifolium	Leaves	Sennoside a and b
Catharanthus roseus	Mainly roots, all parts of plants	Vincaleucoblastine, leucohistine, Ajmalicine
Cinchona ledgeriana	Bark	Quinine
Cymbopogon spp.	Whole plants	Citral, geraniol, citronella
Datua innoxia	Flowers	Hyoscyamine
Digitalis lanata	Leaves	Cetyldigoxin, degitoxin, digoxin, lantosides
Dioscorea deltoidea/floribunda	Tubers	Diosgenin
Glycyrrhiza glabra	Roots and rhizomes	Glycirrhizin
Panax ginseng	Roots	Ginseng
Papaver sominferum	Capsules	Codiene, morphine

Alkaloid	*Source*
Codeine	*Papaver somniferum*
Colchicine	*Colchicum autumnale*
Caffeine	*Coffea arabica, Camellia sinensis*
Camptothecine	*Camptotheca acuminata*
Emetine	*Cephaelis ipecacuanha*
Euipticine	*Ochrosia ellipfica*
Ephedrine	*Ephedra gerardiana*
Morphine	*Papaver somniferum*
Papaverine	*Papaver somniferum*
Quinine, quinidine	*Cinchona ledgeriana*
Reserpine	*Rauwolfia serpentina*
Vinblastine, vincristine	*Catharanthus roseus*

Role and Classification of Common Secondary Metabolites

1. Alkaloid

Alkaloids are naturally occurring chemical compounds containing basic nitrogen atoms. The name derives from the word alkaline and is used to describe any nitrogen-containing base. Alkaloids are produced by a large variety of organisms, including bacteria, fungi, plants and animals and are part of the group of natural products (also called secondary metabolites). Many alkaloids can be purified from crude extracts by acid-base extraction. Many alkaloids are toxic to other organisms. They often have pharmacological effects and are used as medications, as recreational drug, or in entheogenic rituals. Examples are the local anesthetic and stimulant cocaine,stimulant caffeine, nicotine, analgesic morphine or the anti malarial drug quinine. Most alkaloids have a bitter taste.

Alkaloid Classifications

The classification of the alkaloids is complex and may be guided by a set of rules that take into account the structure and other chemical features of the alkaloid molecule, its biological origin, as well as the biogenetic origin where known. For example, where the biosynthesis pathway of an alkaloid is unknown, it may be grouped based on structural similarities with known compounds, including non-nitrogenous compounds or by the organism(s) from which the alkaloid was isolated.

- ☆ **Pyridine group**: Piperine, coniine, trigonelline, arecoline, arecaidine, guvacine, cytisine, lobeline, nicotine, anabasine, sparteine, pelletierine.
- ☆ **Pyrrolidine group**: Hygrine, cuscohygrine, nicotine
- ☆ **Tropane group**: Atropine, cocaine, ecgonine, scopolamine, catuabine
- ☆ **Indolizidine group**: Senecionine, swainsonine
- ☆ **Quinoline group**: Quinine, quinidine, dihydroquinine, dihydroquinidine, strychnine, brucine, veratrine, cevadine

- ✰ **Isoquinoline group**: Opium alkaloids (papaverine, narcotine, narceine), pancratistatin, sanguinarine, hydrastine, berberine, emetine, berbamine, oxyacanthine
- ✰ **Phenanthrene alkaloids**: Opium alkaloids (morphine, codeine, thebaine, oripavine)
- ✰ **Phenethylamine group**: Mescaline, ephedrine, dopamine

1. Indole Group

- ✰ Tryptamines: serotonin, bufotenine, psilocybin
- ✰ Ergolines (the ergot alkaloids): ergine, ergotamine, lysergic acid
- ✰ Beta-carbolines: harmine, harmaline, tetrahydroharmine
- ✰ Yohimbans: reserpine, yohimbine
- ✰ Vinca alkaloids: vinblastine, vincristine
- ✰ Kratom (*Mitragyna speciosa*) alkaloids: mitragynine, 7-hydroxymitragynine
- ✰ Tabernanthe iboga alkaloids: ibogaine, voacangine, coronaridine
- ✰ *Strychnos nux-vomica* alkaloids: strychnine, brucine

2. Purine Group

- ✰ Xanthines: caffeine, theobromine, theophylline
- ✰ Terpenoid group:
- ✰ Acontium alkaloids: aconitine
- ✰ Steroid alkaloids (containing a steroid skeleton in a nitrogen containing structure):
- ✰ Solanum (*e.g.* potato and tomato) alkaloids (solanidine, solanine, chaconine)
- ✰ Veratrum alkaloids (veratramine, cyclopamine, cycloposine, jervine, muldamine)
- ✰ Fire Salamander alkaloids (samandarin)

3. Others: Conessine

- ✰ Quaternary ammonium compounds: muscarine, choline, neurine
- ✰ Miscellaneous: capsaicin, cynarin, phytolaccine, phytolaccotoxin

Biochemistry and Biosynthesis of Secondary Metabolites

Compound	*Species*	*Site of Biosynthesis*
Shikonin	Lithospermum	Roots
Jatrorhizine	Berberies	Roots
Sanguinarine	Papaver	Roots
Nicotine	Nicotiana	Roots
Ajmalicine	Catharanthus	Roots
Harmane alkaloids	Peganum	Roots
Glycirrhizine	Glycirrhiza	Roots

Alkaloids: Function and Biosynthesis

Secondary metabolites of plants are naturally occurring products that appear to have no direct function to growth and development.

Physiological Role: They are used by an organism (a) for defense against herbivores, pathogens and (b) as attractants of pollinators.

Insecticides of Plant Origin

Active Principle	*Species*	*Plant Source*
Amorpholone	*Tephrosia candida*	Stem and leaves
	Delbergia monetaria	Seeds
Azadiractin	*Azadirachta indica*	Seeds
Rotenoids	*Derris elliptica*	Seeds
	Tephrosia vogelii	Roots
Pyrethrins	*Chrysanthemum cinerariaefolium*	Floral heads
Nicotine	*Nicotiana tabacum*	Leaves
Quassin	*Quassia amara*	wood

Types of Alkaloids

The monoterpenoid indole alkaloids comprise a large family of alkaloids, with over 1800 members of rich structural diversity. Many of these natural products are physiologically active in mammals. Among the monoterpenoid indole alkaloid pharmaceuticals that are still commercially isolated from plant material are the anti malarial drug quinine from *C. officinalis*, rat poison and homeopathic drug strychnine from *Strychnos nuxvomica* and the antineoplastic chemotherapeutic agents vincristine and vinblastine from *Catharanthus roseus* (periwinkle).

The tropane class of alkaloids, which are found mainly in the Solanaceae, contains the anticholinergic drugs hyoscyamine and scopolamine. Solanaceous plants have been used traditionally for their medicinal, hallucinogenic and poisonous properties, which are due to their tropane alkaloids. The narcotic topical anesthetic and central nervous system stimulant cocaine is a tropane alkaloid which is a useful medicinal in small, controlled doses is potentially lethal when abused. *Datura* leaves are smoked for the hallucinogenic effects of this alkaloid.

The benzylisoquinoline alkaloids are very large and diverse class of alkaloids. This family contains such varied physiologically active members as emetine (an antiamoebic), colchicines (a microtubule disrupter and gout suppressant), berberine (an antimicrobial against eye and intestinal infections), morphine (a narcotic analgesic), codeine (a narcotic analgesic and antitussive) and sanguinarine (an antimicrobial used in oral hygiene). As with the pharmacologically active members of indole and tropane classes of alkaloids, those of pharmaceutical interest from the benzylisoquinoline class are largely still isolated from plants, again due to the complexity of their structures.

Eco-chemical Function of Alkaloids

Their medicinal and toxicological properties clearly make alkaloids important to the people, but their role in plants has been a longstanding question. Why should a plant invest so much nitrogen in synthesizing such a large number of alkaloids of such diverse structure? *Catharanthus roseus* alone contains over 100 different monoterpenoid indole alkaloids. That many alkaloids are cytotoxic provides insight into their potential function in the chemical defense arsenal of the plant. It needs to be systematically addressed whether alkaloids function as phytoalexins, as nitrogen storage forms, as UV protectants, or in any combination of these and other potential functions. The question of alkaloid function in plants and the relevance of that function have been reviewed. One alkaloid whose eco-chemical functions have been thoroughly analyzed in recent years is nicotine, a highly toxic alkaloid that has been identified in the leaves, stems and roots of a number of *Nicotiana* species. Nicotine sulfate, a byproduct of the tobacco industry, serves commercially as a very effective insecticide and fumigant. To date, no insect has evolved a resistance mechanism against nicotine. In mammals, ingestion of nicotine results in nausea, vomiting, diarrhea, mental confusion, convulsions and respiratory paralysis. With such strong and varied physiological responses, nicotine would seem to be an effective deterrent against insect attack and herbivore grazing.

Cyanogenic glycosides are found in the fruits (and wilting leaves) of the rose family including cherries, apples, plums etc. Role in plant defense against herbivores due to bitter taste and release of toxic hydrogen cyanide upon tissue disruption Herbivores have acquired the ability to metabolize cyanogenic glycosides them for use in their predator defense. A large group of compounds found in thousands of plant species and stored in leaves, seeds or roots. Enzymatic release of cyanide gas following herbivore Glucosinolates (mustard oil glycosides) which is principally found in plants from the Brassicaceae Enzymatic release following herbivore and also which can act as volatile defensive compounds (toxins, deterrents).

The terpenoids, sometimes called isoprenoids are a large and diverse class of naturally-occurring organic chemicals similar to terpenes, derived from five-carbon isoprene units assembled and modified in thousands of ways. Most are multicyclic structures that differ from one another not only in functional groups but also in their basic carbon skeletons. These lipids can be found in all classes of living things and are the largest group of natural products. Plant terpenoids are used extensively for their aromatic qualities. They play a role in traditional herbal remedies and these are under investigation for antibacterial, antineoplastic and other pharmaceutical functions. Terpenoids contribute to the scent of eucalyptus, the flavors of cinnamon, cloves and ginger and the color of yellow flowers. Well-known terpenoids include citral, menthol, camphor, Salvinorin A in the plant *Salvia divinorum* and cannabinoids found in Cannabis.

The steroids and sterols in animals are biologically produced from terpenoid precursors. Sometimes terpenoids are added to proteins, *e.g.*, to enhance their attachment to the cell membrane; this is known as isoprenylation.

Examples

- ☆ Resins of conifers are monoterpenes
- ☆ Essential oils – pepper mint, limon, basil, sage may be in glandular hairs on epidermis
- ☆ **Pyrethrum** (from chrysanthemums) and **azadirachtin** (from Neem tree) can be used as "natural" insecticides in agricultural practices or in horticulture

Monoterpine as Feeding Deterrent

Verpenone feeding deterrent to spruce bark beetle Cucurbitins in cucurbits acts as feeding detterent. Iridoids glycoside present in 57 plant families of Dicotyledoneae is bitter substance act as feeding detterent. The essential oil of *Cymbopogan citratus* shows fumigant activity in the management of cereals in storage without mammalian toxicity.

Monoterpene as Growth and Development Inhibitor

Citronellal acts as oviposition deterrent against *Amrasca devastans* and also *Limonene* inhibit embryonic development of *Spodoptera litura* and Essential oil of lamiaceae affect larval development fruit flies.

- ☆ Diterpine ryanodine from *Ryania speciosa* active against caterpillars.
- ☆ 16-kaurin-19-oic acid in sunflower used to protect from *Helicoverpa armigera.*
  - Azadiractin acts as antifeedent affect growth and development, oviposition deterrent, reduce fecundity
  - Active components are nimbin, nimbinin and nimbidin.

Sesqueterpenoids

Gossypol

It is a polyphenol derived from the cotton plant (genus *Gossypium*, family *Malvaceae*). Gossypol is a polyphenolic aldehyde that permeates cells and acts as an inhibitor for several dehydrogenase enzymes. Protect cotton from *Earias insulana* and *Earias vittella.*

Phenols

In organic chemistry, phenols, sometimes called phenolics, are a class of chemical compounds consisting of a hydroxyl group (-OH) bonded directly to an aromatic hydrocarbon group. The simplest of the class is phenol.

Phenol is the simplest of the phenols. Although similar to alcohols, phenols have unique properties and are not classified as alcohols (since the hydroxyl group is not bonded to a saturated carbon atom). They have higher acidities due to the aromatic ring's tight coupling with the oxygen and a relatively loose bond between the oxygen and hydrogen. The acidity of the hydroxyl group in phenols is commonly intermediate between that of aliphatic alcohols and carboxylic acids (their pKa is usually between 10 and 12). Loss of a positive hydrogen ion (H+) from the hydroxyl group of a phenol forms a corresponding negative phenolate ion or phenoxide ion and the corresponding salts are called phenolates or phenoxides, although the term aryloxides is preferred according to the IUPAC Gold Book. Phenols can have two or more hydroxy groups bonded to the aromatic ring(s) in the same molecule. The simplest examples are the three benzenediols, each having two hydroxy groups in benzene ring. Some phenols are germicidal and are used in formulating disinfectants. Others possess estrogenic or endocrine disrupting activity.

Phenolics vs Plant growth

- ✰ Phenolics affecting seed germination and dormancy and these substances, found in both seed coats and embryos, have been identified as phenolic acids, hydroxycinnamic acids and coumarins.
- ✰ Non-germinating seeds of *Melilotus alba* were found to possess a large amount of free coumarin, while in rapidly germinating seeds coumarinic acid β-glucoside was more prevalent.
- ✰ Another naturally occurring phenolic compound inhibiting the germination of seeds of the same or of other species is ferulic acid
- ✰ Pure solution of ferulic acid gave strong inhibition of *Raphanus sativus* seed germination at concentration of 10^{-4} M.
- ✰ It has been suggested that phenolics may be active as germination inhibitors by inhibiting phenolics and plant defense 33 transport of amino acids and the formation of proteins in the seeds.

Phenolics vs Fungal Pathogens

- ✰ It plays an important role in non-host resistance to filamentous fungi.
- ✰ Some antibiotic phenolics are stored in plant cells as inactive bound forms but are readily converted into biologically active antibiotics by plant hydrolysing enzymes (glycosidases) in response to pathogen attack.
- ✰ First demonstrated example of phenolics providing disease resistance - catechol (I) and protocatechuic acid (II) to prevent onion smudge disease *Colletotrichum circinans*.
- ✰ The coloured outer onion scales of resistant onion varieties contain enough of these two phenols to reduce spore germination of *C. circinans* to below 2 per cent, while susceptible varieties lack these compounds and the germination rate is over 90 per cent.
- ✰ Adequate levels of chlorogenic acid (III) account for the resistance of potato tubers against *Streptomyces scabies*, *Verticillium alboatrum* and *Phytophthora infestans*, while at low concentrations it stimulates the growth of *P. infestans* and *Fusarium solani* var. *coeruleum*.

Antifungal Phenolics

Examples of Anti-fungal Phenolics

Sl.No.	Fungus	Compound
1.	*Colletotrichum circinans*	Catechol, Protocatechuic acid
2.	*Botrytis cinerea*	2, 5 – Dimethoxybenzoic acid
3.	*Eutypa lata*	Salicylic acid
4.	*Phytophthora infestans*	Vanillic acid
5.	*Verticillium albo-atrium*	Chlorogenic acid
6.	*Phytophthora infestans*	Chlorogenic acid
7.	*Fusarium oxysporum*	Chlorogenic acid, Rutin
8.	*Penicillium glabrum*	3- and 7- Hydroxyflavone
9.	*Fusarium* spp.	Dihydroquercetin
10.	*Aspergillus* sp.	Flavone, Flavanone
11.	*Fusarium oxysporum*	5,8-Dihydroxy-6,7-dimethoxy flavan
12.	*Helminthosporium oryzae*	5,8-Dihydroxy-6,7-dimethoxy flavan
13.	*Cercospora bieticola*	Genistein, Biochanin
14.	*Helminthosporium sativum*	Hordatine A and B

Phenolics vs Pest Resistance

- Proximate and ultimate determinants of host-plant range in phytophagous insects.
- A well-known example of simple phenolics being feeding barriers to insect herbivores is represented by salicylates in *Salix* leaves and their role in the feeding and growth of the polyphagous larvae of *Operophtera brumata.*
- It has been observed that the levels of salicylates correlated negatively with growth: larvae exposed to leaves rich in these compounds grew slowly and consumed less material and this suggests that salicylates could be considered as antifeedants for *O. brumata.* Four isoflavonoids (judaicin, judaicin-7-*O*-glucoside, 2-methoxyjudaicin and maackiain) isolated from wild relatives of chickpea (*Cicer arietinum*), were shown to deter larval feeding by *Heliocoverpa armigera* at 100 ppm. All four isoflavonoids showed a dose- dependent decrease in activity, with judaicin and maackiain retained their antifeedant activity at 50 ppm and 10 ppm, respectively.
- The isoflavonoids are tested in combinations and with chlorogenic acid. The combinations containing judaicin and maackiain are most active and chlorogenic acid enhanced the antifeedant activity of all four isoflavonoids. *H. armigera* is the only one of four noctuids to be deterred by all four isoflavonoids. *Spodoptera littoralis* was deterred by judaicin alone and *S. frugiperda* by maackiain alone. *Heliothis virescens* and *S. exigua* are not deterred from feeding by any of the isoflavonoids.

- ✰ Positive relationship between resistance/susceptibility characteristics against aphids (*Aphis craccivora* Koch) and flavonoid glycoside content of cowpea lines was found. The resistant lines showed a flavonoid content higher than susceptible ones.

Coumarine

Coumarin is a chemical compound (specifically, a benzopyrone) found in many plants. It has a sweet scent, readily recognised as the scent of newly-mown hay and has been used in perfumes since 1882. It has clinical medical value as the precursor for several anticoagulants, notably warfarin and it is used as a gain medium in some dye lasers.

- ✰ They affect oviposition in insects Furanocoumarine oviposition detterent and toxic to *H. armigera, Spodoptera litura.* Coumarin is a potent rodenticide: rats and other rodents largely metabolize it to 3, 4-coumarin epoxide, a toxic compound that can cause internal hemorrhage and death.

Tannin

Tannins are astringent, bitter plant polyphenols that either bind and precipitate or shrink proteins. The astringency from the tannins which causes the dry and puckery feeling in the mouth following the consumption of unripened fruit or red wine. Likewise, the destruction or modification of tannins with time plays an important role in the ripening of fruit and the ageing of wine.

- ✰ Tannin affects the reproduction in gypsy moth
- ✰ Tannin extract of sorghum BSK 5 and BTX 625 affect of bioactivity of rice weevil.
- ✰ Tannin in cotton causes resistance to bollworm complex.

Functions of Tannins on Insects

- ✰ They have an astringent taste, which affects palatability and decreases feed consumption, they form complexes of reduced digestibility with proteins and they act as enzyme inactivators.
- ✰ Tannins are present at high levels in most of the plant seeds and grains, where they negatively impact the use of seeds and grain in insect feeding.
- ✰ In cowpea, condensed tannins (proanthocyanidins) contribute to resistance to infestation by insects such as the cowpea weevil *Callosobruchus maculatus.*

Seed Storage vs Secondary Metabolites

Alkaloids derived from neem products are NIMBIN, NIMBIZENE, NIMBANDIOL were isolated but active compounds have been isolated from Neem this include "AZADIRACTIN" Neem kernel powder protect the grains in wheat storage against *Rhizopertha dominica, Sitophilus oryzea* and *Tribolium granarium* in paddy storage against insect pests and in storage of jowar and mung and against stored product pests. Plant products are naturally evolved ingredients of the biosphere, they not only have an edge over the synthetic alien molecules, but also rate at least a prior acceptance from a view point an environmental stability, greater biodegradability and lesser hazards were expected in plant products.

In recent years attempts have been made to replace synthetic pesticides by natural pesticides of plant origin which are safe, less persistent and more specific. Plant products are easy availability and safe application to seeds. green plants appear to be the reservoirs of effective chemotherapeutant and would constitute an inexhaustible source of harmless pesticide. Many of the botanicals possess antifungal and anti bacterial activity but few reports were available for antifungal and anti bacterial activity. To overcome this, use of botanicals to reduce the process of seed deterioration, insect infestation and infection have decisive role to play.

Botanicals

Trees	*Botanical Name*	*Parts Used*
Neem	*Azadirachta indica*	Leaves, flower, seed, kernels
Custard apple	*Annona squamosa*	Seeds, leaves
Arappu	*Albizia amara*	Leaves
Eucalyptus	*Eucalyptus globulus*	Leaves
Pungam	*Pongamia pinnata*	Leaves
Prosopis	*Prosopis juliflora*	Leaves
Notchi	*Vitex negundo*	Leaf
Bael	*Aegle marmelos*	Oil from leaves

Custard Apple

Leaf, unripe fruits, seeds and roots have exhibited insecticidal properties. Seeds containing anolobine, anninine artabotrine, municine and souamosin and also contain glycerides of hydroxylated unsaturated acids which are toxic to insects.

Effect of different Forms of Custard Apple Seed Kernal Powder on the Adult Survival and Seed Damage

Sl.No.	*Treatments*	*Mean per cent Survival Adult Beetles after Release (3rd day)*	*7th day*	*Mean per cent of Seed Damage Progressive in Second Generation*
1.	Custard apple seed kernel powder @ 1 gm	56.66	10.66	0.00
2.	Custard apple seed kernal powder @ 2 gm	25.00	0.00	0.00
3.	Custard apple seed kernal aqueous extract	85.00	51.66	13.33
4.	Custard apple seed kernal powder paste @ 1 gm	58.33	6.66	0.00
5.	Custard apple seed kernal powder paste @ 2 gm	28.33	1.66	0.00
6.	Control	86.66	45.00	100.00

Result: Custard apple seed powder and paste each at the rate of 1-2 g/100 g of seed could be used as seed protectant against pulse beetle. Only aqueous extract of seeds was ineffective.

Arappu

- ☆ Leaves are a source of triterpenic saponins and sapogenins.leaves contain a methyl ester of a fatty acid and a triterpene saponin. Leaves are repellant in nature. Leaf powder containing the nutrients especially the micronutrients might synergistically interact with the amino acid, tryptophan to form the indole acetic acid.
- ☆ Ragi, panivaragu, kuthiraivali treated with arapu leaf powder was better in preserving viability and vigour potential in storage.

Effect of Seed Treatments, Period (Month) of Storage on Germination per cent in Thenai Cv. Co.5

Sl.No.	*Treatments*	*Months*				
		0	*2*	*4*	*6*	*8*
1.	Control	93	96	90	94	80
2.	Bavistin (2 g kg^{-1})	93	90	83	82	84
3.	Captan (2 g kg^{-1})	93	90	83	86	84
4.	Aappu powder (10 g kg^{-1})	93	98	96	94	86
5.	Thiram (2 g kg^{-1})	93	98	92	84	79

Result: Arappu recorded 93 per cent germination and after 12 months of storage, germination per cent was slightly decreased to 86 per cent

Pungam

- ☆ Leaves are used for storage purpose. Seed contain 27-39 per cent of fatty oil, called pungam oil. Pungam oil contain karanjin possess insecticidal and antifungal properties.

Storability of Seeds in Various Treatments after 5 Months

Treatments	*0.25*	*0.50*	*0.75*	*1.00*
Neem oil	37.67	19.00	0.00**	0.00**
Karanj oil	40.67	25.33	7.33*	0.00**
Control	87.00	–	–	–

Result: Neem oil at 0.75 and 1 per cent levels and pungam oil at 1 per cent level completely prevented adult Emergence.

Conclusion

Secondary metabolites involved in the defense mechanisms of plants against fungal pathogens and insect herbivores. Plants currently bred for increased levels of

SM might well influence different aspects of plant/pathogen and/or plant/insect interactions. Studies on these interactions are needed in order to have a better understanding of how increasing the expression of an endogenous SM compounds in plants could influence plant resistance characteristics against pathogens and phytophagous insects. Standardization and commercialisation of extraction methods for different Secondary metabolites compounds could be a welcome future.

References

Agosta, William. *Bombardier Beetles and Fever Trees: A Close-up Look at Chemical Warfare and Signals in Animals and Plants.* Reading, MA: Addison-Wesley, 1996.

Bell E.A., Charlwood B.V. (1980). Secondary plant products. *Encyclopedia of Plant Physiology* 8, Springer, Heidelberg, D.

Bidlack, Wayne R. *Phytochemicals as Bioactive Agents.* Lancaster, PA: Technomic Publishers, 2000.

Byrde, R.J.W., Fielding, A.H. and Williams, A.H. 1960, Phenolics in Plants in Health and Disease, J.B. Pridham (Ed.), Pergamon Press, London, 95.

Constabel, C.P. 1999, Induced Plant Defenses Against Herbivores and Pathogens: Biochemistry, Ecology and Agriculture, Agrawaal, A.A., Tuzun, S. and Bent, E.(Eds), APS Press, St. Paul, MN (USA), 137.

Friend, J. 1979, Recent Advances in Phytochemistry, Biochemistry of Plant Phenolics, Swain, T., Harborne, J.B., Van Sumere, C.F. (Eds), Plenum Press, NewYork, 557.

Grain Legumes No.15 – Dec. 1996–Jan., Feb. 1997

Hahlbrock, K. and Scheel, D. 1989, Ann. Rev. Plant Physiol. Plant Mol. Biol. 40,347.

Harborne, J.B. and Grayer, R.J. 1993, Flavonoids: Advances in research since 1986, Harborne, J.B. (Ed.), Chapman and Hall, London, 589.

Karban, Richard and Ian T. Baldwin. *Induced Responses to Herbivory.* Chicago: University of Chicago Press, 1997.

Makanfuola,W.A.(1990) evaluation of extracts of neem (*Azadiracta indica*) for the control of some stored product pests. *J. Stored Product. Res.*, 25(4):231-237.

Michael (1996) Resistance to insects and other pests: role of plant secondary metabolites

Overeem, J.C., 1976, Biochemical Aspects of Plant Parasite Relationships, Friend, J. and Threlfall, D.R. (Eds), Academic Press, London, 195.

Ramawat, K.G and Merillon, J.M (1999) Biotechnology technology secondary metabolites, Mohan Rimlaini for Oxford and IBH Publishing Co. Pvt. Ltd., New Delhi.

Rawat R.R. and Singh O.P.(1981).effectiveness of various forms of custard apple seed powder as seed protectant against the pulse beetle *Callosobruchus chinensis. Bulletin of Grain Technology.* 19(1); 53-56.

Rosenthal, Gerald A. and May R. Berenbaum. *Herbivores, Their Interactions with Secondary Plant Metabolites*. San Diego, CA: Academic Press, 1991.

Vincenzo Lattanzio *etal.* (2006)Role of phenolics in the resistance mechanisms of plants against fungal pathogens and insects, Phytochemistry: Advances in Research, 23-67

Vonnie *etal.* (2007) Feeding responses to selected alkaloids by gypsy moth larvae, Lymantria dispar (L.), Department of Biological Sciences, Towson University, 8000 York Road.

Chapter 9

Seed Fortification and Germination of Seed

Seed Gortification and Germination Studies in Barnyard Millet

The study conducted with barnyard millet indicated that the on set of germination occurred at 7 DAS in all the treatments including control.

Germination

The seeds treated with *Moringa oleifera, Annona squamosa* and *Aegle marmellos* recorded 100 per cent germination. The lowest germination per cent (92) was recorded in *Morinda tinctoria* and control(seed treatment with distilled water).

Shoot and Root Length

The seeds treated with *Annona squamosa* recorded the longest root length of 12.86 cm and it was on par with *Aegle marmellos, Moringa oleifera* and *Morinda tinctoria* with 12.41, 12.22, 12.21 cm, respectively. The lowest root length of 11.21 cm was recorded in control while the longest shoot of 11.20 cm was recorded in the seeds treated with *Annona squamosa* followed by *Moringa oleifera* with 9.94 cm and the shortest shoot of 7.22 cm was recorded in control.

Vigour Index

Among the different seed treatments the seeds treated with *Annona squamosa* recorded the highest vigour index of 2406.2. It was followed by *Moringa oleifera* with 2215.9 vigour index and the lowest vigour index of 1630.2 was recorded in control.

Table 9.1: Germination Per cent of Vrikshyurveda Fortified Seeds of Barnyard Millet

GLM Incorporation	Moringa	Aegle	Morinda	Annona	Mean
Albizia	66.91 (84.61)	67.85 (85.76)	65.17 (82.31)	66.93 (84.62)	**67.18 (84.90)**
Gliricidia	69.13 (87.30)	67.59 (85.35)	67.20 (85.00)	69.13 (87.30)	**68.50 (86.54)**
Pongamia	68.129 (86.15)	67.26 (85.00)	65.42 (82.69)	67.84 (85.76)	**67.33 (85.09)**
Delonix	68.47 (86.54)	68.16 (86.15)	66.91 (84.61)	69.47 (87.69)	**68.57 (86.63)**
Peltophorum	68.81 (86.92)	65.17 (82.31)	65.42 (82.69)	68.48 (86.54)	**66.97 (84.61)**
MEAN	**68.30 (86.31)**	**67.21 (84.92)**	**66.02 (83.46)**	**68.37 (86.38)**	
	M	***S***	***M at S***	***S at M***	
SEd	0.97	0.13	1.44	1.24	
CD (0.05)	NS	0.28	NS	NS	
Zero input	66.00 (83.46)				
Fertiliser application	70.14 (88.46)				

Data in the parentheses indicate observed values.

Seed Fortification and Germination Studies in Finger Millet (Ragi)

Rapid and uniform field emergence is the essential pre requisite to increase the yield, quality and ultimately the profits in annual crops. Pre-sowing invigoration is one such technological highlight focused to the above goal. The study was carried out with five different freshly prepared leaf extracts *viz., Moringa oleifera, Aegle marmellos, Morinda tinctoria* and *Annona squamosa* and *Vitex negundo* in five percent concentration for 6 hrs and seed treatment with distilled water served as control.

Germination

In a laboratory experiment seeds of finger millet were fortified with leaf extracts of five tree species *viz., Aegle marmellos, Annona squamosa, Moringa oleifera, Morinda tinctoria* and *Vitex negundo* and the effect of seed fortification on germination of seeds studied. Treating seeds with *Annona squamosa* leaf extract recorded 92 per cent germination while fortification of seeds with distilled water registered 90 per cent germination. But one hundred per cent germination was registered in the rest of the tree species under lab conditions.

Shoot and Root Length

The seeds treated with *Vitex negundo* leaf extracts produced longest root of 14.90 cm and it was on par with *Aegle marmellos, Moringa oleifera, Annona squamosa* and *Morinda tinctoria* with 14.86 cm, 14.22 cm, 14.21 cm and 14.41 cm respectively. The shortest root of 12.50 cm was produced in control while the maximum shoot length of 9.98 cm was recorded in the seeds treated with *Vitex negundo*, which was on par with *Moringa oleifera* and *Aegle marmellos*. The smallest shoot of 7.22 cm was produced in control.

Table 9.2: Germination Per cent of Vrikshyurveda Fortified Seeds of Finger Millet (Lab Study)

GLM Incorporation	*Germination DAS*	*Germination (Per cent)*	*Root Length (cm)*	*Shoot Length (cm)*	*Vigour Index*
Moringa oleifera	7	100 (88.83)	14.22	9.94	2416.00
Morinda tinctoria	7	96 (78.46)	14.41	8.49	2198.40
Annona squamosa	7	92 (73.57)	14.21	8.37	2077.36
Aegle marmellos	7	100 (88.83)	14.86	9.87	2473.02
Vitex negundo	7	100 (88.83)	14.90	9.98	2488.00
Distilled water	7	90 (71.57)	12.50	7.22	1774.80
SEd			0.34	0.14	37.91
CD (P≤0.05)			0.74	0.31	82.60

Table 9.3: Germination Per cent of Vrikshyurveda Fortified Seeds of Finger Millet (Field Study)

GLM Incorporation	*Moringa*	*Aegle*	*Morinda*	*Annona*	*Mean*
Albizia	81.61 (64.61)	72.42 (58.33)	83.95 (66.38)	73.70 (59.15)	**77.92 (62.12)**
Gliricidia	79.27 (62.92)	83.02 (65.67)	80.86 (64.06)	79.73 (63.25)	**80.72 (63.97)**
Leucaena	85.36 (67.51)	76.45 (60.99)	76.20 (60.82)	79.45 (63.05)	**79.37 (63.09)**
Delonix	95.17 (78.03)	93.71 (75.49)	93.69 (75.75)	93.90 (76.12)	**94.12 (76.35)**
Peltophorum	68.95 (56.14)	69.42 (56.43)	69.41 (56.43)	73.82 (59.24)	**70.40 (57.06)**
Mean	**82.07 (69.41)**	**79.01 (63.38)**	**80.82 (64.69)**	**80.12 (64.16)**	
	M	***S***	***M at S***	***S at M***	
SEd	1.37	0.75	2.00	1.68	
CD (P≤0.05)	3.82	1.60	4.87	3.57	
Fertiliser application			82.79		
Zero input			61.75		

Vigour Index

Among different seed treatments, the seeds treated with *Vitex negundo* recorded the highest vigour index of 2488. It was on par with *Aegle marmellos* and *Moringa oleifera* with the value of 2473 and 2416 respectively. The lowest vigour index of 1774 was recorded in control.

Points for Discussion

Rapid and uniform field emergence is the essential pre requisite to increase the yield, quality and ultimately the profits in annual crops. Pre-sowing invigoration is one such technological highlight focused to the above goal. Pre-sowing seed invigoration treatments are numbered many by the researchers and all are claimed to have invigorative effect at field for enhancing the yield of a crop to a tune of 10-15 per cent. Some of the widely followed presowing seed treatment techniques are seed fortification with growth regulators, botanicals, seed pelleting, osmotic priming and seed infusion. Adoption of any of these techniques for a particular crop required standardization work as the responses of the seed to the pre-sowing treatment vary with chemical concentration and duration of the treatment. Standardization is even stressed, based on the lot, as ageing is one another factor that modifies or alters the response of seed to any physical or physiological management technique. The results expressed that seed fortification with *Vitex negundo* leaf extract at 5 per cent found to maximize the germination percentage, root length, shoot length and vigour index of seed, which was on par with *Moringa oleifera* and *Aegle marmellos* leaf extract and control had the lowest germination. The increase in germination by these botanical leaf extracts is not very much known due to lack of information on the active principle involved in it. However, it is presumed that these botanicals contain some of the micro nutrients which are conducive for seed invigoration and presence of saponin like substance which acts as a precursor of GA_3 and invigorated the seed at a particular concentration.

The study also revealed that root length was maximum in the treatment of seed soaking with *Vitex negundo*. The presence of phenols in the leaf extracts would have promoted the root length which is reported by several workers. The control (Seeds are soaked with distilled water) recorded the lowest root length in finger millet. This might be due to low availability of nutrients in the water.

Increased emergence and vigorous plant development in crop seeds primed with *Moringa* leaf extract due to the presence of calcium, potassium, ascorbic acid and Cytokinin has been observed. Seed fortification with plant growth regulators and nutrient solution are effective in increasing seed germination, growth and development of plants. The probable reason for higher germination in fortification treatments could be due to greater hydration of colloids, increase in bound water content, lower water deficit, more efficient root system and increased metabolic activities. The efficiency of *Moringa* leaf extract as a seed fortification agent has been documented. It may help in increasing antioxidant status, chlorophyll content, leaf area and yield of crops.

Studies at Tamil Nadu Agricultural University indicated that leaf extracts of *Vitex negundo* caused significant changes in the germination percentage. As compared to control, the aqueous leaf extracts of *V.negundo* at 5 per cent concentration levels exhibited promotory effect on seed germination and seedling growth in both green gram and black gram. The differential degree of inhibitory and stimulatory effects on germination and growth of seedlings may be due to the presence of allelochemicals at different level in the leaf extracts of *Vitex negundo*. The higher seed germination percent and enhanced seedling vigour were noticed after seed treatment with *Coleus aromaticus, Adathoda vasica* and *Vitex negundo* extracts

As far as the growth parameters are concerned, the majority of leaf extracts showed a phytostimulatory effect by way of enhanced seed germination percentage and improved emergence of the seedlings.

Seed Fortification and Germination Studies in Maize

The study conducted with maize indicated that the on set of germination occurred at 3 DAS in all the treatments including control.

Germination

The germination percent of all treatments and control recorded 100 per cent germination value and the Germination energy was calculated and it indicated that among the seed treatment the higher germination value was obtained in *Annona squamosa* (2.38) followed by *Morinda tinctoria* (1.96) and control (1.96) while among the seed treatments the maximum germination value was recorded in *Annona squamosa* (433) followed by *Morinda tinctoria* and control which were on par, with others barring *Alangium salvifolium*. The Emergence energy value (EEV) was also worked out and it presented that among the seed treatment the maximum Emergence energy value was recorded in *Annona squamosa* (43.30) followed by *Morinda tinctoria* and control which were on par, recorded the value of 34.10 and minimum in *Alangium salvifolium* (22.50). As regards Germination relative index (GRI), among the seed treatments the maximum germination relative index (GRI) maximum were obtained in *Annona squamosa* (44.00) followed by *Aegle marmellos, Morinda tinctoria* and control which were on par (34.00) and minimized *Alangium salvifolium* (22) and also emergence energy value was maximized in *Annona squamosa* (43.30) followed by *Morinda tinctoria* and control which were on par (34.10) and minimized in *Alangium salvifolium* (22.50).

Table 9.4: Germination of Vrikshyurveda Fortified Seeds of Maize (Lab Study)

GLM Incorporation	*Germination DAS*	*Germination (Per cent)*	*Germination Energy (GE)*	*Germination Value (GV)*	*Emergence Energy Value (EEV)*	*Germination relative Index(GRI)*
Alangium	3	100	1.30	225.00	22.50	22
Aegle	3	100	1.71	341.66	31.60	34
Morinda	3	100	1.96	341.66	34.10	34
Annona	3	100	2.38	433.33	43.30	44
Control	3	100	1.96	341.66	34.10	34
SEd			0.213	40.148	4.554	2.923
CD (P≤0.05)			**0.444**	**83.747**	**9.500**	**6.097**

Table 9.5: Germination Per cent of Vrikshyurveda Fortified Seeds of Maize (Field Study)

GLM Incorporation	*Alangium*	*Aegle*	*Morinda*	*Annona*	*Mean*
Albizia	58	68	74	79	**69.75**
Gliricidia	68	68	58	63	**64.25**
Leucaena	53	63	68	74	**64.50**
Delonix	53	74	63	74	**66.00**
Peltophorum	74	63	63	68	**67.00**
Mean	**61.20**	**67.20**	**65.20**	**71.60**	
	M	***S***	***M at S***	***S at M***	
SEd	0.60	0.43	1.04	0.98	
CD ($P \leq 0.05$)	NS	0.86	2.07	1.94	
Fertiliser application	65				
Zero input (C_1)	58				

Discussion Points

The increase in germination by these botanical leaf extracts is not very much known due to lack of information on the active principle involved in it. However, it is presumed that these botanicals contain some of the micro nutrients as well as saponin like substance which acts as a precursor of GA_3 which are conducive for seed invigouration. Germination value combines as it does both total germination capacity and germination energy and hence is a better measure of seed performance. In terms of this seed germination value also, *Annona squamosa* showing higher value (433.33) followed by *Morinda tinctoria* and control which were on par, recorded (341.66) than *Alangium salvifolium*, which is indicated to be superior where the germination value, as an integrated measure of seed quality. Thus the study highlighted that Seed fortification with 5 per cent per cent *Annona squamosa* leaf extract enhanced the seed and seedling quality characters and hence could be adopted as a pre sowing seed treatment for maize crop.

Seed Fortification and Germination Studies in Black Gram

The similar study conducted with black gram indicated that the on set of germination occurred at 3 DAS in all the treatments including control.

Germination

The higher germination per cent was recorded in *Moringa oleifera* (94). It was followed by *Annona squamosa* (92) and the lowest germination per cent (82) was recorded in control.

Shoot and Root Length

Among the seed treatments the longest root was recorded in *Moringa oleifera* (28.03). It was followed by *Annona squamosa* (26.63) and the shortest root was recorded in *Aegle marmellos* (21.67) while among the seed treatment the maximum shoot length was recorded in *Moringa oleifera* (24.57). It was followed by *Annona squamosa* (23.20) and the minimum shoot length were recorded in *Aegle marmellos* (18.85).

Vigour Index

Among the seed treatment the higher vigour index was recorded in *Moringa oleifera* (4944). It was followed by *Annona squamosa* (4584) and the lower vigour index was recorded in *Aegle marmellos* (3647).

Table 9.6: Germination Per cent of Vrikshyurveda Fortified Seeds of Black Gram (lab study)

Treatments	*Germination DAS*	*Germination (Per cent)*	*Root Length (cm)*	*Shoot Length (cm)*	*Vigour Index*
Moringa	3	94	28.03	24.57	4944.09
Aegle	3	90	21.67	18.85	3647.40
Morinda	3	92	23.81	20.78	4102.28
Annona	3	92	26.63	23.20	4584.97
Control	3	82	24.00	22.67	3826.94
SEd			0.77	0.06	
CD (P≤0.05)			**2.15**	**0.18**	

Table 9.7: Germination Per cent of Vrikshyurveda Fortified Seeds of Black Gram (Field Study)

GLM Incorporation	*Foliar Applications*				*Mean*
	Moringa	*Aegle*	*Morinda*	*Annona*	
Albizia	78	73	73	80	**76**
Gliricidia	75	73	69	75	**73**
Pongamia	83	74	78	73	**77**
Delonix	73	78	68	64	**70**
Peltophorum	64	62	73	75	**68**
Mean	**74**	**72**	**72.2**	**73.4**	
	M	***S***	***M at S***	***S at M***	
SEd	0.32	0.54	1.10	1.22	
CD (P≤0.05)	0.89	1.16	2.42	2.61	
Fertiliser application	64				
Zero input	52				

Discussion

The cry on organic farming elaborates the need of user friendly and eco friendly technique for seed management techniques. The present study carried out with four different freshly prepared leaf extracts *viz., Moringa oleifera, Aegle marmellos, Morinda tinctoria* and *Annona squamosa* in five percent concentration for 12 h soaking along with untreated seeds. The results expressed that seed fortification with *Moringa oleifera* leaf extract at 5 per cent found to maximize germination percentage, root length, shoot length and vigour index of seed, followed by *Annona squamosa* leaf extract and control. The increase in germination by these leaf extracts is not very much known due to lack of information on the active principle involved in it. However, it is presumed that these botanicals contain some of the micro-nutrients as evidenced from this study which are conducive for seed invigoration.

Seed Fortification and Germination Studies in Onion

Sprouting Percentage

The soil incorporation of green leaf manures had significant effect on sprouting of aggregatum onion at 3 and 6 DAS. The highest percentage of sprouting (69.98) was recorded in *Senna siamea*, which was on par with *Albizia lebbek* and *Gliricidia sepium* and the lowest percentage of sprouting (67.03) was recorded in *Leucaena leucocephala* at 3 DAS. While at 6 DAS the highest percentage of sprouting (85.37) was recorded in *Leucaena leucocephala*, which was on par with *Gliricidia sepium* and *Delonix regia* the lowest percentage of sprouting (82.79) was recorded in *Senna siamea*. The absolute control zero input recorded 42.01 and 55.23 per cent sprouting at 3 and 6 DAS while recommended dose of inorganic fertilizers recorded 70.1 and 85.5 percent of sprouting at 3 and 6 DAS respectively.

Points for Discussion

The study revealed that sprouting percentage was highest in the treatment with *Gliricidia sepium* as green leaf manure and bulb soaking with *Aegles marmellos*. This might be due to the presence phenols in the leaf extracts, which would have promoted the sprouting in onion. The recommended dose of inorganic fertilizers recorded lower percentage of sprouting. The absolute control recorded the lowest percentage of sprouting in aggregatum onion, since this plot did not receive any application of organic and inorganic nutrient. Hence this might have recorded the lowest sprouting.

Seed Fortification and Germination Studies in Bhendi

Germination

The soil incorporation of green leaf manures had significant effect on germination percentage of Bhendi at 5 DAS. The highest percentage of germination (85.5) was recorded in *Albizia lebbek* follwed by *Leucaena leucocephala* and the lowest percentage of germination (81.0) was recorded in *Delonix regia* at 5 DAS. The seed treatment with leaf extracts had significant effect on Bhendi. Among the seed treatments, *Aegles marmellos* registered the highest percentage of germination (85.9), while *Alangium salvifolium* recorded the lowest (81.8) at 5 DAS. The combination of soil application of *Albizia lebbek* and seed fortification with leaf extract of *Annona squamosa* recorded the highest germination percentage 92.2 followed by *Gliricidia sepium* +*Aegles marmellos* and the lowest was recorded in *Gliricidia sepium* + *Annona squamosa* with 77.2 per cent at 5 DAS.

The increase in germination percentage of *Albizia lebbek* + *Annona squamosa* over absolute control and inorganic fertilizer alone was 27.2 and 6.9 respectively.

Table 9.8: Germination Per cent of Vrikshyurveda Fortified Seeds of Bhendi

GLM Incorporation	*Alangium*	*Annona*	*Aegles*	*Morinda*	*Ocimum*	*Mean*
Albizia	82.27	92.25	81.72	86.03	85.73	85.59
Delonix	81.24	79.58	84.33	80.69	79.32	81.03
Gliricidia	78.72	77.21	90.52	80.01	82.71	81.83
Leucaena	85.37	85.75	87.27	81.94	83.64	84.79
Mean	81.89	83.69	85.96	82.16	82.85	83.31
Zero input						65.00
Fertiliser application						85.28
	M	***S***	***M x S***			
SEd	0.620	0.739	1.460			
CD (P≤0.05)	1.518	1.505	3.083			

Discussion

The germination percentage revealed that it was high in the treatment *Albizia lebbek* as green leaf manure and seed soaking with *Annona squamosa* leaf extract. The increased level of germination in the present study may be due to the synergistic effect of leaf extracts and green leaf manures. This effect may be because of the presence of the saponins, GA_3 in traces and micronutrients especially zinc. The recommended dose of inorganic fertilizers recorded low percentage of seed germination. The absolute control recorded the lowest percentage of germination in bhendi, since this plot did not receive any application of organic and inorganic nutrient and hence the absence of boosting mechanism resulted in the lowest germination.

Chapter 10
Soil Fertility and Nutrient Management

Fertility of a soil is defined by its ability to provide all essential nutrients in adequate quantities and in the proper balance for the growth of plants in dependent of direct application of nutrients when other growth factors like light, temperature and water are favorable. This ability depends on the nutrient content of the soil and also on its efficiency in transforming nutrients within the farm's nutrient cycle. The break down of biomass from crop residues, green manures and mulch, these contribute to build up of soil organic matter, including humus. They also play an essential role in transferring nutrients from the soil organic matter to the mineral stage, which is available to plants. Soil organisms also protect plants from disease and make the soil crumbly. A fertile soil is easy to work, absorbs rain water well and it is robust against siltation and erosion. It neutralizes acids, which pass through contaminated air to the soil surface and decomposes pollutants such as pesticides rapidly. And last but not least a fertile soil is an efficient storage for nutrients and CO_2. In this way a fertile soil prevents the eutrophication of rivers, lakes and oceans and contributes to the reduction of global warming. It is the responsibility of farmers to understand soil ecology to the point that they can create or restore the conditions for a robust balance in the soil. If a soil does not regularly bring good yields, farmers should investigate the reasons for it. Soil organisms play a vital role in transformation of nutrients in soil.

Properties of a Fertile Soil

a) It is rich in nutrients necessary for basic plant nutrition (including nitrogen, phosphorus, potassium, calcium, magnesium and sulphur);

b) It contains sufficient micronutrients for plant nutrition (including boron, copper, iron, zinc, manganese, chlorine and molybdenum);

c) It has optimum amount of soil organic matter;

d) It has a neutral pH in a suitable range for crop production (between 6.0 and 7.5);

e) It has crumbly structure and is biologically active; and

g) It has good water holding capacity.

The plants require 16 essential nutrients to grow to its full potential. Out of the 16 essential elements hydrogen, carbon and oxygen are obtained mainly from the air and from water. The other essential elements come from the soil and are generally managed by the farmers. Some of these nutrients are required in large quantiy known as Macro (major) nutrients such as nitrogen (N), phosphorus (P) potassium (K); Some others are required in small quantity known as (minor) nutrients such as boron (B), copper (Cu), iron (Fe), chloride (Cl), manganese (Mn), molybdenum (Mo) and zinc (Zn). A few others are essential as they are limiting in soil such as calcium (Ca), magnesium (Mg) and sulphur (S). Recycling organic matter such as crop residues and tree leaves is an excellent way of providing micronutrients to growing plants.

Conditions for Uptake of Nutrients from the Soil

a) The soil must be sufficiently moist to allow the roots to take up and transport the nutrients. Sometimes supplying water to plants will eliminate nutrient deficiency symptoms.

b) The pH of the soil must be within a certain range for nutrients to be releasable from the soil particles.

c) The temperature of the soil must fall within a certain range for nutrient uptake to occur.

d) The nutrients must be within the root zone in order for the roots to access them.

The optimum range of temperature, pH and moisture is different for different crops. Thus, nutrients may be physically present in the soil, but not available to plants. Knowledge of soil pH, texture and history can be very useful for predicting which nutrients may become deficient.

Neutral Soil pH

Soil pH, its acidity or alkalinity, is highly relevant to how nutrients become readily available in soil, known as solubility of nutrients. Plants differ in their sensitivity to a low or high pH level. Some plants can withstand or even prefer a somewhat low pH level. Soils with pH less than 6.5 and which respond to liming may be considered as acid soils. When potassium, calcium and magnesium leach from the soil, it becomes acidic. This may happen if there is a lot of rain (or irrigation water) that washes nutrients away, or if too much mineral nitrogen fertilizers are applied. Phosphorous gets immobilized and its availability is reduced. Most of the activities of beneficial organisms like *Azotobacter* and nodule forming bacteria of legumes are adversely affected as acidity increases under acidic conditions, the bacteria fix less nitrogen and decompose less organic matter, which results in fewer available nutrients. Addition of lime or compost will help to neutralize acidity and to increase the pH, so that the availability of nutrients will be increased. Alkaline soils are formed due to concentration of exchangeable sodium and high pH. The pH of alkaline soil can be corrected through application of gypsum: Application of molasses or growing of green manure crops and their incorporation in the field can also help to correct an alkaline soil.

Plant roots prefer soil with a crumbly structure, like well-made bread. Such soil is well-aerated and the plant roots are able to penetrate easily. This allows them to grow both wide and deep allowing them to access more nutrients to support good growth. Soil aggregation is also an important indicator of the workability of the soil. Soils that are well aggregated are said to have "good tilth". A good soil structure also contributes to reduction of erosion of topsoil, as water infiltrates more easily into the soil and the aggregates resist to the raindrops.

Organic Agriculture for Soil Fertility Management

Intensively managed soils require good soil fertility management to ensure long term sustainability. That is why proper soil fertility management is central importance in organic crop production and farming. Basically organic farmers approach soil fertility management by conserving and protecting their soils from sun, rain and wind and feeding it with organic material in an appropriate way, so as to allow it to feed the plants in a balanced way. When the soil is fertile in the organic sense, it can produce good crop yields for several years.

Organic soil fertility management can be done through strategic approach to manage soil fertility and plant nutrition.

Strategy 1

Soil and moisture conservation and availabilty- Conservation of soil, soil organic matter and soil water from loss which can be achieved through

- ✫ Formation of contour ridges and bunds, grass strips and terraces, application of mulch to the soil surface prevents soil erosion and water loss.
- ✫ Use of mulch and raising cover crops protect the soil.
- ✫ Construction of small water harvesting structures like saucer basins, micro catchment, soil pits, farm pond, etc help in water harvesting.
- ✫ Minimum tillage to minimize soil disturbance.

Strategy 2

Soil health and Soil Organic Carbon- Improving organic matter content and enhancing biological activity in the soil through addition of organic sources that can build an active soil with good structure which can hold water and supply plant nutrients. It includes practices like

- ✫ Application of composts and other organic materials to the soil improves soil structure and water holding capacity and contributing to soil organic matter content.

- ✰ Growing of green manure crops and incorporation into the soil facilitate growth and multiplication of soil organisms and thus mineralize rapidly to provide nutrients to the crop that follows.
- ✰ Application of green leaf manures of nitrogen fixing trees.
- ✰ Recycling of valuable animal manures for composting.
- ✰ Soil fertility is best improved through integrated and combined implementation of the different practices as practicing any one of the techniques though it helps to maintain the soil fertility but it takes a long time and hence the combination of techniques is ideal to speedy improvement of soil fertility.

Strategy 3

Soil chemical balance and supplementation-Application of soil amendments. These supplementary measures include

- ✰ Application of soil amendments such as lime and gypsum to correct soil pH and microbial inoculations to enhance biological activity of the soil and nitrogen fixation in the soil.
- ✰ Artificial application of water through different types and modes to supplement water requirements.

A holistic approach is essential to get optimum soil fertility and soil productivity for crop growth.

Use of Commercial Organic and Selected Mineral Fertilizers to Satisfy Specific Nutrient Needs

Organic soil fertility management principally relies on natural sources and biological processes and seeks for a long-term improvement of soil fertility based on optimization of farm-own nutrient management. In some conditions though, it may be necessary to supply organic materials from out- side of the farm to build up soil fertility and produce reasonable harvests in a short time. In certified organic agriculture mineral fertilizers shall be used as a supplement to biologically-based fertility methods only. Their use shall be justified by appropriate soil and leaf analysis.

Commercial Organic Fertilizers

Commercial organic fertilizers are mostly by-products of agro-processing and food industrial waste like seed oil cakes (soybean, sunflower, neem, peanut), pelleted chicken manure and agro-processing by-products such as fruit peels, coffee husks, wood shavings and dust, rice husks and plant ashes. Others include bone meal, fish meal as well as commercially produced composts.

Biofertilizers

Microorganisms play an important role in the soil in providing nutrients to plants. Some microbes add nutrients to the soil through mineralization. Others add nitrogen by fixing it from the atmosphere. These include *Rhizobium, Azospirillum* and

Azotobacter. Other microbes, such as *Mycorrhizal* fungi, help to supply plants with phosphorus. Pseudomonas species are a diverse group of bacteria that can use a wide range of compounds that plants give off when their roots leak or die. They are able to solubilize phosphorus and may help to suppress soil-borne plant diseases.

Some farmers and companies may recommend the application of microorganisms to the soil to enhance decomposition processes and control diseases. These biofertilizers are usually sold as ready-to-use products for application as sprays or with irrigation water, or for mixing with compost. These products contain living microorganisms and need to be stored and applied cautiously. Microbial products should be used before their expiry date. It is recommended to find out the effect of these products first by testing them on a small-scale and comparing result with the untreated plots. Microbial fertilizers cannot substitute appropriate soil management practices on the farm. Most bacteria, fungi and other microorganism are naturally present in the soil and can be enhanced by proper application of compost.

Nitrogen Dynamics in Soil

Nitrogen (N) is the most frequently deficient nutrient in most non legume cropping systems. Many inorganic and organic sources are available to supply N to crops. Understanding the chemistry and biology of soil N is essential for maximizing productivity while reducing the impacts of N inputs on the environment.

The ultimate source of all N used by plants is N_2, which constitutes 78 per cent of the earth's atmosphere. Unfortunately, higher plants cannot metabolize N_2 directly into protein. N_2 must first be converted to plant available N by:

- Microorganisms that live symbiotically on legume roots,
- Free-living or non-symbiotic soil microorganisms,
- Atmospheric electrical discharges forming N oxides, or
- The manufacture of synthetic N fertilizers.

The rate of NO_3^- uptake is usually high, causing an increase in rhizosphere pH. When plants absorb high levels of NO_3^-, there is an increase in anion (HCO_3^-, OH^-, organic anions) transport out of cells and an increase in cation absorption (Ca^{2+}, Mg^{2+}, K^+). Plants metabolize NO_3^- to NH_4^+ to amino acids and to proteins. NO_3^- reduction is an energy-requiring process that uses 2 nitrate reductase (NADH) molecules for each NO_3^- reduced in protein synthesis. Thus, NH^{4+} is the preferred N source since energy is conserved compared to NO_3^- (one less step in the reduction process). Plants supplied with NH_4^+ may have increased carbohydrate and protein levels compared with NO_3^-.

Plant uptake of NH_4^+ proceeds best at neutral pH values and is depressed by increasing acidity. Absorption of NH_4^+ reduces Ca^{2+}, Mg^{2+} and K^+ uptake while increasing $H_2PO_4^-$, SO_4^{-2} and Cl^- absorption. Rhizosphere pH decreases with NH_4^+ uptake, caused by H^+ exuded by the root to maintain electro neutrality or charge balance inside the plant. Differences in 2 pH units have been observed for NH_4^+ versus NO_3^- uptake. The acidification can affect both nutrient availability and biological activity in the vicinity of roots.

NH_4^+ tolerance limits are narrow. High levels of NH_4^+ can retard growth and restrict K^+ uptake. In contrast, plants can accumulate and tolerate comparatively high NO_3^- levels in tissues.

Preference of plants for either NH_4^+ or NO_3^- is determined by plant age and type, environment and other factors. Cereals, corn, sugar beets, pineapple, rice and ryegrass use either form of N. Kale, celery, bush beans and squash grow best when provided with some NO_3^-. Some plants, such as blueberries, *Chenopodium album* and certain rice cultivars, cannot tolerate NO_3^-. Solanaceous crops, such as tobacco, tomato and potato, prefer a high NO_3^-/NH_4^+ ratio.

Plant growth is often improved when the plants are nourished with both NO_3^- and NH_4^- compared with NO_3^- or NH_4^+ alone. Mixtures are beneficial at certain growth stages for some genotypes of corn, sorghum, soybeans, wheat and barley; increased non legume yields with NH_4^+ nutrition are associated with greater tillering. Corn yields increased from 8 to 25 per cent with $NH_4^+ + NO_3^-$, compared with yields with NO_3^- alone, which was related to increased numbers of kernels/plant and not to heavier kernels.

Before NO_3^- can be used in the plant, it must be reduced to NH_4^+ or NH_3. Nitrate reduction involves 2 enzyme-catalyzed reactions that occur in roots and/or leaves, depending on the plant species. Both reactions occur in series so that toxic nitrite (NO_2^-) does not accumulate.

	Reduction Reaction	*Enzyme*	*Reaction Site*
Step 1	$NO_3^- \rightarrow NO_2^-$	Nitrate reductase	Cytoplasm
Step 2	$NO_2^- \rightarrow NH_3$	Nitrite reductase	Chloroplast

The NH_3 produced is assimilated into amino acids that are subsequently combined into proteins and nucleic acids. Proteins provide the framework for chloroplasts, mitochondria and other structures in which most biochemical reactions occur. The type of protein formed is controlled by a specific genetic code found in nucleic acids, which determines the quantity and arrangement of amino acids in each protein. One of these nucleic acids, deoxyribonucleic acid (DNA), present in the nucleus and mitochondria of the cell, duplicates the genetic information in the chromosomes of the parent cell to the daughter cell. Ribonucleic acid (RNA), present in the nucleus and cytoplasm of the cell, executes the introductions coded within DNA molecules. Most enzymes controlling these metabolic processes are also proteins and are continually metabolized and resynthesized.

In addition to the formation of proteins, N is an integral part of chlorophyll, which converts light into chemical energy needed for photosynthesis. An adequate supply of N is associated with high photosynthetic activity, vigorous vegetative growth and a dark green colour. An excess of N in relation to other nutrients, such as P, K and S can delay crop maturity. Stimulation of heavy vegetative growth early in the growing season can be a disadvantage in regions where soil moisture limits plant growth. Early-season depletion of soil moisture without adequate replenishment before the grain-filling period can depress yields. If N is used properly in conjunction with other needed inputs, it can speed the maturity of crops and reduce the energy required to dry grain to 15.5 per cent moisture or permit earlier harvest.

Supplying sufficient NH_4^+ and NO_3^- to meet plant requirement depends on the quantity of mineralized from soil organic N with the remainder provided through fertilizer or organic N applications. The amount of plant available N released from organic N depends on many factors affecting N mineralization, immobilization and losses of NH_4^+ and NO_3^- from the soil.

N Mineralization and Immobilization in Soil

N mineralization is the conversion of organic N to NH_4^+ through two reactions, aminization and ammonification. Mineralization occurs through the activity of heterotrophic microorganisms that require organic C for energy. Heterotrophic bacteria dominate the breakdown of proteins in neutral and alkaline environments, with some involvement of fungi, while fungi predominate in acid soils. The end products of the activities of one group furnish the substrate for the next and so on until the material is decomposed. Aminization converts proteins in residues to amino acids, amines and urea. These are organic N compounds that are further converted to inorganic NH_4^+ by ammonification. A diverse population of aerobic and anaerobic bacteria, fungi and actinomycetes is capable of converting the products of aminization to NH_4^+.

The NH_4^+ produced through ammonification is subject to several fates. NH_4^+ can be:

- ☆ Converted to NO_2^- and NO_3^- (nitrification);
- ☆ Absorbed directly by higher plants (N uptake);
- ☆ Utilized by heterotrophic bacteria to decompose residues (immobilization)
- ☆ Fixed as biologically unavailable N in the lattice of certain clay minerals (NH_4^+ fixation), or
- ☆ Converted to NH_3 and slowly released back to the atmosphere (volatilization).

Soil moisture content regulates the proportions of aerobic and anaerobic microbial activity. Maximum aerobic activity and N mineralization occur between 50 and 70 per cent water-filled pore space. Soil temperature also influences microbial activity and N mineralization. Optimum soil temperature for microbial activity ranges between 25 and 35°C. Soil OM contains about 5 per cent N and during a growing season 1 to 4 per cent of organic N is mineralized to inorganic N. As total soil N content increases, the quantity of N mineralized from soil organic N increases. Therefore, soil and crop management strategies that conserve or increase soil OM will result in a greater contribution of mineralizable N to N availability to crops. The quantity of N mineralized during the growing season can be estimated. For example, if a soil contained 4 per cent OM with 2 per cent mineralization rate, thus, each year, 80 lbs N/a as NH_4^+ are mineralized, which can be utilized by plants or other soil N processes.

N immobilization is the conversion of inorganic N (NH_4^+ and NO_3^-) to organic N and is basically the reverse of N mineralization. If decomposing OM contains low N, microorganisms will immobilize NH_4^+ and NO_3^- in the soil solution. The microbes need N in a C:N ratio of about 8:1' therefore, inorganic N in the soil is utilized by the rapidly growing population. N immobilization during crop residue decomposition can reduce NH_4^+ and NO_3^- to very low levels. Microorganisms compete effectively with plants for NH_4^+ and NO_3^- during immobilization and plants can readily become N deficient. Fortunately, in most cropping systems, sufficient fertilizer N is applied to compensate for immobilization and crop requirements. If added organic material contains high N, immobilization will not proceed because the residue contains sufficient N to meet microbial demand and inorganic N will increase from mineralization of organic N in the residue.

The ratio of per cent C to per cent N (C:N ratio) defines the relative quantities of these two elements in crop residues and other fresh organic materials, soil OM and soil microorganisms. The N content of humus or stable soil OM ranges from 5.0 to 5.5 per cent, whereas C ranges from 50 to 58 per cent, giving a C:N ratio ranging between 9 and 12. Whether N is mineralized or immobilized depends on the C:N ratio of the OM being decomposed by soil microorganisms.

Generally, when organic residues with C:N > 20:1 are added to soil, soil N is immobilized during the initial decomposition process. For residues with C:N < 20:1 there is a release of mineral N early in the decomposition process. Soil OM may also increase, depending on the quantity and type of residue added and the quantity of OM loss through oxidation (mineralization) or physical soil loss. The time required for residue decomposition depends on the quantity of OM added, inorganic soil N supply, resistance of the residue to microbial attack (a function of the amount of lignins, waxes and fats present) and soil temperature and moisture.

Phosphorus Dynamics in Soil

Phosphorus (P) is less abundant in soils than N and K. Total P in surface soils varies from 0.005 to 0.15 per cent. The average total P content of soils is lower in the humid Southeast than in the Prairie Provinces and Western states. Unfortunately, the quantity of total P in soils has little or no relationship to the availability of P to plants. Although prairie soils are often high in total P, many of them are characteristically low in plant available P. Therefore, understanding the relationships and interactions of the various forms of P in soils and the numerous factors that influence P availability is essential to efficient P management.

The P Cycle

The decrease in soil solution P with absorption by plant roots is buffered by both inorganic and organic P fractions in soils. Primary and secondary P minerals dissolve to resupply $H_2PO_4^-$ and HPO_4^{-2} in solution. Inorganic P adsorbed on mineral and clay surfaces as $H_2PO_4^-$ and HPO_4^{-2} (labile inorganic P) can also desorb to buffer solution P. Soil microorganisms digest plant residues containing P and produce organic P compounds that are mineralized through microbial activity to supply solution P.

Water-soluble fertilizer P applied to soil readily dissolves and increases P in soil solution. In addition to P uptake by roots, inorganic and organic P fractions buffer the increase in solution P through P adsorption on mineral surfaces, precipitation as secondary P minerals and immobilization as microbial or organic P. Maintaining solution P concentration (intensity) for adequate P nutrition depends on the ability of labile P (quantity) to replace soil solution P taken up by the plant. The ratio of quantity to intensity factors defines buffer capacity (BC) or the relative ability of the soil to buffer changes in soil solution P. The larger the BC, the greater the ability to buffer solution P. The P cycle can be simplified to the following relationship.

$$\text{Soil solution P} \leftrightharpoons \text{labile P} \leftrightharpoons \text{nonlabile P}$$

where labile and nonlabile P represent both inorganic and organic fractions. Labile P is the readily available portion of the quantity factor that exhibits a high dissociation rate and rapidly replenishes solution P. Depletion of labile P causes some non labile P to become labile, but at a slow rate. Thus, the quantity factor comprises both labile and non labile P fractions. Understanding the dynamics of P transformations in soils will provide the basis for sound management of soil and fertilizer P to ensure adequate P availability to plants.

Forms and Functions of P in Plants

P concentration in plants ranges between 0.1 and 0.5 per cent, considerably lower than N and K. Plants absorb either H_2PO_4 or HPO_4^{-2} (orthophosphate) depending on soil pH. Plants also absorb soluble, low-molecular weight organic P compounds (*i.e.*, nucleic acid and phytin) that are products of soil OM decomposition. Because of the instability of many organic P compounds in the presence of active microbes, their importance as sources of P for higher plants is limited.

Functions

The most essential function is in energy storage and transfer. Adenosine di- and triphosphates (ADP and ATP) act as "energy currency" within plants. P is an essential element in deoxyribonucleic acid (DNA) and ribonucleic acid (RNA) that contain the genetic code of the plant to produce proteins and other compounds essential for plant structure, seed yield and genetic transfer. Phospholipids, phosphoproteins, coenzymes and nucleotides are important structural components of membrane chemistry and related functions. Thus, P is essential for vigorous growth and development of reproductive parts (fruits, seeds, etc.).

Adequate P is associated with increased root growth. When soluble $H_2PO_4^-$ is applied in a band, plant roots proliferate extensively in P-treated soil.

Therefore, if a soil contains 2 per cent OM in the surface 6 in. (assume 1.5 per cent OM degradation rate), the organic P content is:

2×10^6 lbs soil/afs x 0.015 x 0.02 = 600 lbs organic P/afs

Soil organic P generally increases with increasing organic C and/or N; however, the C:P and N:P ratios are more variable among soils than C:N ratio. Soils are characterized by C:N:P:S ratio, which also varies among soils. Average C:N:P:S ratio in soil is 140:10:1.3:1.3.

Most soil organic P compounds are esters of orthophosphate ($H_2PO_4^-$) including inositol phosphates (10 to 50 per cent), phospholipids (1 to 5 per cent) and nucleic acids (0.2 to 2.5 per cent). Inositol phosphates represent a series of phosphate esters ranging from monophosphate up to hexaphosphate.

P Mineralization and Immobilization in Soils

In general, P mineralization and immobilization are similar to N in that both processes occur simultaneously in soils and can be depicted as follows:

Organic P

Soil organic P originates from plant and animal residues, which are degraded by microorganisms to produce other organic compounds and release inorganic P. Pholsphatase enzymes catalyze the mineralization reaction of organic P by:

$$\underset{\text{Immobilization}}{\overset{\text{Mineralization}}{\rightleftarrows}} \text{Inorganic P} \; (H_2PO_4^- / HPO_4^{-2})$$

The quantity of P mineralized in soils increases with increasing organic P content. In contrast, the quantity of inorganic P immobilized is inversely related to soil organic P, such that as the ratio of soil organic C:P increases (*i.e.*, decreasing organic P), P immobilization increases. Evidence of organic P mineralization can be provided by measuring changes in soil organic P during the growing season. Organic P content decreases with crop growth and increases again after harvest.

Residue C:P ratio determines the predominance of P mineralization over immobilization, just as C:N influenced N mineralization and immobilization. The following guidelines have been suggested.

Expressed as per cent residue P, net P immobilization occurs when residue P < 0.2 per cent and net mineralization occurs with > 0.3 per cent residue P. When residues are added to soil, net P immobilization occurs during the early stages of decomposition, followed by net P mineralization as the C:P ratio of the residue decreases. P mineralization/immobilization processes are similar to N.

Other factors affecting the quantity of P mineralization/immobilization are temperature, moisture, aeration, pH, cultivation intensity and P fertilization. The environmental effects are similar to those described for N mineralization/immobilization, since both are microbial processes.

Inorganic fertilizer P can be immobilized to organic P by microorganisms. The quantity of P immobilized varies widely, with values of 25 to 100 per cent of applied P reported. Continued fertilizer P applications can increase organic P content and subsequently increase P mineralization. In general, organic P will accumulate with P fertilization when C and N are available in quantities relative to the C:N:P ratio of soil OM. Inorganic P will likely accumulate if C and N are limiting.

Measuring organic P cycling in soils is more difficult than for N because inorganic P produced through mineralization can be removed from solution by i) P adsorption to clay and other mineral surfaces and ii) P precipitation as secondary Al-, Fe-, or Ca-

P minerals. Immobilization and mineralization processes for P, C and N cycling are similar and related. Addition of adequate amounts of N and P to soil to which crop residues are returned then some of the added nutrients may be immobilized. The continued cultivation of soils without the addition of N and P results in depletion of nutrients in soil through mineralization. If N and P are present in insufficient amounts, the synthesis of soil OM may be reduced.

P Sources

Animal and municipal wastes are excellent sources of plant available P, with manure accounting for 98 per cent of organic P applied to crop land. The form and content of P in fresh animal waste varies greatly depending on P content of the feed and animal type. Typically, inorganic P ranges from 0.3 to 2.4 per cent of the dry weight, while organic P ranges from 0.1 to 1 per cent. In fresh manure, organic P represents 30 to 70 per cent of total P. Mineralization of organic P during storage usually increases inorganic P content and decreases organic P.

Microbial P Fertilization

Phosphate-solubilizing bacilli and other rhizobacteria are abundant in soil and may be readily isolated from plant rhizospheres. The rhizobacteria colonizing plant roots and stimulating plant growth are known as plant growth promoting bacteria (PGPR). *Bacillus* spp. are among the most abundant P-solubilizers and have the advantage of forming stress-resistant spores, a necessary requirement for developing seed inoculants.

Potassium Dynamics in Soil

Potassium (K) is absorbed by plants in larger amounts than any other nutrient except N. Although total soil K content exceeds crop uptake during a growing season, in most cases only a small fraction of it is available to plants. Total soil K content ranges between 0.5 to 2.5 per cent and is lower in coarse-textured soils formed from sandstone or quartzite and higher in fine-textured soils formed from rocks high in K-bearing minerals. In tropical soils, total K content is generally low because of greater weathering by high rainfall and temperatures, thus K deficiency frequently occurs after a few years of cropping a virgin soil.

Forms of K

Listed in increasing order of plant availability, soil K exists in four forms:

Mineral	5,000-25,000 ppm (0.5 to 2.5 per cent)
Non-exchangeable	50 – 750 ppm
Exchangeable	40 – 600 ppm
Solution	1 – 10 ppm

Mineral K accounts for 90 to 98 per cent of the total soil K, whereas slowly available (non-exchangeable) and readily available (exchangeable and solution) represents 1 to 10 per cent and 0.1 to 2 per cent, respectively. K transformations among K forms in soils are dynamic. Because of K removal by crop uptake and leaching, there is a continuous but slow transfer of K from primary minerals to exchangeable and slowly available forms. With applications of large amounts of fertilizer K, some reversion to slowly available forms will occur.

Exchangeable and solution K equilibrate rapidly, whereas non exchangeable K equilibrates very slowly with exchangeable and solution forms. Transfer of K from the mineral fraction to more available forms is extremely slow in most soils and this K is considered essentially unavailable to crops during a single growing season.

Functions and Forms of K in Plants

K is absorbed by plant roots as K^+ and its tissue concentration ranges from 0.5 to 6 per cent in dry matter.

Unlike N, P and Most other nutrients, K is not a component of biochemical compounds in the plant. K exists solely in solution or bound to (-) charges on tissue surfaces. As a result K^+ has functions particularly related to solution ionic strength in plant cells. K is involved in water relations, charge balance and osmotic pressure in cells and across membranes, which explains its high mobility in the plant.

K is important for many crop quality characteristics due to its involvement in synthesis and transport of photosynthates to plant reproductive and storage organs (grain, fruit, tubers etc.) and subsequent conversion into carbohydrates, proteins, oils and other products. In fruit and vegetable crops (citrus, bananas, tomatoes, potatoes, onions, etc.), adequate K enhances fruit size, color, taste and peel thickness, which is important for storage and shipping quality.

K deficiency influences metabolic processes primarily related to photosynthesis and synthesis and translocation of enzymes. The increase in dark respiration under K deficiency reduces plant growth and crop quality.

Photosynthesis and Energy Relations

K is essential to photosynthesis through several functions, including:

- ATP synthesis
- production and activity of specific photosynthetic enzymes (*i.e.* RuBP carboxylase)
- CO_2 absorption through leaf stomata
- maintenance of electro neutrality during photophosphorylation in chloroplasts.

Enzyme Activation

K is involved in activation of enzymes important to energy utilization, starch synthesis, N metabolism and respiration. These enzymes are abundant in meristematic tissue at the growing points both above and below ground level, where cell division occurs rapidly and where primary tissues are formed. For example, the starch synthetase enzyme converts sugars into starch, which is critical to starch accumulation in grains, fruits and vegetables. Decreased starch conversion causes soluble sugar accumulation that affects fruit quality. Also, K influences the nitrogenase enzyme required for reduction of atmospheric N_2 to NH_3 in *Rhizobium*. N_2 reduction depends on carbohydrate supply, where K enhances carbohydrate transport to nodules for amino acids synthesis.

Translocation of Assimilates

Once CO_2 is assimilated into sugars during photosynthesis, they are transported from leaves to fruits, roots, tubers, seeds and grains where they are stored or used for growth. Translocation of sugars uses energy from ATP that requires K for its synthesis. Sugar translocation is greatly reduced in K-deficient plants where for example normal translocation rate in sugarcane leaves is = 2.5 cm/min that is reduced to half in K-deficient plants. Under adequate K nutrition, osmotic potential of the phloem sap, water flow rate and sucrose concentration are higher than in K-deficient plants. K is also important as a counter ion (maintaining electrical balance) for NO_3^- transport in the xylem.

Water Relations

K provides much of the osmotic "pull" that draws water into plant roots. K-deficient plants are less able to withstand water stress, mostly because of their inability to fully utilize available water. Maintenance of plant turgor is essential for optimum photosynthetic and metabolic processes. Stomatal opening occurs when there is an increase of turgor pressure in the guard cells surrounding each stoma, which is brought about by an influx of K. Malfunctioning of stomata due to K deficiency is related to lower rates of photosynthesis and less efficient water use. Transpiration, or water loss through stomata, accounts for most plant water use. K can affect the rate of transpiration and water uptake through regulation of stomata openings.

Sources of K

Organic K

K in organic wastes (manures and sewage sludge) occurs predominately as soluble inorganic K^+. In animal waste, K content ranges between 0.2 and 2 per cent (4 to 40 lbs K/5) of dry matter. The average K content in sewage sludge is 10 lbs K/t. Therefore, waste materials can supply sufficient quantities of plant available K, depending on the rate applied. Most waste application rates are governed by the quantity of N or P applied to minimize impacts of land application of waste on surface and groundwater quality. If low waste rates are utilized on K-deficient soils, additional K may be needed.

The knowledge gained from the various research trails conducted on different crops on the uptake pattern of major nutrients such as nitrogen, phosphorus and potassium are presented and discussed hereunder for better understanding.

Crop : Barnyard Millet

Nitrogen

The incorporation of green leaf manures had significant effect on nitrogen in barnyard millet field on available nitrogen at 45 days after incorporation of the green leaf manures nitrogen uptake by plants and soil available nitrogen at the post harvest stage. The highest available nitrogen 310.4 kg ha^{-1} was recorded in *Peltophorum ferrugineum*, it was followed by *Delonix regia* recorded (285.3 kg ha^{-1}) and the lowest available nitrogen of 260.28 kg ha^{-1} was recorded in *Gliricidia sepium* at 45 days after incorporation of the green leaf manures. The highest plant nitrogen uptake of 74.08 kg ha^{-1} was recorded in *Peltophorum ferrugineum* and the lowest nitrogen uptake 59.5 kg ha^{-1} was recorded in *Gliricidia sepium*. The highest available nitrogen of 219.52 kg ha^{-1} at post harvest stage was recorded *Albizia lebbek*, it was on par with *Pongamia pinnata* (216.3 kg ha^{-1}). All other treatments were on par with each other. The absolute control, zero input recorded plant nitrogen uptake of 42.5 kg ha^{-1}. When compared to zro input, *Peltophorum ferrugineum* showed an increased nitrogen uptake of 21.5 kg ha^{-1}. The recommended dose of fertilizers recorded (61.75 kg ha^{-1}) nitrogen uptake. When compared to this *Peltophorum ferrugineum* showed a decreased nitrogen uptake of 7.67 kg ha^{-1} in barnyard millet uptake. The absolute control zero input recorded available nitrogen of 183.3 kg ha^{-1} at post harvest stage and recommended dose of fertilizers recorded 170.2 kg ha^{-1} of available nitrogen at post harvest stage.

Phosphorus

The incorporation of green leaf manures had significant effect on phosphorus in barnyard millet field at 45 days after incorporation of the green leaf manures, plant phosphorus uptake and soil available phosphorus at post harvest stage. The highest available phosphorus 23.15 kg ha^{-1} was recorded in *Delonix regia*, it was followed by *Peltophorum ferrugineum* (22.01 kg ha^{-1}) and the lowest available phosphorus of 16.93 kg ha^{-1} was recorded in *Gliricidia sepium* at 45 days after incorporation of the green leaf manures. The highest phosphorus uptake by plants of 12.16 kg ha^{-1} was recorded in *Delonix regia* and the lowest phosphorus uptake of 8.69 kg ha^{-1} was recorded in *Gliricidia sepium* and it was on par with *Albizia lebbek* with 9.2 kg ha^{-1}. The highest available phosphorus (8.97 kg ha^{-1}) at post harvest stage was recorded in *Peltophorum ferrugineum*, it was followed by *Albizia lebbek* (8.62 kg ha^{-1}). The lowest available phosphorus was recorded (6.46 kg ha^{-1}) in *Gliricidia sepium* at post harvest stage. The absolute control zero input recorded phosphorus uptake of 6.05 kg ha^{-1}. When compared to zero input, *Delonix regia* showed an increased phosphorus uptake of 6.11 kg ha^{-1}. The recommended dose of fertilizers recorded of phosphorus uptake of 13.65 kg ha^{-1}. When compared to this *Delonix regia* showed a decreased phosphorus uptake of 1.48 kg ha^{-1} in barnyard millet at post harvest stage. The absolute control zero input recorded available phosphorus of 7.80 kg ha^{-1} at post harvest stage and recommended dose of fertilizers recorded 12.05 kg ha^{-1} of available phosphorus at post harvest stage

Potassium

The incorporation of green leaf manures had significant effect on potassium in barnyard millet field at 45 days after incorporation of the green leaf manures, potassium uptake by plants and soil available potassium at post harvest stage. The highest available potassium content 348.46 kg ha^{-1} was recorded *Peltophorum ferrugineum*, it was on par with *Delonix regia* (347.97 kg ha^{-1}) and the lowest available potassium content of 300.75 kg ha^{-1} was recorded in *Gliricidia sepium* at 45 days after incorporation of the green leaf manures. The highest potassium uptake of 64.03 kg ha^{-1} was recorded in *Delonix regia* and the lowest potassium uptake of 43.45 kg ha^{-1} was recorded in *Gliricidia sepium*. The highest soil available potassium 262.75 kg ha^{-1} at post harvest stage was recorded in *Delonix regia*, it was followed by *Peltophorum ferrugineum* (M_5) recorded 253.38 kg ha^{-1}. The lowest available potassium was recorded 240.71 kg ha^{-1} *Albizia lebbek* at post harvest stage. The absolute control zero input recorded potassium uptake of 40.71 kg ha^{-1}. When compared to zero input, *Delonix regia* showed an increased potassium uptake of 23.32 kg ha^{-1}. The recommended dose of fertilizers recorded 67.05 kg ha^{-1} potassium uptake. When compared to this *Delonix regia* showed a decreased potassium uptake of 3.02 kg ha^{-1} in barnyard millet at after harvest. The absolute control zero input recorded available potassium of 220.30 kg ha^{-1} at post harvest stage and recommended dose of fertilizers recorded 224.85 kg ha^{-1} of available potassium at post harvest stage.

Table 10.1: Influence of Vrikshyurveda Farming on Plant Nutrients Uptake at Harvest in Barnyard Millet

GLM Incorporation	*Plant N Uptake kg ha^{-1}*	*Plant P Uptake kg ha^{-1}*	*Plant K Uptake kg ha^{-1}*
Albizia	64.57	9.20	48.70
Gliricidia	59.55	8.69	43.45
Pongamia	67.20	10.46	51.93
Delonix	71.05	12.16	64.03
Peltophorum	74.08	10.41	55.46
Mean	**67.30**	**10.19**	**53.07**
SEd	0.94	0.28	2.05
CD ($p \leq 0.05$)	2.05	0.60	3.23
Zero input (C_1)	52.50	6.05	40.71
Crop production guide recommendations	81.75	13.65	67.05

The uptake of nutrients *viz.*, nitrogen, phosphorus and potassium was high in the treatment combination *Pheltophorum ferrugineum* and *Annona squamosa*. Higher nutrient uptake may be due to the addition of leaf manures might be ascribed to the release of different organic acids on their decomposition, better microbial activity and chelation with Fe and Al which helps in solubilization of unavailable forms of phosphorus and release the mono calcium phosphate into the labile pool of soil, thus making it available to the growing crops. Apart from this, the release of nutrients at a slower rate during the decomposition favours the availability of nutrients for the entire crop growth period. The high nutrient uptake stimulated the microbial growth and root growth due to improvement in soil physical conditions created by addition of organics. The increased nutrient uptake in fertilized plots was due to the adequate supply of nutrients and well developed root system resulting in better absorption of water and nutrient. The efficient absorption and translocation of nutrients from soil to foliage may be probably due to favourable and possible interaction of nitrogen and potassium on the availability of more nutrients at critical stages of crop growth. Potassium performs many functions in plant metabolism *viz.*, promoting photosynthesis, harnessing interaction between potassium and nitrogen could have increased the nutrient uptake. The absolute control registered the lowest uptake of nutrients due to low availability of nutrients in the soil and also non addition of external inputs.

Table 10.2: Influence of Vrikshyurveda Farming on Soil Fertility after Harvest of Barnyard Millet

GLM Incorporation	*Soil Nitrogen kg ha^{-1}*	*Soil P_2O_5 kg ha^{-1}*	*Soil K_2O kg ha^{-1}*
Albizia	219.52	8.62	240.71
Gliricidia	201.68	6.46	241.50
Pongamia	216.38	7.39	247.38
Delonix	207.27	7.74	262.75
Peltophorum	210.11	8.97	253.38
Mean	**210.99**	**7.84**	**249.11**
SEd	1.75	0.15	1.48
CD ($p \leq 0.05$)	3.87	0.29	3.21
Zero input	183.31	7.80	224.85
Crop production guide recommendations		12.05	220.30

Note: Time zero values for Nitrogen 250.8 kg ha^{-1}, Phosphorus 14.5 kg ha^{-1} and Potassium 283.40 kg ha^{-1}.

Crop : RAGI

Nitrogen

The incorporation of green leaf manures had significant effect on available nitrogen at soil 45 days after incorporation of GLM, nitrogen uptake by plants and available nitrogen at soil after harvest. The available nitrogen status in the soil was at the rate of 244.6 kg ha^{-1} before green leaf manure incorporation. After 45 DAI, the highest available nitrogen of 303.3 kg ha^{-1} was registered in *Delonix regia*. It was followed by two treatments such as *Leucaena leucocephala* and *Gliricidia sepium*, while both were on par with each other. The lowest available nitrogen of 254.3 kg ha^{-1} was registered in *Peltophorum ferrugineum* in soil after 45 days incorporation of green leaf manures. Among the treatments, nitrogen uptake was higher (71.28 kg ha^{-1}) in *Delonix regia* and *Peltophorum ferrugineum* registered lower nitrogen uptake (61.05 kg ha^{-1}) at harvest stage. The highest available nitrogen of 242.66 kg ha^{-1} in postharvest soil was recorded in *Delonix regia*, which was followed by *Gliricidia sepium* and *Leucaena leucocephala*, while both were on par with each other. *Peltophorum ferrugineum* recorded lowest available nitrogen in post harvest soil with value of 178.07 kg ha^{-1}. The absolute control zero input recorded 56.26 kg ha^{-1} of nitrogen uptake by plants. The recommended dose of fertilizers recorded the nitrogen uptake with value of 80.72 kg ha^{-1}. When compared to this, *Delonix regia* showed an decreased nitrogen uptake of 9.44 kg ha^{-1} in finger millet. In post harvest soil, the zero input recorded available nitrogen of 178.56 kg ha^{-1} and recommended dose of fertilizers recorded 188.35 kg ha^{-1} of available nitrogen.

Phosphorus

The incorporation of green leaf manures had significant effect on available phosphorus at 45 days after incorporation of GLM, phosphorus uptake by plants and phosphorus availability in soil after harvest. The available phosphorus in the soil was of 16.50 kg ha^{-1} before green leaf manure incorporation. After 45 DAI, The highest available phosphorus with value of 26.40 kg ha^{-1} was recorded in *Delonix regia*, which was on par with *Leucaena leucocephala* and *Gliricidia sepium*. The lowest available phosphorus of 18.97 kg ha^{-1} was registered in *Peltophorum ferrugineum* at 45 days after incorporation of green leaf manures.Phosphorus uptake (13.26 kg ha^{-1}) by finger millet was higher in *Delonix regia*). It was on par with *Gliricidia sepium* and *Leucaena leucocephala* and the lowest phosphorus uptake of 9.48 kg ha^{-1} was registered in *Peltophorum ferrugineum*) at harvest stage. The available phosphorus was more (10.50 kg ha^{-1}) in the treatment of *Delonix regia* in soil after harvest. It was comparable with *Leucaena leucocephala* and *Albizia lebbek*. The treatment *Peltophorum ferrugineum* registered low available phosphorus of 7.59 kg ha^{-1} in post harvest soil. The absolute control zero input recorded phosphorus uptake of 7.42 kg ha^{-1}. When compared to zero input, *Delonix regia* showed an increased 5.84 kg ha^{-1} of phosphorus uptake by plants. While recommended dose of fertilizers showed 13.53 kg ha^{-1} of phosphorus uptake by plants. When compared to this, *Delonix regia* showed a lesser uptake of 0.27 kg ha^{-1} phosphorus in finger millet at harvest stage. In post harvest soil, the zero input recorded available phosphorus of 8.25 kg ha^{-1} and recommended dose of fertilizers recorded 12.37 kg ha^{-1} of available phosphorus.

Potassium

The incorporation of green leaf manures had significant effect on available potassium at 45 days after incorporation of green leaf manures, potassium uptake by plants and potassium availability in soil after harvest. The available potassium in the soil was 299.7 kg ha^{-1} before green leaf manure incorporation. After 45 DAI, the highest available potassium with a value of 368.6 kg ha^{-1} was recorded in *Leucaena leucocephala*, which was on par with *Delonix regia*. The lowest available potassium of 317.6 kg ha^{-1} was registered in *Albizia lebbek* at 45 days after incorporation of green leaf manures. Potassium uptake (69.75 kg ha^{-1}) was higher in *Delonix regia*. It was followed by *Leucaena leucocephala* and the lowest potassium uptake of 47.65 kg ha^{-1} was registered in *Albizia lebbek* at harvest stage. The available potassium was more (275.3 kg ha^{-1}) in the treatment of *Delonix regia* in soil after harvest. It was followed by *Leucaena leucocephala*. The treatment *Peltophorum ferrugineum* registered less amount phosphorus available (224.7 kg ha^{-1}) in post harvest soil. The absolute control zero input recorded potassium uptake of 44.95 kg ha^{-1}. When compared to zero input, *Delonix regia* showed an increased 24.80 kg ha^{-1} potassium uptake by plants. The recommended dose of fertilizers recorded 74.92 kg ha^{-1} of potassium uptake. When compared to this, *Delonix regia* showed an decreased uptake of 5.17 kg ha^{-1} potassium by finger millet at harvest stage. In post harvest soil, the zero input had available potassium of 224.5 kg ha^{-1} and recommended dose of fertilizers recorded 239.7 kg ha^{-1} of available potassium. The uptake of nutrient was highest in the treatment for nitrogen, phosphorus and potassium in the treatment with *Delonix regia* as green leaf manure and foliar sprays of *Moringa oleifera*. This might be due to the timely decomposition of green leaf manures applied due to microbial activity in the organic matter would have made the nutrients available to the crop and better use efficiency would have resulted in better uptake of nitrogen, phosphorus and potassium. The presence of secondary metabolites in the leaf extracts would have promoted the microbial activity leading to better availability and there by the uptake of nutrient was better. The absolute control recorded lower uptake of nutrient. Since there was no inputs of organic and organic nutrients. Hence the lower uptake of nutrients by the crop.

Table 10.3: Influence of Vrikshyurveda Farming on Plant Nutrients Uptake and Soil Fertility during Harvest Stage in Finger Millet

GLM Incorporation	*Plant Nutrients Uptake*			*Soil Nutrient Status After Harvest*		
	Nitrogen (kg ha^{-1})	*Phosphorus (kg ha^{-1})*	*Potassium (kg ha^{-1})*	*N (kg ha^{-1})*	*P_2O_5 (kg ha^{-1})*	*K_2O (kg ha^{-1})*
Albizia	64.7	11.39	47.65	199.96	9.24	254.14
Gliricidia	66.52	12.37	53.46	217.22	8.66	257.30
Leucaena	66.65	11.90	58.98	216.60	10.03	265.41
Delanix	71.28	13.26	69.75	242.66	10.50	275.35
Peltophorum	61.05	9.48	49.90	178.07	7.59	224.77
Mean	**66.14**	**11.68**	**55.95**	**210.90**	**9.20**	**255.39**
SEd	2.13	0.55	0.63	0.70	0.59	0.58
CD ($p \leq 0.05$)	5.92	1.53	1.75	1.94	1.64	1.62
Zero input	80.72	13.53	74.92	188.35	12.37	239.76
Fertiliser applications	**56.26**	**7.42**	**44.95**	**178.56**	**8.25**	**224.59**

Time zero values for soil Nitrogen 244.61 kg ha^{-1}, phosphorus 16.50 kg ha^{-1} and Potassium 299.7 kg ha^{-1}.

Application of organic manure could reduce N loss and conserve soil N by mineralization, thus maintaining a continuous availability of N in the entire life cycle of rice plant which would result in the increase of total N uptake. Available phosphorus and potassium content also increased in repeated use of green manure either alone or in combination of NPK. Increase of available phosphorus might be due to the decomposition of organic matter accompanied by the release of appreciable quantities of carbon-di-oxide, as carbon-di-oxide production plays an important role in increasing phosphate availability. Increase of potassium might be attributed due to direct addition of potassium to the available pool due to the increase of K uptake after incorporation of green leaf manures. Integrated use of farm yard manure with green manure enhanced nutrient uptake by preventing the leaching and other losses.

Crop : Maize

Nitrogen

The incorporation of green leaf manures had significant effect on nitrogen in maize at 45 days after incorporation of the green leaf manures on available nitrogen, nitrogen uptake and available nitrogen at the post harvest stage. The highest available nitrogen (212.60 kg ha^{-1}) was recorded in *Albizia lebbek,* it was followed by *Peltophorum ferrugineum* recorded (205.60 kg ha^{-1}) and the lowest available nitrogen (180.1 kg ha^{-1}) was recorded in *Delonix regia* at 45 days after incorporation of the green leaf manures. The highest nitrogen uptake (103.8 kg ha^{-1}) was recorded in *Albizia lebbek* and the lowest nitrogen uptake (69.71 kg ha^{-1}) was recorded in *Leucaena leucocephala.* The highest available nitrogen at post harvest stage was (121.9 kg ha^{-1}) recorded *Leucaena leucocephala,* it was followed by *Gliricidia sepium* recorded 117.6 kg ha^{-1}. The lowest available nitrogen was recorded (94.65 kg ha^{-1}) *Delonix regia* at post harvest stage. The absolute control zero input recorded nitrogen uptake of 71.58 kg ha^{-1}. When compared to control, *Albizia lebbek* showed increased nitrogen uptake of 32.24 kg ha^{-1}. The recommended dose of fertilizers recorded (208.9 kg ha^{-1}) nitrogen uptake. When compared to this, *Albizia lebbek* showed a decreased nitrogen uptake of 105.10 kg ha^{-1} in maize. The absolute control zero input recorded available nitrogen of 103.5 kg ha^{-1} at post harvest stage and recommended dose of fertilizers recorded 98.95 kg ha^{-1} of available nitrogen at post harvest stage.

Phosphorus

The incorporation of green leaf manures had significant effect on phosphorus in maize at 45 days after incorporation of the green leaf manures, phosphorus uptake and available phosphorus at post harvest stage. The highest available phosphorus (15.50 kg ha^{-1}) was recorded in *Delonix regia,* it was followed by *Albizia lebbek* recorded (13.90 kg ha^{-1}) and the lowest available phosphorus (11.70 kg ha^{-1}) was recorded in *Gliricidia sepium* at 45 days after incorporation of the green leaf manures. The highest phosphorus uptake (9.30 kg ha^{-1}) was recorded in *Delonix regia,* it was followed by *Albizia lebbek* recorded 8.34 kg ha^{-1}. And the lowest phosphorus uptake (7.02 kg ha^{-1}) was recorded in *Gliricidia sepium.* The highest available phosphorus at post harvest stage was (5.98 kg ha^{-1}) recorded *Delonix regia,* it was followed by *Albizia lebbek* recorded 5.25 kg ha^{-1}. The lowest available phosphorus was recorded (4.37 kg ha^{-1}) *Gliricidia sepium* at post harvest stage. The absolute control zero input recorded phosphorus uptake of 4.90 kg ha^{-1}. When compared to control, *Delonix regia* showed an increased phosphorus uptake of 4.40 kg ha^{-1}. The recommended dose of fertilizers recorded 36.5 kg ha^{-1} of phosphorus uptake. When compared to this, *Delonix regia* showed a decreased phosphorus uptake of 27.20 kg ha^{-1} in maize at post harvest stage. The absolute control zero input recorded available phosphorus of 5.32 kg ha^{-1} at post harvest stage and recommended dose of fertilizers recorded 34.75 kg ha^{-1} of available phosphorus at post harvest stage.

Potassium

The incorporation of green leaf manures had significant effect on potassium in maize at 45 days after incorporation of the green leaf manures, potassium uptake and available potassium at post harvest stage. The highest available potassium (299.7 kg ha^{-1}) was recorded in *Albizia lebbek*, it was followed by *Gliricidia sepium* recorded (297.2 kg ha^{-1}) and the lowest available potassium (285.9 kg ha^{-1}) was recorded in *Leucaena leucocephala* at 45 days after incorporation of the green leaf manures. The highest potassium uptake (149.8 kg ha^{-1}) was recorded in *Albizia lebbek* and the lowest potassium uptake (142.9 kg ha^{-1}) was recorded in *Leucaena leucocephala*. The highest available potassium at post harvest stage was (140.2 kg ha^{-1}) recorded *Albizia lebbek*, it was followed by *Peltophorum ferrugineum* recorded 137.67 kg ha^{-1}. The lowest available potassium was recorded (131.1 kg ha^{-1}). *Leucaena leucocephala* at post harvest stage. The absolute control zero input recorded Potassium uptake of 115.2 kg ha^{-1}. When compared to control, *Albizia lebbek* showed an increased Potassium uptake of 34.73 kg ha^{-1}. The recommended dose of fertilizers recorded (168.9 kg ha^{-1}) Potassium uptake. When compared to this, *Albizia lebbek* showed a decreased Potassium uptake of 19.06 kg ha^{-1} in maize at after harvest. The absolute control zero input recorded available potassium of 164.7 kg ha^{-1}at post harvest stage and recommended dose of fertilizers recorded 159.1 kg ha^{-1} of available potassium at post harvest stage. The uptake of nitrogen, phosphorus and potassium were highest in the treatment that received *Albizia lebbek* as green leaf manure and foliar spraying of *Annona squamosa*. Higher nutrient uptake due to the addition of organic leaf manures might be ascribed to the release of different organic acids and their influence on decomposition, better microbial activity and chelation with Fe and Al which helps in solubilization of unavailable forms of phosphorus and release the mono calcium phosphate into the labile pool of soil, thus making it available to the growing crops. Apart from this, the release of nutrients at a slower rate during the decomposition favours the availability of nutrients for the entire crop growth period and high nutrient uptake stimulated the

Table 10.4: Influence of Vrikshyurveda Farming on Plant Nutrients Uptake and Soil Fertility during Harvest Stage in Maize

GLM Incorporation	*Plant Nutrients Uptake*			*Soil Nutrient Status After Harvest*		
	Nitrogen (kg ha^{-1})	*Phosphorus (kg ha^{-1})*	*Potassium (kg ha^{-1})*	*N (kg ha^{-1})*	*P_2O_5 (kg ha^{-1})*	*K_2O (kg ha^{-1})*
Albizia	103.82	8.34	149.86	105.52	5.25	140.25
Gliricidia	79.90	7.02	148.60	117.60	4.37	136.90
Leucaena	69.71	7.86	142.98	121.92	4.98	131.12
Delonix	83.28	9.30	145.51	94.65	5.98	135.34
Peltophorum	87.91	8.10	145.45	114.79	5.15	137.67
Mean	**84.92**	**8.12**	**146.48**	**116.70**	**5.15**	**136.26**
SEd	**2.65**	**0.17**	**0.58**	1.37	0.12	**0.71**
CD ($p \leq 0.05$)	7.36	0.48	1.62	3.80	0.33	1.98
Fertiliser applications	208.92	36.5	168.92	98.95	34.75	159.14
Zero input	71.58	4.90	115.13	103.56	5.32	164.75

Time zero vales for soil N 178.45 kg ha^{-1}, phosphorus 10.50 kg ha^{-1} and 287.8 potassium kg ha^{-1}.

microbial growth and root growth due to improvement in soil physical conditions created by addition of organics. The increased uptake in recommended dose of fertilizers was due to the supply of readily available nutrients and well developed root system under balanced nutrient application resulting in better absorption of water and nutrient. The efficient absorption and translocation of nutrients from soil to foliage may be probably due to favourable and possible interaction of nitrogen and potassium on the availability of more nutrients at critical stages of crop growth. Potassium performs many functions in plant metabolism *viz.*, promoting photosynthesis, harnessing interaction between potassium and nitrogen could have increased the nutrient uptake. The absolute control registered the lowest uptake of nutrients due to low availability of nutrients in the soil.

Crop : Blackgram

Nitrogen

The incorporation of green leaf manures had significant effect on nitrogen in black gram at 45 days after incorporation of the green leaf manures on available nitrogen, nitrogen uptake and available nitrogen at the post harvest stage. The highest available nitrogen (299.8 kg ha^{-1}) was recorded in *Pongamia pinnata*, it was followed by *Albizia lebbek* recorded (294.8 kg ha^{-1}) and the lowest available nitrogen (290.9 kg ha^{-1}) was recorded in *Peltophorum ferrugineum* at 45 days after incorporation of the green leaf manures. The highest nitrogen uptake (68.3 kg ha^{-1}) was recorded in *Pongamia pinnata* and the lowest nitrogen uptake (38.4 kg ha^{-1}) was recorded in *Delonix regia*. The highest available nitrogen at post harvest stage was (200.47 kg ha^{-1}) recorded *Gliricidia sepium*; it was followed by *Peltophorum ferrugineum* recorded 196.42 kg ha^{-1}. The lowest available nitrogen was recorded (187.3 kg ha^{-1}) *Delonix regia* at post harvest stage. The absolute control zero input recorded nitrogen uptake of 47.66 kg ha^{-1}. When compared to control, *Pongamia pinnata* showed increased nitrogen uptake of 20.70 kg ha^{-1}. The recommended dose of fertilizers recorded (70.59 kg ha^{-1}) nitrogen uptake. When compared to this, *Pongamia pinnata* showed a decreased nitrogen uptake of 2.23 kg ha^{-1} in black gram uptake. The absolute control zero input recorded available nitrogen of 176.0 kg ha^{-1} at post harvest stage and recommended dose of fertilizers recorded 152.8 kg ha^{-1} of available nitrogen at post harvest stage.

Phosphorus

The incorporation of green leaf manures had significant effect on phosphorus in black gram at 45 days after incorporation of the green leaf manures, phosphorus uptake and available phosphorus at post harvest stage. The highest available phosphorus (24.41 kg ha^{-1}) was recorded in *Delonix regia*, it was followed by *Pongamia pinnata* recorded (23.16 kg ha^{-1}) and the lowest available phosphorus (16.50 kg ha^{-1}) was recorded in *Gliricidia sepium* at 45 days after incorporation of the green leaf manures. The highest phosphorus uptake (12.45 kg ha^{-1}) was recorded in *Delonix regia*, it was followed by *Pongamia pinnata* recorded 10.35 kg ha^{-1}. And the lowest phosphorus uptake (6.38 kg ha^{-1}) was recorded in *Gliricidia sepium*. The highest available phosphorus at post harvest stage was (12.01 kg ha^{-1}) recorded *Gliricidia sepium*, it was followed by *Albizia lebbek* recorded 11.03 kg ha^{-1}. The lowest available phosphorus was recorded (9.83 kg ha^{-1}) *Delonix regia* at post harvest stage. The absolute control zero input recorded phosphorus uptake of 5.34 kg ha^{-1}. When compared to control, *Delonix regia* showed increased phosphorus uptake of 7.11 kg ha^{-1}. The recommended dose of fertilizers recorded 16.45 kg ha^{-1} of phosphorus uptake. When compared to this, *Delonix regia* showed decreased phosphorus uptake of 4.0 kg ha^{-1} in black gram at post harvest stage. The absolute control zero input recorded available phosphorus of 8.26 kg ha^{-1} at post harvest stage and recommended dose of fertilizers recorded 14.75 kg ha^{-1} of available phosphorus at post harvest stage.

Potassium

The incorporation of green leaf manures had significant effect on potassium in black gram at 45 days after incorporation of the green leaf manures, potassium uptake and available potassium at post harvest stage. The highest available potassium (353.7 kg ha^{-1}) was recorded in *Delonix regia*, it was followed by *Pongamia pinnata* recorded (346.4 kg ha^{-1}) and the lowest available potassium (332.9 kg ha^{-1}) was recorded in *Peltophorum ferrugineum* at 45 days after incorporation of the green leaf manures. The highest potassium uptake (25.65 kg ha^{-1}) was recorded in *Pongamia pinnata* and the lowest potassium uptake (17.96 kg ha^{-1}) was recorded in *Peltophorum ferrugineum*. The highest available potassium at post harvest stage was (266.8 kg ha^{-1}) recorded *Pongamia pinnata*, it was followed by *Albizia lebbek* recorded 264.6 kg ha^{-1}. The lowest available potassium was recorded (258.5 kg ha^{-1}) *Peltophorum ferrugineum* at post harvest stage. The absolute control zero input recorded Potassium uptake of 15.85 kg ha^{-1}. When compared to control, *Pongamia pinnata* showed increased Potassium uptake of 9.80 kg ha^{-1}. The recommended dose of fertilizers recorded (28.25 kg ha^{-1}) Potassium uptake. When compared tothis, *Pongamia pinnata* showed a decreased Potassium uptake of 2.60 kg ha^{-1} in black gram at after harvest. The absolute control zero input recorded available potassium of 269.35 kg ha^{-1} at post harvest stage and recommended dose of fertilizers recorded 267.15 kg ha^{-1} of available potassium at post harvest stage. The uptake of nitrogen, phosphorus and potassium was high in the treatment that received *Pongamia pinnata* as green leaf manure and foliar spraying of *Moringa oleifera*. The biological nitrogen fixation is considered to be the main source of nitrogen in pulses. However, with the onset of reproductive phase, the developing seed become the primary sink for material synthesized in the plant and this diversion of photosynthesis to seeds in turn slows down and eventually stops the nitrogen fixation by rhizobia. Although, nutrient uptake by plant continues, yet the plant is deprived by its major nitrogen nutrient at this critical stage pod development. The organic manures increased the adsorption power of soil for cations and anions, particularly phosphate and nitrate and these were released slowly for the benefit of crop during entire crop growth period leading to higher yield. Higher nutrient uptake due to the addition of organic leaf manures might be ascribed to the release of different organic acids on their decomposition, better microbial activity and chelation with Fe and Al which helps in solubilization of unavailable forms of phosphorus and release the mono calcium phosphate into the labile pool of soil, thus making it available to the growing crops. Apart from this, the release of nutrients at a slower rate during the decomposition favours the availability of nutrients for the entire crop growth period and high nutrient uptake stimulated the microbial growth and root growth due to improvement in soil physical conditions created by addition of organics. The increased uptake in recommended dose of fertilizers was due to the supply of nutrients and well developed root system under balanced nutrient application resulting in better absorption of water and nutrient. The efficient absorption and translocation of nutrients from soil to foliage may be probably due to favourable and possible interaction of nitrogen and potassium on the availability of more nutrients at critical stages of crop growth. Potassium performs many functions in plant metabolism *viz.*, promoting photosynthesis, harnessing interaction between potassium and nitrogen could have increased the nutrient uptake. The absolute control registered the lowest uptake of nutrients due to low availability of nutrients in the soil.

Table 10.5: Influence of Vrikshyurveda Farming on Plant Nutrients Uptake and Soil Fertility during Harvest Stage in Blackgram

GLM Incorporation	*Plant Nutrients Uptake*			*Soil Nutrient Status After Harvest*		
	Nitrogen (kg ha^{-1})	*Phosphorus (kg ha^{-1})*	*Potassium (kg ha^{-1})*	*N (kg ha^{-1})*	*P_2O_5 (kg ha^{-1})*	*K_2O (kg ha^{-1})*
Albizia	56.94	9.21	23.21	194.27	11.03	264.65
Gliricidia	50.27	6.38	20.54	200.47	12.01	262.67
Leucaena	68.36	10.35	25.65	191.51	10.24	266.82
Delonix	38.45	12.45	19.68	187.30	9.83	262.80
Peltophorum	45.74	8.85	17.96	196.42	11.00	258.50
Mean	**51.95**	**9.44**	**21.40**	**193.99**	**10.82**	**263.09**
SEd	0.95	0.96	**0.42**	1.38	0.01	**1.38**
CD (p≤0.05)	**2.65**	**2.66**	1.16	**3.82**	**0.28**	3.82
Fertiliser applications	70.59	16.45	28.25	152.81	14.75	267.15
Zero input	**47.66**	**5.34**	**15.85**	**176.01**	**8.26**	**269.35**

Time zero vales for soil N **226.9** kg ha^{-1}, phosphorus 14 kg ha^{-1} and 291.2 Potassium kg ha^{-1}.

Crop : Onion

Nitrogen

The incorporation of green leaf manures had significant effect on nitrogen uptake in aggregatum onion at 30, 60 and 90 DAS. The highest nitrogen uptake (12.72 kg ha^{-1}) was recorded in *Gliricidia sepium*, which was on par with *Senna siamea* recorded (12.29 kg ha^{-1}), *Leucaena leucocephala* recorded (12.26 kg ha^{-1}) and the lowest nitrogen uptake (11.50 kg ha^{-1}) was recorded in *Albizia lebbek* at 30 DAS, While at 60 DAS, the highest nitrogen uptake (17.39 kg ha^{-1}) was recorded in *Albizia lebbek* and the lowest nitrogen uptake (15.96 kg ha^{-1}) was recorded in *Leucaena leucocephala*. The application of leaf extract through seed treatment and foliar sprays had a significant effect on nitrogen uptake in aggregatum onion. Among the treatments, *Annona squamosa* recorded the nitrogen uptake (12.35 kg ha^{-1}) which was on par with *Morinda tinctoria* (12.24 kg ha^{-1}), *Aegle marmelos* recorded (12.19 kg ha^{-1}) and the lowest nitrogen uptake

(11.69 kg ha^{-1}) was recorded in *Alangium salvifolium* at 30 DAS, while at 60 DAS the highest nitrogen uptake (16.93 kg ha^{-1}) was recorded in *Aegle marmelos*, which was on par with *Morinda tinctoria* and the lowest nitrogen uptake (16.16 kg ha kg ha^{-1}) was recorded in *Annona squamosa*. The interaction effect was significant, the treatment combination *Gliricidia sepium* + *Aegle marmelos* recorded the highest nitrogen uptake (13.42 kg ha^{-1}), which was on par with *Leucaena leucocephala* + *Morinda tinctoria combination* (12.80 kg ha^{-1}), *Delonix regia* + *Annona squamosa* (12.65 kg ha^{-1}) and *Leucaena leucocephala* + *Annona squamosa* (12.65 kg ha^{-1}) at 30 DAS. The combination of *Gliricidia sepium* + *Aegle marmelos*) recorded highest nitrogen uptake (18.95 kg ha^{-1}), which was on par with *Gliricidia sepium* + *Morinda tinctoria* (18.58 kg ha^{-1}) and the lowest *Gliricidia sepium* + *Annona squamosa* (13.26 kg ha^{-1}) of nitrogen uptake at 60 DAS. The absolute control recorded nitrogen uptake of 12.25 kg ha^{-1} and 14.28 kg ha^{-1} at 30 and 60 DAS. When it was compared to control, *Gliricidia sepium* + *Aegles marmellos* combination showed increased nitrogen uptake of 1.17 kg ha^{-1} and 4.67 kg ha^{-1}. The recommended dose of fertilizers recorded nitrogen uptake of 18.25 and 22.25 kg ha^{-1}at 30 and 60 DAS. When compared to this, *Gliricidia sepium* + *Aegle marmelos* showed decreased nitrogen uptake of 4.83 and 3.3 kg ha^{-1}. At 90 DAS the highest nitrogen uptake (23.37 kg ha^{-1}) was recorded in *Gliricidia sepium* and the lowest nitrogen uptake (20.54 kg ha^{-1}) was recorded in *Senna siamea*. Among the treatments, *Aegle marmelos* recorded the highest nitrogen uptake (22.37 kg ha^{-1}) and the lowest nitrogen uptake (20.27 kg ha^{-1}) recorded in *Alangium salvifolium* at 90 DAS. The interaction effect was significant, treatments *Gliricidia sepium* + *Aegle marmelos* recorded the highest nitrogen uptake (26.75 kg ha^{-1}) which was on par with *Albizia lebbek* + *Aegle marmelos* with (25.25 kg ha^{-1}) of nitrogen uptake at 90 DAS. The absolute control recorded nitrogen uptake of 18.25 kg ha^{-1}. When it was compared to control, *Gliricidia sepium* + *Aegle marmelos* showed increased nitrogen uptake of 8.5 kg ha^{-1}. The recommended dose of fertilizers recorded (29.12 kg ha^{-1}) nitrogen uptake. When compared to this, *Gliricidia sepium* + *Aegle marmelos* showed decreased nitrogen uptake of 2.37 kg ha^{-1} in aggregatum onion at 90 DAS.

Phosphorus

The incorporation of green leaf manures had significant effect on phosphorus uptake in aggregatum onion at 30, 60 and 90 DAS. The highest phosphorus uptake (7.97 kg ha^{-1}) was recorded in *Gliricidia sepium* and the lowest phosphorus uptake (6.37 kg ha^{-1}) was recorded in *Delonix regia*) at 30 DAS, while at 60 DAS the highest phosphorus uptake (11.15 kg ha^{-1}) was recorded in *Gliricidia sepium* and the lowest phosphorus uptake (9.46 kg ha^{-1}) was recorded in *Senna siamea*. The application of leaf extract through seed treatment and foliar sprays had a significant effect on phosphorus uptake in aggregatum onion. Among the treatments *Annona squamosa* recorded the phosphorus uptake (7.30 kg ha^{-1}) and the lowest phosphorus uptake (6.28 kg ha^{-1}) was recorded in *Aegle marmelos* at 30 DAS, while at 60 DAS the phosphorus uptake of 10.83 kg ha^{-1} was recorded in *Aegle marmelos* and the lowest phosphorus uptake (9.78 kg ha kg ha^{-1}) in *Annona squamosa*. The interaction effect was significant, the treatment *Gliricidia sepium* + *Annona squamosa* recorded the highest phosphorus uptake (8.21 kg ha^{-1}), which was on par with *Gliricidia sepium* + *Aelges marmellos* recorded (8.19 kg ha^{-1}) at 30 DAS and *Gliricidia sepium* + *Aelges marmellos* recorded highest phosphorus uptake (12.00 kg ha^{-1}) at 60 DAS. The absolute control recorded phosphorus uptake of 3.56 kg ha^{-1} and 6.25 kg ha^{-1} at 30 and 60 DAS. When compared to control, *Gliricidia sepium* + *Aegle marmelos* showed increased phosphorus uptake of 4.63 and 5.75 kg ha^{-1}. The recommended dose of inorganic fertilizers recorded phosphorus uptake of 11.05 and 15.60 kg ha^{-1}at 30 and 60 DAS. When it was compared to this, *Gliricidia sepium* + *Aegle marmelos* showed decreased phosphorus uptake of 2.86 and 3.6 kg ha-$^{-1}$. At 90 DAS, the highest phosphorus uptake (14.06 kg ha^{-1}) was recorded in *Gliricidia sepium* and the lowest phosphorus uptake (13.20 kg ha^{-1}) was recorded in *Albizia lebbek*. Among the treatments, *Annona squamosa*) recorded the highest phosphorus uptake (14.14 kg ha^{-1}) and the lowest phosphorus uptake (13.05 kg ha^{-1}) was recorded in *Alangium salvifolium*. The interaction effect was significant the *Gliricidia sepium* + *Aegle marmelos* which recorded the highest phosphorus uptake (15.65 kg ha^{-1}) and the lowest phosphorus uptake of 12.32 kg ha^{-1} was recorded *Albizia lebbek* + *Morinda tinctoria*. The absolute control recorded phosphorus uptake of 7.35 kg ha^{-1}. When compared to control, *Gliricidia sepium* + *Aegle marmelos* showed increased phosphorus uptake of 8.3 kg ha^{-1}. The recommended dose of inorganic fertilizers recorded phosphorus uptake of 18.25 kg ha^{-1}at 90 DAS.When compared to this, *Gliricidia sepium* + *Aegle marmelos* showed an decreased phosphorus uptake of 2.6 kg ha $^{-1}$ in aggregatum onion.

Potassium

The incorporation of green leaf manures had significant effect on potassium uptake in aggregatum onion at 30, 60 and 90 DAS. The highest potassium uptake (12.71 kg ha^{-1}) was recorded in *Gliricidia sepium* and the lowest potassium uptake (9.99 kg ha^{-1}) was recorded in *Albizia lebbek* at 30 DAS, while at 60 DAS the highest potassium uptake (17.89 kg ha^{-1}) was recorded in *Gliricidia sepium* and the lowest potassium uptake (15.57 kg ha^{-1}) was recorded in *Delonix regia*. The application of leaf extract through seed treatment and foliar sprays had a significant effect on potassium uptake in aggregatum onion. Among them highest potassium uptake (11.50 kg ha^{-1}) was recorded in *Aegle marmelos* and the lowest potassium uptake (10.83 kg ha^{-1}) was recorded in *Alangium salvifolium* at 30 DAS. While at 60 DAS the highest potassium uptake (17.35 kg ha^{-1}) was recorded in *Morinda tinctoria*, which was on par with *Aegles marmellos* which recorded (17.13 kg ha^{-1}) and the lowest potassium uptake (16.42 kg ha^{-1}) was recorded in *Alangium salvifolium*. The interaction effect was significant, the treatment *Gliricidia sepium* + *Aegle marmelos* recorded the highest potassium uptake (14.75 kg ha^{-1}) at 30 DAS and *Gliricidia sepium* + *Aegle marmelos* recorded highest potassium uptake (19.25 kg ha^{-1}), which was on par with *Senna siamea* + *Annona squamosa* at 60 DAS. The absolute control recorded potassium uptake of 7.58 and 12.23 kg ha^{-1} at 30 and 60 DAS. When compared to control, *Gliricidia sepium* + *Aegle marmelos* showed increased potassium uptake of 7.17 and 7.02 kg ha^{-1}. The recommended dose of inorganic fertilizers recorded potassium uptake of 16.25 and 25.65 kg ha^{-1} at 30 and 60 DAS. When compared to this, *Gliricidia sepium* + *Aegle marmelos* showed decreased potassium uptake of 1.5 and 6.31 kg ha-$^{-1}$. At 90 DAS the highest potassium uptake of 26.79 kg ha^{-1} was recorded in *Gliricidia sepium* and the lowest potassium uptake of 22.46 kg ha^{-1} was recorded in *Albizia lebbek*. Among the treatments, *Aegle marmelos* recorded the highest potassium uptake (25.55

Table 10.6: Influence of Vrikshyurveda Farming on Plant Nutrients Uptake at 90 Days after Sowing in Aggregatum Onion cv. CO 5

GLM Incorporation	*Plant Uptake of Nutrients (kg ha^{-1})*		
	Nitrogen	*Phosphorus*	*Potassium*
Albizia	21.18	13.20	22.46
Gliricidia	20.54	13.45	25.81
Leucaena	23.37	14.06	26.79
Delonix	20.60	13.22	26.40
Peltophorum	21.63	13.50	24.67
Mean	21.46	13.48	25.23
SEd	0.55	0.30	0.92
CD (p≤0.05)	1.29	0.69	2.14
Zero input	18.25	7.35	15.16
Fertiliser application	29.12	18.25	30.15

Table 10.7: **Influence of Vrikshyurveda Farming on Plant Nutrients Uptake at 90 Days after Sowing in Aggregatum Onion cv. CO 5**

GLM Incorporation	*Plant Uptake of Nutrients (kg ha^{-1})*		
	Nitrogen	*Phosphorus*	*Potassium*
Albizia	21.18	13.20	22.46
Gliricidia	20.54	13.45	25.81
Leucaena	23.37	14.06	26.79
Delonix	20.60	13.22	26.40
Peltophorum	21.63	13.50	24.67
Mean	21.46	13.48	25.23
SEd	0.55	0.30	0.92
CD (p≤0.05)	1.29	0.69	2.14
Zero input	18.25	7.35	15.16
Fertiliser application	29.12	18.25	30.15

kg ha^{-1}), which was on par with *Annona squamos* recorded (25.53 kg ha^{-1}) and the lowest potassium uptake (24.70 kg ha^{-1}) recorded in *Alangium salvifolium* at 90 DAS. The interaction effect was significant treatment *Gliricidia sepium + Aegle marmelos* recorded the highest potassium uptake (28.90 kg ha^{-1}) which was on par with *Gliricidia sepium + Morinda tinctoria,* recorded (28.89 kg ha^{-1}). The absolute control recorded potassium uptake of 15.16 kg ha^{-1} at 90 DAS. When compared to control, *Gliricidia sepium + Aegle marmelos* showed increased potassium uptake of 13.74 kg ha^{-1}. The recommended dose of inorganic fertilizers recorded potassium uptake of 30.15

kg ha^{-1} at 90 DAS. When compared to this, Gliricidia *sepium* + *Aegle marmelos*) showed decreased potassium uptake of 1.25 kg ha^{-1}. The uptake of nutrient was highest in the treatment for nitrogen phosphorus and potassium in the treatment with *Gliricidia sepium* as green leaf manure and bulb soaking along with foliar sprays of *Aegles marmellos*. This might be due to the timely decomposition of green leaf manures applied due to microbial activity in the organic matter would have made the nutrients available to the crop and better use efficiency would have resulted in better application of green leaf manures and uptake of nitrogen, phosphorus and potassium. The presence of secondary metabolites in the leaf extracts would have promoted the microbial activity leading to better availability and there by the uptake of nutrient was better. The recommended dose of inorganic fertilizers recorded highest uptake of nutrient. This might be due to the available form of inorganic fertilizers, which could have resulted in higher uptake while absolute control recorded lower uptake of nutrient. Since there was no inputs of organic and organic nutrients. Hence, the lower availability of nutrient in the soil.

Crop : Bhendi

Nitrogen

The incorporation of green leaf manures had significant effect on nitrogen uptake in Bhendi at 30, 60, 90 and 120 DAS. The highest nitrogen uptake (13.89 kg ha^{-1}) was recorded in *Albizia lebbek* and the lowest (11.65 kg ha^{-1}) was recorded in *Leucaena leucocephala* at 30 DAS, While at 60 DAS, the highest nitrogen uptake (33.92 kg ha^{-1}) was recorded in *Albizia lebbek*) and the lowest nitrogen uptake (30.84 kg ha^{-1}) was recorded in *Leucaena leucocephala*. The application of leaf extract through seed treatment and foliar sprays had a significant effect on nitrogen uptake in Bhendi. Among the treatments, *Aegles marmellos* recorded the highest nitrogen uptake (13.61 kg ha^{-1}), which was on par with *Annona squamosa* (13.16 kg ha^{-1}) and the lowest (12.05 kg ha^{-1}) was recorded in *Ocimum sanctum* at 30 DAS, while at 60 DAS the highest nitrogen uptake (33.44 kg ha^{-1}) was recorded in *Annona squamosa* and the lowest (30.79 kg ha kg ha^{-1}) was recorded in *Ocimum sanctum*. The interaction effect was significant, the treatment combinations *Albizia lebbek+ Annona squamosa* recorded the highest nitrogen uptake (16.45 kg ha^{-1}) and the lowest (10.84 kg ha^{-1}) nitrogen uptake was recorded in *Leucaena leucocephala* + *Annona squamosa* during 30 DAS. At 60 DAS*Albizia lebbek+ Annona squamosa* recorded the highest nitrogen uptake (41.44 kg ha^{-1}) and lowest nitrogen uptake (29.52 kg ha^{-1}) was recorded in *Gliricidia sepium* + *Ocimum sanctum*. The absolute control recorded nitrogen uptake of 8.25 and 19.28 kg ha^{-1} at 30 and 60 DAS respectively. When compared to control, *Albizia lebbek+ Annona squamosa*, showed increased nitrogen uptake of 8.20 kg ha-$^{-1}$ and 22.16 kg ha-$^{-1}$. The recommended dose of inorganic fertilizers recorded nitrogen uptake of 18.25 and 43.23 kg ha^{-1} at 30 and 60 DAS. When compared to this, *Albizia lebbek+ Annona squamosa* showed decrease in nitrogen uptake of 1.8 and 1.79 kg ha-$^{-1}$. At 90 DAS, the highest nitrogen uptake (53.63 kg ha^{-1}) was recorded in *Albizia lebbek* and the lowest nitrogen uptake (50.01 kg ha^{-1}) was recorded in *Leucaena leucocephala*. *Albizia lebbek* recorded the highest nitrogen uptake (65.17 kg ha^{-1}) and the lowest nitrogen uptake (61.23 kg ha^{-1}) was recorded in *Leucaena leucocephala* at 120 DAS. Among the treatments, *Aegles marmellos* recorded the highest nitrogen uptake (53.32 kg ha^{-1}), which was on par (53.26 kg ha^{-1}) with *Annona squamosa* and the lowest nitrogen uptake (50.13 kg ha^{-1}) was recorded in *Alangium salvifolium* at 90 DAS. During 120 DAS, the highest nitrogen uptake (65.79 kg ha^{-1}) was recorded in *Annona squamosa* and the lowest nitrogen uptake (60.97 kg ha^{-1}) was recorded in *Ocimum sanctum*. The interaction effect was significant, the treatment combinations *Albizia lebbek+ Annona squamosa* recorded the highest nitrogen uptake (61.98 kg ha^{-1}) and the lowest nitrogen uptake (49.28 kg ha^{-1}) was recorded in *Delonix regia + Ocimum sanctum* at 90 DAS. While, *Albizia lebbek+ Annona squamosa* recorded the highest nitrogen uptake (77.77 kg ha^{-1}) and the lowest nitrogen uptake (60.04 kg ha^{-1}) was recorded in *Leucaena leucocephala+ Morinda tinctoria* at 120 DAS. The absolute control recorded nitrogen uptake of 25.75 kg ha^{-1} and 35.27 kg ha^{-1} at 90 and 120 DAS. When compared to control, *Albizia lebbek+ Annona squamosa* showed an increase in nitrogen uptake of 36.23 kg ha^{-1} and 42.50 kg ha^{-1}. The recommended dose of inorganic fertilizers recorded (63.20 kg ha^{-1} and 79.67 kg ha^{-1}) nitrogen uptake. When compared to this, *Albizia lebbek+ Annona squamosa* showed a decrease in nitrogen uptake of 1.22 kg ha^{-1} and 1.90 kg ha^{-1} in Bhendi at 90 and 120 DAS.

Phosphorus

The incorporation of green leaf manures had significant effect on phosphorus uptake in bhendi at 30, 60, 90 and 120 DAS. The highest phosphorus uptake (2.71 kg ha^{-1}) was recorded in *Gliricidia sepium* and the lowest (2.48 kg ha^{-1}) in *Delonix regia* at 30 DAS, while at 60 DAS the highest phosphorus uptake (3.81 kg ha^{-1}) was recorded in *Albizia lebbek* and the lowest (3.36 kg ha^{-1}) *Leucaena leucocephala*. The application of leaf extract through seed treatment and foliar sprays had a significant effect on phosphorus uptake in bhendi. Among the treatments, *Annona squamosa* recorded the highest phosphorus uptake (2.71 kg ha^{-1}) and the lowest (2.47 kg ha^{-1}) was recorded in *Albizia lebbek* at 30 DAS, while at 60 DAS, the phosphorus uptake of 3.65 kg ha^{-1} was recorded in *Aegles marmellos* and the least (3.43 kg ha kg ha^{-1}) was found in *Ocimum sanctum*. The interaction effect was significant, the treatment combinations, *Albizia lebbek + Annona squamosa* recorded the highest phosphorus uptake (3.40 kg ha^{-1}) and the lowest (2.11 kg ha^{-1}) was recorded in *Albizia lebbek + Morinda tinctoria* at 30 DAS and *Albizia lebbek + Annona squamosa* recorded the highest phosphorus uptake (4.12 kg ha^{-1}) and *Leucaena leucocephala+ Morinda tinctoria* was found to have the lowest phosphorus uptake (3.12 kg ha^{-1}) at 60 DAS. The absolute control recorded phosphorus uptake of 1.01 kg ha^{-1} and 2.05 kg ha^{-1} at 30 and 60 DAS. Where as, the *Albizia lebbek + Annona squamosa* showed an increase in phosphorus uptake of 2.39 kg ha^{-1} and 2.07 kg ha-$^{-1}$ over control. The recommended dose of inorganic fertilizers recorded phosphorus uptake of 4.02 kg ha^{-1} and 5.26 kg ha^{-1} at 30 and 60 DAS. When compared to this, *Albizia lebbek + Annona squamosa* showed a decrease in phosphorus uptake of 0.62 kg ha^{-1} and 1.14 kg ha^{-1} respectively. At 90 DAS, the highest phosphorus uptake (4.55 kg ha^{-1}) was recorded in *Albizia lebbek* and the lowest phosphorus uptake (4.19 kg ha^{-1}) was recorded in *Leucaena leucocephala*, while at 120 DAS, the highest phosphorus uptake (5.88 kg ha^{-1}) was recorded in *Albizia lebbek* and the lowest (5.59 kg ha^{-1}) phosphorus uptake in *Leucaena leucocephala*. Among the treatments, *Annona squamosa* recorded the highest phosphorus uptake (4.67 kg ha^{-1}) and the lowest phosphorus uptake (4.28 kg ha^{-1}) were recorded in *Aegles marmellos* and *Morinda tinctoria* at 90 DAS, while at 120 DAS, the highest phosphorus uptake (5.95 kg ha^{-1}) was recorded in *Annona squamosa* and the lowest phosphorus uptake (5.58 kg ha^{-1}) was recorded in *Ocimum sanctum*. The interaction effect was significant, *Albizia lebbek* + *Annona squamosa* recorded the highest phosphorus uptake (5.62 kg ha^{-1}) and the lowest phosphorus uptake of 4.13 kg ha^{-1} was recorded *Leucaena leucocephala* + *Aegles marmellos* at 90 DAS. At 120 DAS, the highest phosphorus uptake (6.78 kg ha^{-1}) was recorded in *Albizia lebbek* + *Annona squamosa* and the lowest phosphorus uptake of 5.38 kg ha^{-1} was recorded in *Albizia lebbek* + *Ocimum sanctum*. The absolute control recorded phosphorus uptake of 3.25 kg ha^{-1} and 4.65 kg ha^{-1} at 90 and 120 DAS. When compared to this, *Albizia lebbek* + *Annona squamosa* showed an increase in phosphorus uptake of 2.37 kg ha^{-1} 2.13 kg ha^{-1} at 90 and 120 DAS respectively. The recommended dose of inorganic fertilizers recorded phosphorus uptake of 6.75 kg ha^{-1} and 7.95 at 90 DAS and 120 DAS respectively. When compared to this, *Albizia lebbek* + *Annona squamosa* showed a decrease in phosphorus uptake of 1.13 kg ha^{-1} and 1.17 kg ha^{-1} in bhendi.

Potassium

The incorporation of green leaf manures had significant effect on potassium uptake in bhendi at 30, 60, 90 and 120 DAS. The highest potassium uptake (21.57 kg ha^{-1}) was recorded in *Gliricidia sepium* and the lowest (20.22 kg ha^{-1}) was recorded in *Leucaena leucocephala* at 30 DAS, while at 60 DAS, the highest potassium uptake (53.32 kg ha^{-1}) was recorded in *Albizia lebbek* and the lowest (50.59 kg ha^{-1}) was recorded in *Leucaena leucocephala*. The application of leaf extract through seed treatment and foliar sprays had a significant effect on potassium uptake in bhendi. Among them, the highest potassium uptake (21.53 kg ha^{-1}) was recorded in *Annona squamosa* and the lowest (20.35 kg ha^{-1}) was recorded in *Ocimum sanctum* at 30 DAS, Where as at 60 DAS, the highest potassium uptake (53.70 kg ha^{-1}) was recorded in *Annona squamosa*, which was on par (53.09 kg ha^{-1}) with *Aegles marmellos* and the lowest potassium uptake (51.04 kg ha^{-1}) was recorded in *Ocimum sanctum*. The interaction effect was significant, the treatment combinations, *Albizia lebbek* + *Annona squamosa* recorded the highest potassium uptake (24.12 kg ha^{-1}) and lowest potassium uptake (19.72 kg ha^{-1}) was recorded in *Delonix regia* + *Morinda tinctoria* at 30 DAS and *Albizia lebbek* + *Annona squamosa* recorded the highest potassium uptake (59.67 kg ha^{-1}) and the lowest potassium uptake (49.93 kg ha^{-1}) was recorded in *Leucaena leucocephala*+ *Alangium salvifolium* at 60 DAS. The absolute control recorded potassium uptake of 15.26 and 23.35 kg ha^{-1} at 30 and 60 DAS. When compared to control, *Albizia lebbek* + *Annona squamosa* showed an increase in potassium uptake of 8.86 and 36.32 kg ha^{-1}. The recommended dose of inorganic fertilizers recorded potassium uptake of 26.50 and 62.23 kg ha^{-1} at 30 and 60 DAS. When compared to this, *Albizia lebbek* + *Leucaena leucocephala* showed a decrease in potassium uptake of 2.38 and 2.56 kg ha^{-1} at 30 and 60 DAS respectively. At 90 DAS the highest potassium uptake of 82.31 kg ha^{-1} was recorded in *Albizia lebbek* and the lowest potassium uptake of 80.22 kg ha^{-1} was recorded in *Leucaena leucocephala*. At 120 DAS, the highest potassium uptake (111.03 kg ha^{-1}) was recorded in *Albizia lebbe*) and the lowest (103.03 kg ha^{-1}) was recorded in *Morinda tinctoria*. Among the treatments, *Annona squamosa* recorded the highest potassium uptake (83.60 kg ha^{-1}) and the lowest (79.99 kg ha^{-1}) was recorded in

Ocimum sanctum at 90 DAS, while at 120 DAS the highest potassium uptake (109.18 kg ha^{-1}) was recorded in *Annona squamosa* and the lowest (102.61 kg ha^{-1}) was recorded in *Ocimum sanctum*. The interaction effect was significant. The treatment combinations, *Albizia lebbek* +*Annona squamosa* registered the highest potassium uptake (91.01 kg ha^{-1}) and the lowest (77.93 kg ha^{-1}) was recorded in *Albizia lebbek* + *Ocimum sanctum* at 90 DAS. At 120 DAS, the highest potassium uptake (124.7 kg ha^{-1}) was recorded in *Albizia lebbek* +*Annona squamosa* and the lowest (100.9 kg ha^{-1}) was recorded in Delonix *regia* +*Morinda tinctoria*. The absolute control recorded potassium uptake of 39.75 kg ha^{-1} and 53.23 kg ha^{-1} at 90 and 120 DAS. When it was compared with *Albizia lebbek* +*Annona squamosa*, showed an increase in potassium uptake of 51.26 and 71.55 kg ha-$^{-1}$ at 90 and 120 DAS respectively. The recommended dose of inorganic fertilizers recorded potassium uptake of 92.15 kg ha^{-1} and 125.27 kg ha^{-1} at 90 and 120 DAS respectively. When compared to this, *Albizia lebbek* +*Annona squamosa* showed a decrease in potassium uptake of kg ha^{-1} 1.14 and 0.49 kg ha^{-1}. The uptake of nitrogen, phosphorus and potassium were high in the treatment received *Albizia lebbek* as green leaf manure and seed soaking along with foliar sprays of *Annona squamosa*. Higher nutrient uptake due to the addition of organic leaf manures might be due to the release of different organic acids on their decomposition and chelation with Fe and Al which helps in solubilization of unavailable forms of phosphorus and release the mono calcium phosphate into the labile pool of soil, thus making it available to the growing crops. Apart from this, the release of nutrients at a slower rate during the decomposition favours the availability of nutrients for the entire crop growth period. High nutrient uptake stimulated the microbial growth and root growth due to improvement in soil physical conditions created by addition of organics. The increased uptake was due to the supply of nutrients and well developed root system under balanced nutrient application resulting in better absorption of water and nutrient. The efficient absorption and translocation of nutrients from soil to foliage may be probably due to favourable and possible interaction of nitrogen and potassium on the availability of more nutrients at critical stages of crop growth. Potassium performs many functions in plant metabolism *viz.*, promoting photosynthesis, harnessing interaction between potassium and nitrogen could have increased the nutrient uptake. The absolute control registered the lowest uptake of nutrients due to low availability of nutrients in the soil.

References

Barber, S.A. 1984. Soil nutrient bioavailability: A mechanistic approach. New York: John Wily and sons

Boswell, F.C., J.J. Meisinger and N.L. Case. 1985. Production, marketing and use of nitrogen fertilizer (pp. 229-292). In O.P. Engelstad (Ed.), fertilizer technology and use. Madison, Wis.: Soil Science Society of America.

Follett, R.F. (Ed.) 1989. Nitrogen management and groundwater protection. In developments in agricultural managed-forest ecology 21. New York: Elsevier.

Hauck, R.D. 1984. Nitrogen in crop production. American Society of Agronomy, Madison, Wis.

Power, J.F. and R.I. Papendick. 1985. Organic source of nitrogen (pp. 503-520). In O.P. Engelstad (Ed.), Fertilizer technology and use. Madison, Wis.: Soil Science Society of America.

Stevenson, F.J. 1986. Cycles of soil. New York: John Wily and Sons.

Barber, S.A. 1984. Soil nutrient bio availability: A magnetic approach. New York. John Wiley and sons.

Khasawneh, F.E. (Ed.). 1980. The role of Phosphorus in agriculture. Madison, Wis: American Society of Agronomy Crop Science Society of America, Soil Science Society of America.

Lindsay, W.L. 1979. Chemical equilipria in soils. New York: John Wily and sons.

Stevenson, F.J. 1986. Cycles of soil: Carbon, nitrogen, phosphorus, sulfur, micronutrients. New York: John Wily and sons.

Barber, S.A., R.D. Munson and W.B. Dancy. 1985. Production, marketing and use of potassium fertilizer (pp. 377-410). In O.P. Englestad (Ed.), fertilizer technology and use. Madison, wis: Soil Science Society of America.

Mortvedt, J.J., L.S. Murphy and R.H. Follett. 1999. Willoughby, Ohio: Fertilizer technology application. Meister Publ. Co.

Munson, R.D. (Ed.). 1985. Potassium in Agriculture. Madison, Wis: Soil Science Society of America.

Chapter II
Soil Microbial Studies

Soil Microorganisms

Soil microbiology emerged as a distinct branch of soil science in 1838 after the French agricultural chemist and farmer, Boussingault, showed that legumes could obtain nitrogen from air when grown in soil which was not heated. Soil is the outer covering of the earth, which consists of loosely arranged layers of materials composed of inorganic and organic compounds in different stages of organization. It is a natural medium in which microbes live, multiply and die. Increasing attention is being drawn to microorganisms because the fertility of soil depends not only on its chemical composition, but also on the qualitative and quantitative nature of microorganisms inhabiting it. Maintenance of viable, diverse populations and functioning microbial communities in the soil is essential for sustainable agriculture. Microorganisms are generally divided into five major taxonomic categories: algae, bacteria, fungi, protists and viruses. In soil, they are closely associated with soil particles, mainly clay organic matter complexes. Eubacteria are prokaryotic microorganisms. They are recognized as the most dominant group of microorganisms among the various kinds of soil. In general, horizon A (soil with organic matter) of a soil profile consists of more microorganisms than horizon B (silicate clay minerals plus organic matter) and C (weathered parent material).

Bacteria

More dominant group of microorganisms in the soil and equal to one half of the microbial biomass in soil. Population 100,000 to several hundred millions for gram of soil. Majority are Heterotrophs. (Common soil bacteria - *Arthrobacter, Bacillus, Clostridium* and *Micrococcus*).

Actinomycetes

Intermediate group between bacteria and fungi. Numerous and widely distributed in soil. Abundance is next to bacteria, 10^4–10^8 g^{-1} soil. Seventy percent of soil actinomycetes are Streptomyces. Many of them are known to produce antibiotics and population increases with depth of soil.

Fungi

More numerous in surface layers of well-aerated and cultivated soils and dominant in acid soils. Common genera in soil are Aspergillus, Mucor, Penicillium Trichoderma, Alternaria, Rhizopus. Algae are found in most of the soils in number ranges from 100 to 10,000 g^{-1}.

Role of Soil Microorganisms

- ✰ They are involved in nutrient transformation process and Decomposition of resistant components of plant and animal tissue
- ✰ The act of surface blooming reduces soil erosion losses
- ✰ They participate in humus formation
- ✰ Play a role in microbial antagonism and are predators to nematodes
- ✰ They help in maintenance of biological equilibrium and Improves soil structure

Bacteria

Bacteria live in soil as cocci (sphere, 0.5µm), bacilli (rod, 0.5–0.3µm) or spiral. The bacilli are common in soil, whereas spirilli are very rare in natural environments. Bacteria have been classified into two broad categories, the autochthonous and the zymogenous organisms. The autochthonous or indigenous populations are more uniform and constant in soil, since their nutrition is derived from native soil organic or mineral matter (*Arthrobacteria* and *Nocardia*). The zymogenous bacteria require the input of an external substrate and their activity in soils is variable. They often produce resting propagules (*Pseudomonas* and *Bacillus*). When specific substrates are added to soil, the number of zymogenous bacteria increases and gradually declines when the added substrate is exhausted (cellulose decomposers, nitrogen utilizing bacteria, *Nitrosomonas, Nitrobacter*). The most common bacteria belong to the genera *Pseudomonas, Arthrobacter, Clostridium, Achromobacter, Bacillus, Micrococcus, Flavobacterium, Corynebacterium, Sarcina, Azosprillium* and *Mycobacteria*.

Bacteria are single-cell organisms and the most numerous denizens of agriculture, with populations ranging from 100 million to 3 billion in a gram. They are capable of very rapid reproduction by binary fission (dividing into two) in favourable conditions. One bacterium is capable of producing 16 million more in just 24 hours. Most soil bacteria live close to plant roots and are often referred to as rhizobacteria. Bacteria live in soil water, including the film of moisture surrounding soil particles and some are able to swim by means of flagella. The majority of the beneficial soil-dwelling bacteria need oxygen (and are thus termed aerobic bacteria), whilst those that do not require air are referred to as anaerobic and tend to cause putrefaction of dead organic matter. Aerobic bacteria are most active in a soil that is moist (but not saturated, as this will deprive aerobic bacteria of the air that they require) and neutral soil pH and where there is plenty of food (carbohydrates and micronutrients from organic matter) available. Hostile conditions will not completely kill bacteria; rather, the bacteria will stop growing and get into a dormant stage and those individuals with pro-adaptive mutations may compete better in the new conditions. Some gram-positive bacteria produce spores in order to wait for more favourable circumstances and gram-negative bacteria get into a "non-culturable" stage. Bacteria are colonized by persistent viral agents (bacteriophages) that determine gene word order in bacterial host.

Nitrification

Nitrification is a vital part of the nitrogen cycle in which bacteria (which manufacture their own carbohydrate supply without using the process of photosynthesis) are able to transform nitrogen into ammonium into nitrates, which are available to growing plants.

Nitrogen Fixation

It is the process of nitrogen fixation constantly puts additional nitrogen into biological network which is done by free-living nitrogen-fixing bacteria in the soil or water such as *Azotobacter*, These are able to convert nitrogen from the atmosphere into nitrogen-containing organic substances.

Denitrification

While nitrogen fixation converts nitrogen from the atmosphere into organic compounds, a series of processes which returns an approximately equal amount of fixed atmospheric nitrogen, back to the atmosphere. Denitrifying bacteria tend to be anaerobes, or facultatively anaerobes. The purification process caused by oxygen-free conditions converts nitrates and nitrites in soil into nitrogen gas or into gaseous compounds such as nitrous oxide or nitric oxide. In excess, denitrification can lead to overall losses of available soil nitrogen and subsequent loss of soil fertility. However, fixed nitrogen may circulate many times between organisms and the soil before denitrification returns it to the atmosphere.

Fungi

Fungi dominate all types of soils and represent the greatest diversity among soil microorganisms. Fungi possess filamentous mycelium com-posed of individual hyphae. The hyphae may be uni-, bi- or multinucleate and nonseptate or septate. All the environmental factors that influence the distribution of bacteria also apply in fungal flora of soils. However, the quality and quantity of organic matter have a direct bearing on fungal numbers in soils since fungi are heterotrophic organisms. Fungi are dominant in acid soils because an acidic environment is not conducive to the existence of either bacteria or actinomycetes. They are also present in neutral or alkaline soils and some can tolerate a pH over 9.0. Arable soils contain abundant fungi since they are strictly aerobic and an excess of soil moisture decreases their numbers. Filamentous fungi in soil degrade organicmatter and help in soil aggregation. Certain fungi like *Alternaria, Aspergillus, Cladosporium, Helminthosporium* produce substances similar to humic substance in soil and, hence, may be important in the maintenance of soil organic matter.

A gram of garden soil can contain around one million fungi, such as yeasts and moulds. Fungi have no chlorophyll and are not able to perform photosynthesis. They cannot use atmospheric carbon dioxide as a source of carbon, therefore they are chemo-heterotrophic.

Mycorrhizae are those fungi that are able to live symbiotically with living plants, creating a relationship that is beneficial to both, are known as Mycorrhizae (from *myco* meaning fungal and *rhiza* meaning root). Plant root hairs are invaded by the mycelia of the mycorrhiza, which lives partly in the soil and partly in the root and may either cover the length of the root hair as a sheath or be concentrated around its tip. The mycorrhiza obtains the carbohydrates that it requires from the root, in return providing the plant with nutrients including nitrogen and moisture. Later the plant roots will also absorb the mycelium into its own tissues.

Beneficial mycorrhizal associations are to be found in many of our edible and flowering crops. The benefits of mycorrhizal relations to their plant partners are not limited to nutrients, but can be essential for plant reproduction. Recent research has shown that arbuscular mycorrhizal fungi produce glomalin, a protein that binds soil particles and stores both carbon and nitrogen. These glomalin-related soil proteins are an important part of soil organic matter

Actinomycetes

Actinomycetes are soil microorganisms with sufficient distinctive features to delimit them into a distinct group within the prokaryotes. The number of actinomycetes increases in the presence of decomposing organic matter. They are intolerant to acidity and their numbers decline below pH 5.0. The most conducive range of pH is between 6.5 and 8.0. Water logging of soil is unfavorable for the growth of actinomycetes, whereas arid and semi-arid zone soils sustain sizeable populations. The percentage of actinomycetes in the total microbial populations increases with the depth of cultures. soil. Actinomycetes can be isolated in sufficient number even from the C horizon (weathered parent material) of soil profiles. The commonest genera of actinomycetes are Streptomyces (nearly 70 per cent).

The observations made on soil microbial population *viz.*, bacteria, fungi and actinomycetes in different crops are presented hereunder and discussed.

Barnyard Millet

In Bacteria (10^6 g^{-1} in soil), significant variation was observed in the microbial population at 45 days after incorporation (DAI) and after harvest of the barnyard millet crop due to different source of green leaf manure incorporation, the highest bacterial population was recorded (74.67×10^6) in *Peltophorum ferrugineum* at 45 DAI. The lowest microbial population was recorded (49.33×10^6) in *Pongamia pinnata* and it was on par with *Gliricidia sepium*. The bacterial population after harvest also significantly influenced. The higher bacterial population was recorded (58.0×10^6) after harvest in *Peltophorum ferrugineum* and it was on par with *Delonix regia* (55.33×10^6). The lower bacterial population was recorded in *Gliricidia sepium* (37.00×10^6) at harvesting of the crop. For fungi (10^4 g^{-1} in soil) population, the incorporation of green leaf manures had significantly influenced the fungi population on 45 DAI and after

Table 11.1: Dynamics of Soil Microbial Population in Barnyard Millet Grown Field

GLM Incorporation	*Soil Microbial Analysis*								
	Initial			*45 Days After Incorporation of GLM*			*At Harvest*		
	Bacteria (x 10^6 CFU g^{-1} of soil)	*Fungi (x 10^3 CFU g^{-1} of soil)*	*Actinomycetes (x 10^4 CFU g^{-1} of soil)*	*Bacteria (x 10^6 CFU g^{-1} of soil)*	*Fungi (x 10^3 CFU g^{-1} of soil)*	*Actinomycetes (x 10^4 CFU g^{-1} of soil)*	*Bacteria (x 10^6 CFU g^{-1} of soil)*	*Fungi (x 10^3 CFU g^{-1} of soil)*	*Actinomycetes (x 10^4 CFU g^{-1} of soil)*
Albizia				59.33	63.66	38.33	43.67	54.00	46.33
Gliricidia				51.33	53.67	32.67	37.00	50.00	37.33
Pongamia	45.05	49.67	28.15	49.33	55.00	35.00	44.66	45.33	42.67
Delonix				63.67	58.00	33.00	55.33	50.35	37.33
Peltophorum				74.67	70.33	40.00	58.00	60.33	44.67
CPG (C_2)								16.33	28.00
MEAN				**59.70**	**60.20**	**35.98**		**52.02**	**41.66**
SEd				2.45	1.45	0.73	1.46	1.41	0.70
CD(0.05)				5.25	3.15	2.03	3.15	3.04	1.57

harvest of the crop. The higher fungi population was recorded in *Peltophorum ferrugineum* (70.33×10^4), it was followed by *Albizia lebbek* (63.66×10^4) at 45 DAI and all other treatments were on par with each other. After harvest of the crop higher fungi population was recorded in *Peltophorum ferrugineum* (60.33×10^4) it was followed by *Albizia lebbek* (54.00×10^4). The lower fungi population was recorded in *Pongamia pinnata* (45.33×10^4) after harvesting of crop. As regards actinomycetes (10^3 g^{-1} in soil) counts, it was significantly influenced at 45 DAI and post harvest stage. The highest actinomycetes population was registered in *Peltophorum ferrugineum* (40.00×10^3), it was on par with *Albizia lebbek* (38.33×10^3). The all other treatments were on par with each other at 45 DAI. After harvesting the crop the highest actinomycetes population was registered in *Albizia lebbek* (46.33×10^3), it was followed by *Peltophorum ferrugineum* (44.67×10^3) and *Pongamia pinnata*. The lowest actinomycetes population (37.30×10^3) in *Delonix regia* and *Gliricidia sepium* was recorded.

Ragi

Significant variation was observed in the microbial population at 45 days after incorporation of green leaf manures and after harvest of the finger millet crop due to different source of green leaf manures, the highest bacterial population (10^6 g^{-1} in soil) was recorded in *Delonix regia* (74.67×10^6) at 45 DAI. The lowest microbial population (49.33×10^6) was recorded in *Peltophorum ferrugineum*. The bacterial population after harvest also got significantly influenced and the higher bacterial population (58.0×10^6) was recorded after harvest in *Delonix regia* and it was on par with *Leucaena leucocephala* (55×10^6). The lower bacterial population (37.00×10^6) was recorded in *Gliricidia sepium* at harvesting of the crop. For fungi population (10^4 g^{-1} in soil), the incorporation of green leaf manures had significantly influenced the fungi population on 45 DAI and after harvest of the crop. The higher fungi population (70.33×10^4) was recorded in *Delonix regia*, it was followed by *Albizia lebbek* (63.66×10^4) at 45 DAI and all other treatments were on par with each other. After harvest of the crop, the highest fungi population (60.33×10^4) was recorded again in *Delonix regia*. The lowest fungi population was recorded in *Leucaena leuocephala* (45.33×10^4) after harvesting of crop. As regards actinomycetes population, it was not significantly influenced by incorporation of green leaf manures and foliar spraying of leaf extracts at 45 DAI and post harvest soil. The highest actinomycetes population was registered in *Albizia lebbek* (46.33×10^3) at 45 DAI. The lowest actinomycetes population of 37.3×10^3 was recorded in *Delonix regia* and *Gliricidia sepium* at 45 DAI. After harvesting the crop, the highest actinomycetes population was registered in *Peltophorum ferrugineum* (40.00×10^3). And then, *Gliricidia sepium* registered lowest population (32.67×10^3) of actinomycetes in the soil after harvesting of the crop.

Maize

The incorporation of green leaf manures had significantly influenced the bacterial population (10^6 g^{-1} in soil) at 45 days after incorporation (DAI) and after harvest of the crop. The highest bacterial population was recorded (70×10^6) in *Albizia lebbek* at 45 DAI. It was followed by *Peltophorum ferrugineum* (34×10^6) and lower population was obtained under *Leucaena leucocephala* (18×10^6). While at harvest of the crop, bacterial

Table 11.2: Dynamics of Soil Microbial Population in Finger Millet Grown Field

GLM Incorporation	*Soil Microbial Analysis*								
	Time Zero Values in Experimental Field			*45 Days After Incorporation of GLM*			*Post-harvest Soil*		
	Bacteria (x 10^6 CFU g^{-1} of soil)	*Fungi (x 10^3 CFU g^{-1} of soil)*	*Actinomycetes (x 10^4 CFU g^{-1} of soil)*	*Bacteria (x 10^6 CFU g^{-1} of soil)*	*Fungi (x 10^3 CFU g^{-1} of soil)*	*Actinomycetes (x 10^4 CFU g^{-1} of soil)*	*Bacteria (x 10^6 CFU g^{-1} of soil)*	*Fungi (x 10^3 CFU g^{-1} of soil)*	*Actinomycetes (x 10^4 CFU g^{-1} of soil)*
Albizia	**46.05**	**50.67**	**29.15**	60.3	64.6	47.4	43.7	54.0	38.3
Gliricidia				52.3	54.7	38.3	37.0	50.0	32.6
Leucaena				64.7	56.0	43.6	55.3	45.3	35.0
Delonix				75.7	71.3	38.34	58.0	60.3	33.0
Peltophorum				50.3	59.0	45.6	44.6	50.5	40.0
SEd				3.35	3.12	NS	3.30	3.27	NS
CD ($P \leq 0.05$)				8.62	8.02	NS	8.49	8.42	NS
CPG							40.12	42.86	21.57

Table 11.3: Dynamics of Soil Microbial Population in Maize Grown Field

GLM Incorporation	*Soil Microbial Analysis*								
	Time Zero Values in Experimental Field			*45 Days After Incorporation of GLM*			*Post-harvest Soil*		
	Bacteria (x 10^6 CFU g^{-1} of soil)	*Fungi (x 10^3 CFU g^{-1} of soil)*	*Actinomycetes (x 10^4 CFU g^{-1} of soil)*	*Bacteria (x 10^6 CFU g^{-1} of soil)*	*Fungi (x 10^3 CFU g^{-1} of soil)*	*Actinomycetes (x 10^4 CFU g^{-1} of soil)*	*Bacteria (x 10^6 CFU g^{-1} of soil)*	*Fungi (x 10^3 CFU g^{-1} of soil)*	*Actinomycetes (x 10^4 CFU g^{-1} of soil)*
Albizia	19	9.5	7	70	25	25	**40.00**	2	17
Glyricidia				25	16	17	**19.33**	1	10
Pongamia				18	11	16	**15.00**	7	25
Delonix				30	19	17	**22.00**	7	11
Peltophorum				34	25	18	**25.66**	13	10
CPG (C_2)								2	6
MEAN				**70**	**25**	**25**	**40.00**	**2**	**17**
SEd				25	16	17	**19.33**	1	10
CD ($P \leq 0.05$)				18	11	16	**15.00**	7	25

population recorded in *Peltophorum ferrugineum* was 13 × 10^6. It was followed by *Leucaena leucocephala* and *Delonix regia* which recorded (7 × 10^6) and lower level population was obtained under *Gliricidia sepium* (1 × 10^6). As regards fungi (10^3 g^{-1} in soil) population, the incorporation of green leaf manures had significantly influenced the fungi population on 45 DAI and after harvest of the crop. The highest fungi population was recorded (25 × 10^3) in *Albizia lebbek*, it was on par with *Peltophorum ferrugineum* (25 × 10^3) and it was followed by *Delonix regia* which recorded 19 × 10^3 and lower level population was obtained under *Leucaena leucocephala* (11 × 10^3). While at harvest of the crop fungi population recorded in *Leucaena leucocephala* was 25 × 10^3. It was followed by *Albizia lebbek* which recorded 17 × 10^3 and lower level population was obtained under *Gliricidia sepium* and *Peltophorum ferrugineum* (10 × 10^3). For actinomycetes (10^4 g^{-1} in soil), the incorporation of green leaf manures had significantly influenced the actinomycetes population at 45 DAI and after harvest of the crop. The highest actinomycetes population (25 × 10^4) was recorded in *Albizia lebbek*, it was followed by *Peltophorum ferrugineum* (18 × 10^4) and lower level population was obtained under *Leucaena leucocephala* (16 × 10^4). While at harvest of the crop actinomycetes population (69 × 10^4) was recorded in *Gliricidia sepium*. It was followed by *Albizia lebbek* (35 × 10^4) and lower level population was obtained under *Peltophorum ferrugineum* (20 × 10^4).

Black Gram

As regards, Bacteria (10^6 g^{-1} in soil) counts the incorporation of green leaf manures had significantly influenced the bacterial population on 45 days after incorporation (DAI) and after harvest of the crop. The highest bacterial population was recorded (62 × 10^6) in *Pongamia pinnata* at 45 DAI. It was followed by *Albizia lebbek* recorded (47 × 10^6) and the lower level population was obtained under *Gliricidia sepium* (27 × 10^6). While at harvest of the crop (12 × 10^6) of bacterial population was recorded in *Delonix regia*. It was followed by *Peltophorum ferrugineum* which recorded (9× 10^6) and the lower level population was obtained under *Pongamia pinnata* (3 × 10^6) while for fungi (10^3 g^{-1} in soil) population, the incorporation of green leaf manures had significantly influenced the fungi population on 45 DAI and after harvest of the crop. The highest fungi population was recorded (27 × 10^3) in *Pongamia pinnata*, it was followed by *Albizia lebbek* (23 × 10^3) and lower level population was obtained under *Peltophorum ferrugineum* (13 × 10^3). While at harvest of the crop fungi population recorded in *Albizia lebbek* was 18 × 10^3. It was followed by *Pongamia pinnata* (15 × 10^3) and the lower level population was obtained under *Gliricidia sepium* and *Peltophorum ferrugineum* (8 × 10^3). For actinomycetes (10^4 g^{-1} in soil), the incorporation of green leaf manures had significantly influenced the actinomycetes population on 45 DAI and after harvest of the crop. The highest actinomycetes population was recorded (30 × 10^4) in *Pongamia pinnata*, it was followed by *Albizia lebbek* (20 × 10^4) and lower level population was obtained under *Peltophorum ferrugineum* (15 × 10^4). While at harvest of the crop the higher actinomycetes population was recorded (40 × 10^4) in *Albizia lebbek*. It was followed by *Pongamia pinnata* (30 × 10^4) and lower level population was obtained under *Delonix regia* (21 × 10^4).

Table 11.4: Dynamics of Soil Microbial Population in Black Gram Field

GLM Incorporation	*Soil Microbial Analysis*											
	Time Zero Values in Experimental Field				*45 Days After Incorporation of GLM*				*Post-harvest Soil*			
	Bacteria (x 10^6 CFU g^{-1} of soil)	*Fungi (x 10^3 CFU g^{-1} of soil)*	*Actinomycetes (x 10^4 CFU g^{-1} of soil)*	*Mean*	*Bacteria (x 10^6 CFU g^{-1} of soil)*	*Fungi (x 10^3 CFU g^{-1} of soil)*	*Actinomycetes (x 10^4 CFU g^{-1} of soil)*	*Mean*	*Bacteria (x 10^6 CFU g^{-1} of soil)*	*Fungi (x 10^3 CFU g^{-1} of soil)*	*Actinomycetes (x 10^4 CFU g^{-1} of soil)*	*Mean*
Albizia	22	11	10	**14.33**	47	23	20	**30.00**	5	18	40	**21.00**
Gliricidia					27	18	17	**20.67**	6	8	37	**17.00**
Pongamia					62	27	30	**39.67**	3	15	30	**16.00**
Delonix					30	21	17	**22.67**	12	8	21	**13.67**
Peltophorum					27	13	15	**18.33**	9	12	24	**15.00**
CPG									2	5	11	6.00
Mean					**38.60**	**20.40**	**19.80**		**7**	**12.2**	**30.4**	
SEd					0.64	0.65	0.64		0.62	0.64	0.64	
CD (P≤0.05)					1.80	1.80	1.80		1.71	1.80	01.80	

Points for Comprehension

The regulated application of organics as a practice will invigorate the living micro organism of the soil, involved in biochemical activity of importance to soil fertility and plant nutrition. The better availability of nutrients and more amount of organic carbon content of the soil may help to improve the soil microbial population. Further, the population of bacteria, fungi, actinomycetes free living nitrogen fixers, P-solubilizers and enzymes such as dehydrogenase and phosphatase were found significantly higher when the crop was grown organically and the biological activity in soil also increases with the application of organics which in turn increases microbial activity.

References

Alexander, 1977, Introduction to Soil Microbiology, 2nd edition, John Wiley

Alexander, 1994, Biodegradation and Bioremediation, Academic Press

Bardgett, R.D., 2005, The Biology of Soil: A Community and Ecosystem Approach, Oxford University Press

Coleman D.C., 2004, Fundamentals of Soil Ecology, 2nd edition, Academic Press

Coyne, 1999, Soil Microbiology: An Exploratory Approach, Delmar

Doran, J.W., D.C. Coleman, D.F. Bezdicek and B.A. Stewart. 1994. Defining soil quality for a sustainable environment. Soil Science Society of America Special Publication Number 35, ASA, Madison Wis.

Paul, P.A. and F.E. Clark. 1996, Soil Microbiology and Biochemistry, 2nd edition, Academic Press

Richards, 1987, The Microbiology of Terrestrial Ecosystems, Longman Scientific and Technical

Sylvia *et al.*, 1998, Principles and Applications of Soil Microbiology, Prentice Hall

Soil and Water Conservation Society, 2000, Soil Biology Primer.

Tate, 2000, Soil Microbiology, 2nd edition, John Wiley

Van Elsas, 1997, Modern Soil Microbiology, Marcel Dekker.

Wood, 1995, Environmental Soil Biology, 2nd edition, Blackie A and P.

Chapter 12

Plant Extracts for Pest and Disease Management

Introduction

Agricultural crops are under constant assault by insect pests, forcing the cultivators to apply insecticides essential to reduce losses. Synthetic insecticides such as organophosphates are important, effective tools in modern crop management. However, they pose serious threats to the human and environment.Humans come in contact with dangerous pesticides on food, in water and in the air near farms. This "pesticide drift" occurs when pesticide dust and spray travel by wind to places unexposed to pesticides. Almost 80-90 percent of sprayed insecticides/pesticides do not reach their targets. They penetrate to groundwater, pollute streams and harm wildlife, including natural predators of the targeted pests. Mass production farms rely on synthetic pesticides, however, because they are cheaper than organic ones. Many plant species produce substances that protect them by killing or repelling the insects that feed on them. Neem trees produce oil that alters the hormones of bugs so that they cannot fly, breed or eat. It is possible to create effective, natural insecticides from these substances to protect crops that, unlike wild plants, may have lost their capability through cultivation to cope with pests. Plant extracts/botanicals/Natural pesticides have many advantages over synthetic ones and may be more cost-effective as a whole, considering the environmental cost of chemical alternatives. Natural pesticides are biodegradable, barely leave residues in the soil and are less likely to harm humans or animals. In addition, they are cheaper and more accessible in less developed countries.

Plants have a number of chemical constituents known as secondary metabolites and they do not have any special functions in plant physiology. They have become a chemical weapon against crop pests which is accomplished by harmful effects on behaviour, physiology, growth, reproduction and other functions of the phytophagous insects. Plant products being indigenous resources are in use for over a century in India for minimizing losses in stored grain due to insect pests. Research designed to discover the defensive mechanism has led to applications, which have been integrated with existing pest management programmes. Bioactivity of these allelochemicals extended to groups of hexapods other than herbivore-omnivore. A meaningful integration of compounds singly or in combination with other tactics would provide potential for quick application in field practice and commercial exploitation.

Botanicals are the pesticides derived from the plant materials. They are very effective in controlling the insect- pest because they possess repellant, knock down, anti-feedant, broad spectrum properties and also they are important because they are less hazardous, biodegradable and maintain balanced bio-diversity of insect-pest. Use of botanicals is an age old practice where people used to utilize the extracts from the plant parts to control the insects. There are many such examples as use of neem dust to control the stored grain pest, use of various plant extracts like *Acorus* sp., *Melia* sp, *Acacia* spp. etc.

Plant extracts or botanicals have been used for many centuries in India among limited resource farmers to control insect pests of both field crops and stored produce, but their potential was initially limited and ignored. Nicotine, neem products and other plant based extracts were popular for controlling storage pests and other pests in the field. Some of these plant species possess one or more useful properties such as repellency, antifeedant, fast knock down, flushing action, biodegradability, broad-spectrum of activity and ability to reduce insect resistance. However, most of them are either weak insecticidally or may require other plant species with different mode of action (depending on the ratio and rate of application) to increase their efficiency. Extracts of chilli pepper in mixture with garlic (*Allium sativum* L.), onion (*Allium cepa* L.) bulbs extracts and lemon grass (*Cymbopogon citrates* Staph.) leaf extract were found very effective against some leaf eating insect pests of crops. In India rice farmers are said to use a mixture of chilli pepper, dried tobacco leaves and *Croton* species against stem borers.

The farmers mix chilli powder, pepper and wood ash of *Eucalyptus* spp./ *Azadirachta indica* to control diseases and pests of bhendi. An admixture of water from fermented cassava (*Manihot esculentus* Crantz) tubers and bitter leaf has shown potency against ants. Since there is paucity of documented information on the use of plant extract mixtures in pest control, this study was aimed at assessing the efficacy of mixtures of plant extracts for management of pests of different crops.

Threats from Agrochemicals

The first report of pesticide poisoning was reported from Kerala as early in 1958, where over 100 people died after consuming wheat flour contaminated with parathion. This prompted the Indian Council of Agricultural Research to focus attention on the problems caused by pesticides to human life. Measurement of chemicals in the total diet provides the best estimates of human exposure and of the potential risk. The risk of consumers may then be evaluated by comparison with toxicologically acceptable intake levels. Fatty food was the main source of contaminants like DDT and BHC. Pesticides can contaminate soil, water, turf, and other vegetation. In addition to killing insects or weeds, pesticides can be toxic to a host of other organisms including birds, fish, beneficial insects, and non-target plants. Insecticides are generally the most acutely toxic class of pesticides, but herbicides can also pose risks to non-target organisms.

Groundwater pollution and Surface water contamination due to agrochemicals and pesticides is a worldwide problem. Pesticides can reach surface water through runoff from treated plants and soil. Contamination of water by pesticides is widespread. Further contamination of soil is another threat. A large number of transformed products of a wide range of pesticides have been traced from soil. The residues of the banned agrochemicals are still present in soil and these agrochemicals move from soil by runoff and leaching, thereby constituting a problem for the supply of drinking water. The most problematic agrochemicals are undoubtedly pesticides and herbicides. The most influential soil characteristic is the organic matter content. The larger the organic matter content, the greater the adsorption of pesticides. Further heavy treatment of soil with pesticides and herbicides can cause populations of beneficial soil microorganisms to decline. The reduction in population count of soil borne bacteria and fungi leads to soil degradation. Overuse of chemical fertilizers also have the same effects on the soil organisms. Pesticide sprays can directly hit non-target vegetation and/or can drift from sprayed area and contaminate soil, air and beneficial plants.

Integration of Botanicals in Pest Management

Research on botanicals has been carried out in our country since the 1960's. Acquisition of information was only on the role of plant products in pest management. Rapid strides were made in bioprospecting of compounds and their regulatory role in insect plant relationship. During this period, several biocontrol agents namely entomophages and entomopathogens were studied for their role in pest management. Though, biocontrol agents offered promise in the management of crop pests, for a variety of reasons they have to be integrated with other methods. Pesticides used in India are generally broad spectrum in nature and hence their utility in the conservation of entomophages are restricted. Systematic exploration of botanicals revealed the possibilities for rational and judicious integration with many of the entomophages and entomopathogens as the botanicals in general do not alter the diversity of entomophages in the ecosystem.

Botanicals and Entomopathogens

Research and development on the use of plant products and microbials has shown that botanicals can be used to increase the susceptibility of the target pest. Biologically active compounds from plants penetrate the gut wall and allow easy entry of the entomopathogens into the haemocoel. Allelochemicals are known to target sensitive processes, common to all living organisms including humans like respiration, protein synthesis and cell membrane activities while the entomopathogens which are self replicative, multiply in the susceptible insects. In this interactive process the response is generally synergistic. The general principle followed in integration of botanicals and microbials is to apply the agents simultaneously or sequentially for a synergistic response. A wide variety of plant species in the category of weeds, ornamental plants, fruit trees, spices and condiments and medicinal and aromatic plants have been screened for integration with microbial pathogens.

Viruses and Botanicals

Recently laboratory experiments were conducted at TNAU to find out if aqueous extracts of leaves of *T.terrestis, C. roseus* and *V. negundo* would enhance the efficacy of GV applied at a reduced concentration of $1x10^5$ OB/ml against the diamond back moth *Plutella xylostella*. GV at a concentration of 10^5 POB/ml caused 67.5 per cent larval mortality. The three botanicals tested caused a larval mortality of less than 50 per cent. However, when they were combined with GV, the mortality was higher. GV + *V. negundo* caused 80 per cent followed by *T. terrestris* (77.5 per cent). The addition of the extract of *C. roseus* did not increase the larval mortality. A marginal reduction in the adult emergence was also noted, when GV was combined with the botanicals (unpublished data). A close perusal of information on these lines indicates that aqueous extracts from several plant parts were found to enhance mortality of certain pests when combined with Nuclear polyhedrosis virus. Crude aqueous leaf extract (5-20 per cent) of non edible plants are known to increase the efficacy of NPV. A report says that 20 per cent leaf extract of *Vitex negundo* increased the mortality of second instar *H.armigera* by 46.3 times. Whereas 20 per cent aqueous extracts of *Prosopis juliflora* and *Catharanthus roseus* increased the efficacy of HaNPV about 20 times.

Bacteria and Botanicals

One of the studies on use of bacteria and botanicals indicated that addition of guava leaf powder (5 per cent), orange peel powder (5 per cent), lemon grass leaf powder (5 per cent), lemon grass leaf oil (1 per cent), *Euphorbia lathyrus* leaf powder 5 per cent and *E. lathyrus* acetone extract (1 per cent) enhanced the potency of *Bacillus thuringiensis* to *Spodoptera littoralis*.

Fungus and Botanicals

Aqueous extracts of *Ocimum sanctum, Allium sativum, Acorus calamus, Tribulus terrestris*, NSKE and neem oil at one and two per cent inhibited the conidial germination in *Beauvaria bassiana*; neem oil followed by NSKE being the most deleterious.

Botanicals and Entomophagous

The dependency of entomophages for healthy insects for their biological survival has been documented in several situations. The potential of the entomophages could be exploited to the maximum when healthy individuals of prey/host species are available in the field. Hence safer plant products are needed while integrating botanicals with entomophages. On the other hand, in many instances of crop pest control, the prescriptions for the use of entomophages revolved around inundative releases wherein a horizontal control alone is expected. Moreover sequential releases of entomophages are being recommended in many of the biosensitive pest management schedules. The core issue of Integrated Pest Management centers around conservation of bio-diversity. It will be prudent to choose bioactive compounds that are not directly or indirectly deleterious to the beneficial arthropods and non-target organisms.

Predators and Botanicals

Information on the efficacy of botanicals with predators are scanty. The results are either antagonistic or synergistic in general. One report suggested that there was no significant interaction between the predator and the pesticide indicating that the chemical and biological control agents may not work in synergy

The effect of botanicals pesticides on non –target organism has picked up only in recent years. The application of NSKE for the control of *S. litura* did not affect the emergence of the egg parasitoid, *Telenomus remus* Nixon and predator, *Brinckochrysa sceletes* (Banks). However, longevity of the parasitoid was reduced incase of oviposition on pretreated egg masses. Topical application of seed oil of neem, chinaberry and custard apple showed no adverse effect on predatory spider, *Lycosa pseudoannulata* (bosenburg and strand) and was only slightly toxic to the mired bug predator, *Cyrtorhinus lividipennis* Reuter at and above 10 microgram/female.

An active neem seed kernel fraction evaluated against sorghum pests was found to be safe to midge, *Contarina sorghicola* (coquillett); parasitoid, *Tetrastichus sp.* and predator, *Orius sp.* It was, however, toxic to *Apanteles ruficrus* (Haliday), a larval parasite of M. seperata. Field trials with neem oil for the control of sorghum aphid, *Melanapsis sacchari* (Zehntner) did not show any adverse effect on syrphids and coccinellids.

The plant extracts of some rice varieties were found to be highly toxic to important rice pests including *Chilo suppresaalis* (walker), *Nilaparvata lugens* (stal) and *Sogatella furcifera* (Horvath) but were safe to predators of hoppers complex.

Some observations include

- ☆ Application of botanicals - increase the susceptibility of the target pest – increase efficiency of predators and parasitoids
- ☆ Botanicals penetrate gut wall - Pathogens - easy entry
- ☆ Allelochemicals - target sensitive processes on respiration, DNA transcription, protein synthesis and cell membrane activities
- ☆ Entomopathogens - self replicative, multiply in the weakened insect. **The response is synergistic**
- ☆ Integration of botanicals and microbial - simultaneously
- ☆ Uses of chemical insecticides - temporary relief from insect pests disrupt - ecological balance by eliminating natural enemies – overcome by botanicals and BCA.
- ☆ Use of botanicals and BCA – avoids toxic residues, elimination of natural enemies, environmental disharmony and development of resistance.

The experiences and observations on the pest and disease incidences in various crops raised under vrikshauyrvedic farming are presented and discussed in this chapter to get an understanding and contemplation.

RAGI

The study revealed that foliar spraying of leaf extracst protects the crop of finger millet from pest and disease incidence. This might be due to the decomposition green leaf manures which would have released polyphenols due to this the fungal growth would have been suppressed. The presence of antifungal, antibacterial and biopesticidal properties in the leaf extracts which would have reduced the pest and disease in the crop. Most of the common angiospermic plants show a high degree of resistance to fungal infection. Many reports indicated the antimicrobial activity and medicinal property of these plants as they contain both inhibitory as well as phytostimulatory compounds. Most of the extracts proved their efficacy against the majority of fungi. This is due to antifungal compounds present in the extracts, which are responsible for suppression of the fungi. The phytostimulatory properties of the plant extracts may probably be due to growth promoting factors such as IAA, IBA, GA3, or their precursors.

The extract of *Vitex negundo* leaves was found to be effective in the reduction of the bacterial and fungal incidence on seeds. Thus, the plant extracts were found to be effective not only at reducing the incidence of the majority of seedborne fungal pathogens, but also improving the seed germination percentage as well as the seedling vigour and ultimately the yield. It is reported that leaf extracts of *Moringa oleifera* accelerate growth of tomato, peanut, corn and wheat at early vegetative growth, improve resistance to pest and diseases and enhanced more and larger fruits.

MAIZE

Pest Population

The incorporation of green leaf manures had significant effect on pest incidence at 75 DAS. *Albizia lebbek* recorded the lowest mean population of aphids 4.3 nos. plant^{-1}. It was followed by *Peltophorum ferrugineum* recorded the aphids population of 8.3 nos. plant^{-1}. The highest aphids population was recorded in *Delonix regia* (15.7 nos. plant^{-1}). The application of leaf extract through foliar sprays had significant effect on the incidence of aphids at 75 DAS. *Annona squamosa* recorded the lowest mean population of aphids 10.5 nos. plant^{-1}. It was followed by *Aegle marmellos* recorded the aphid's population of 10.6 nos. plant^{-1}. The highest aphid's population was recorded in *Alangium salvifolium* (11.2 nos. plant^{-1}). The interaction effect was significant on pest incidence and *Albizia lebbek* + *Annona squamosa* combination recorded the lowest mean pest population (3.1 nos. plant^{-1}). It was followed by *Albizia lebbek* + *Aegle marmellos* combination recorded the aphids population of 4.5 nos. plant^{-1}. The highest aphids population was recorded in *Delonix regia* + *Aegle marmellos* (16.1 nos. plant^{-1}).

The absolute control zero input recorded aphids population of 22.1 nos. plant^{-1}. When compared to control, *Albizia lebbek* + *Annona squamosa* showed decreased aphids population of 18.9 nos. plant^{-1}. The recommended dose of fertilizers recorded (33.0 nos. plant^{-1}) aphids population. When compared to this, *Albizia lebbek* + *Annona squamosa* showed decreased aphids population of 30.1 nos. plant^{-1} at 75 DAS.

Table 12.1: Pest Incidence (Numbers/plant) in Maize Grown Under Vrikshyurveda Farming

GLM Incorporation	*Pest Population per Plant*				
	Alangium	*Aegle*	*Morinda*	*Annona*	*Mean*
Albizia	4.83 (2.20)	4.50 (2.12)	5.00 (2.23)	3.16 (1.78)	**4.37**
Gliricidia	12.66 (3.56)	12.00 (3.46)	9.83 (3.13)	10.66 (3.26)	**11.29**
Leucaena	14.33 (3.79)	14.33 (3.79)	15.33 (3.92)	14.83 (3.85)	**14.70**
Delonix	15.83 (3.98)	16.17 (4.02)	15.16 (3.89)	15.83 (3.98)	**15.74**
Peltophorum	8.66 (2.94)	8.83 (2.97)	7.83 (2.80)	8.17 (2.86)	**8.37**
MEAN	**11.26**	**11.16**	**10.63**	**10.53**	
	M	***S***	***M at S***	***S at M***	
SE d	0.32	0.22	0.53	0.49	
CD (P≤0.05)	0.90	0.47	1.27	10.05	
Crop Production Guide recommendations (C2)	33.33				
Zero Input (C_1)	22.12				

Points for Discussion

Occurrence of pest incidence as well as weeds evidenced from this observation is also very low in the combination *Albizia lebbek* + *Annona squamosa,* which also contributed higher yield. Leaf and seed extract of *Annona squamosa* in water and also their powders were reported to show insecticidal, antifeedant and repellent activity against number of insects and pests (Vijayalakshmi *et al.,* 2002) that supports present finding.

ONION

Thrips Incidence

The incorporation of green leaf manures had significant effect on thrips incidence percentage of aggregatum onion. The highest thrips incidence of 21.69 per cent was recorded in *Albizia lebbek* and the lowest percentage of thrips incidence of 19.98 was recorded in *Gliricidia sepium.* The application of leaf extract through seed treatment and foliar sprays had a significant effect on thrips incidence percentage of aggregatum onion. Among the treatments, *Alangium salvifolium* recorded 21.54 per cent of thrips incidence and the lowest percentage of thrips incidence of 20.68 was recorded in *Annona squamosa.* The interaction effect was significant. Among the treatments the highest percent of thrips incidence 22.50 was recorded in *Albizia lebbek* + *Alangium salvifolium* and the lowest percentage of thrips incidence 18.19 was recorded in

Gliricidia sepium + *Aegle marmelos*, which was on par with *Senna siamea* + *Annona squamosa* recorded 18.66 per cent of incidence.

The absolute control recorded 35.55 per cent of thrips incidence and the recommended dose of inorganic fertilizers recorded 50.25 per cent of thrips incidence. When it was compared to these, *Gliricidia sepium* + *Aegle marmelos* combination showed decreased thrips incidence per cent of 17.23 and increased incidence per cent 32.

Onion Rotting Percent

The incorporation of green leaf manures had significant effect on rotting percentage of aggregatum onion. The highest percentage of rotting 25.09 was recorded in *Albizia lebbek* and the lowest percentage of rotting 20.01 was recorded in *Senna siamea*. The application of leaf extract through seed treatment and foliar sprays had a significant effect on rotting percentage of aggregatum onion. Among the treatments, *Morinda tinctoria* was recorded 24.46 of per cent of rotting and the lowest percent of rotting 23.70 was recorded in *Ageless marmellos*.

The interaction effect was significant the highest per cent of rotting 25.95 was recorded in *Albizia lebbek* + *Alangium salvifolium* and lowest percent of rotting 20.31 was recorded in *Gliricidia sepium* + *Aegle marmelos*. The absolute control recorded 49.35 per cent of rotting and the recommended dose of inorganic fertilizers recorded 45.25 percentage of rotting. When it was compared these *Gliricidia sepium* + *Aegle marmelos* showed decreased rotting of 23.4 per cent and 19.32 per cent of decreased respectively.

Bhendi

Shoot and Fruit Borer Incidence

The incorporation of green leaf manures had significant effect on shoot and fruit borer incidence percentage of bhendi. The highest shoot and fruit borer incidence of 37.86 per cent was recorded in *Leucaena leucocephala*) and the lowest 30.16 per cent was recorded in *Albizia lebbek*. The application of leaf extract through seed treatment and foliar sprays had a significant effect on percentage of shoot and fruit borer incidence in Bhendi. Among the treatments, *Annona squamosa* recorded the highest incidence (35.16 per cent) and the lowest (29.36 per cent) was recorded in *Ocimum sanctum*. The interaction effect was significant. Among the treatment combinaions, the highest per cent of shoot and fruit borer incidence (42.61 per cent) was recorded in *Gliricidia sepium* + *Annona squamosa* combination and the lowest (23.99 per cent) was recorded in *Albizia lebbek* +*Annona squamosa*.

The absolute control recorded 48.00 per cent of shoot and fruit borer incidence in bhendi. When compared to this, *Albizia lebbek* +*Annona squamosa* showed decrease in shoot and fruit borer incidence per cent of 24.01. The recommended dose of inorganic fertilizers recorded 55.25 per cent of shoot and fruit borer incidence. When compared to this, *Albizia lebbek* +*Annona squamosa* showed decrease in shoot and fruit borer incidence per cent of 31.26.

Table 12.2: Shoot and Fruit Borer Incidence (Per cent) in Bhendi Grown Under Vrikshyurveda Farming

GLM Incorporation	*Alangium*	*Annona*	*Aegle*	*Morinda*	*Ocimum*	*Mean*
Albizia	32.65	23.99	32.59	32.54	29.05	30.16
Delonix	37.59	33.87	32.45	33.35	27.08	32.87
Gliricidia	31.52	42.61	25.12	31.70	26.14	31.42
Leucaena	34.34	40.19	39.39	40.19	35.16	37.86
Mean	30.02	35.16	32.39	34.45	29.36	33.08
Zero input						48.00
Fertiliser application						55.25
	M	***S***	***M x S***			
SEd	1.789	1.748	3.602			
CD (P≤0.05)	4.377					
		3.561	7.702			

Yellow Vein Mosaic Virus

The incorporation of green leaf manures had significant effect on yellow vein mosaic virus percentage in bhendi. The highest percentage of yellow vein mosaic virus (24.67 per cent) was recorded in *Delonix regia*) and the lowest (22.75 per cent) was recorded in *Albizia lebbek*. The application of leaf extract through seed treatment and foliar sprays had a significant effect on yellow vein mosaic virus infection in Bhendi. Among the treatments, (*Morinda tinctoria*) recorded 24.54 per cent of yellow vein mosaic virus and the lowest per cent (22.94) was recorded in *Annona squamosa*. The interaction effect was significant, the highest per cent of yellow vein mosaic virus (25.73 per cent) was recorded in *Gliricidia sepium* + *Morinda tinctoria* and the lowest (19.21 percent) was recorded in *Albizia lebbek* +*Annona squamosa* combination.

Table 12.3: Yellow Mosaic Virus Incidence (per cent) in Bhendi Grown Under Vrikshyurveda Farming

GLM Incorporation	*Alangium*	*Annona*	*Aegle*	*Morinda*	*Ocimum*	*Mean*
Albizia	22.77	19.21	24.06	24.05	23.66	22.75
Delonix	24.49	25.14	24.88	25.39	23.44	24.67
Gliricidia	24.72	24.20	21.46	25.73	22.99	23.82
Leucaena	22.55	23.22	22.55	22.99	22.77	22.82
Mean	**23.63**	**22.94**	**23.24**	**24.54**	**23.22**	**23.52**
Control plots						
Fertiliser application						33.00
Zero input						28.00

Table 12.4: Pest Incidence (Numbers/Plant) in Onion Grown Under Vrikshyurveda Farming

GLM Incorporation	*Thrips*					*Rotting*				
	Alangium	*Annona*	*Aegles*	*Morinda*	*Mean*	*Alangium*	*Annona*	*Aegles*	*Morinda*	*Mean*
Albizia	22.50	22.10	20.86	21.28	21.69	25.95	23.66	24.99	25.62	25.09
Senna	20.86	18.66	21.28	22.90	20.93	24.89	23.48	24.14	23.53	20.01
Gliricidia	20.44	20.01	18.19	21.28	19.98	24.66	23.99	20.31	22.51	22.80
Leucaena	20.86	22.10	21.69	21.28	21.49	23.97	25.62	23.58	25.80	24.74
Delonix	20.01	20.57	21.43	20.97	20.74	24.21	23.56	25.47	25.11	24.24
Mean	**20.93**	**20.68**	**20.69**	**21.54**	**20.96**	**24.14**	**24.06**	**23.70**	**24.46**	**23.37**
	M	***S***	***M X S***			***M***	***S***	***M X S***		
SED	0.12	0.09	0.22			0.23	0.24	0.52		
CD(0.05)	0.29	0.19	0.47			0.54	0.49	1.09		
Control plots										
Control-1	35.55					49.35				
Control-2	50.25					45.25				

The absolute control recorded 33.00 per cent of virus infection and the recommended dose of inorganic fertilizers recorded 28.00 percentage of virus infection. When compared with these *Albizia lebbek* +*Annona squamosa* showed decrease in virus incidence of 13.79 and 8.79 per cent respectively.

Sorghum

With an objective of finding out the influence of bio pesticides from leaves of tree species on germination, yield and aphid incidence in sorghum a study was undertaken at Agricultural College and Research Institute, TNAU, Madurai. Leaf Extracts of tree species *viz., Azadirachta indica, Prosopis juliflora, Pongamia glabra, Pithecellobium dulce, Vitex negundo* were used for prophylactic pest control measure in sorghum. They were given the following treatment methods *viz.*, Seed treatment alone, two protective sprays at 30 DAS and 60 DAS and the combination of treatment 1 + treatment 2. The experiment was set up in a factorial RBD with two replications. Totally there were 15 treatments in all. The observations on plant height, pest incidences, yield attributes and yield were recorded. Two crops were grown in successive years. This experiment showed a mixed response and there was no significant difference in any of the parameter studied. However there was significance among the treatment for Aphid incidence, which is the key to select the best treatment in this study, it inferred that application of Pungam leaf extract @ 3 per cent either during 30 and 45 DAS (2.23 t ha^{-1}) or following seed treatment before sowing and spraying of 3 per cent leaf extract during 30 and 45 DAS (2.19 t ha^{-1}) is the best to control pest incidence and to get higher yields in sorghum. From the study it is recommended that Spraying of leaf extracts of *Pongamia pinnata* L. @ 5 per cent during 30 and 60 DAS registered an average yield of 2.2 t ha^{-1} in Sorghum and also had lower aphid incidence.

Ponits for Discussion

Application of green leaf manure dampens excessive growth forces, thus reducing the crops susceptibility to fungal. Apart from this, the protein fractions of the tree species were highly effective against the spore germination and mycelial growth which might have reduced the disease incidence. The present studies revealed that the percentage of yellow vein mosaic virus, shoot and fruit borer was lowest in the treatment with *Albizia lebbek* as green leaf manures and seed soaking along with foliar sprays of *Annona squamosa*. This might be due to the release of polyphenols during the decomposition of green leaf manures which acts as the suppressors of disease causing organisms The presence of antifungal, antibacterial and biopesticidal properties in the leaf extracts of custard apple in the form of phenolic and non-volatile substances may be attributed to this. Apart from the phenolic compounds, the volatile properties of several plant extracts and compounds were also reported to inhibit fungal growth and spore germination by several workers like spraying with the leaf extracts that has higher PO (peroxidase), PPO (polyphenol oxidase) and PAL (Phenylalanine ammonia lyase) activity considerably reduce the pest and disease menace. The enzymes PO and PPO which are more fungitoxic than phenolics may be responsible for disease resistance. Further application of plant extracts resulted in significant induction of resistance inducing enzymes in the host and thereby reducing the severity of diseases has also been reported elsewhere. The superiority of *Annana squamosa* over other sprays may be due to the fact that plants can absorb the active principles more rapidly or in other words there may be better compatibility between the host components when compared to other sprays. However these hypothesis needs confirmation through means of biochemical and cytological studies. The recommended dose of inorganic fertilizers recorded the highest pest and disease incidence. This might be due to the higher level of nitrogen application, which would have caused excessive vegetative growth, succulency coupled with absence of replellant action favouring the incidence of pest and diseases. The absolute control also recorded higher incidence of pest and diseases due to the reasons that the plants were weak and there was no mechanism to manage them.

Chapter 13
Yield and Yield Parameters

Yield and Yield Attributes of Barnyard Millet

Inflorescence Length

The incorporation of green leaf manure significantly influenced the inflorescence length. The maximum length (23.66 cm) was registered in *Peltophorum ferrugineum* and it was on par with *Delonix regia* (23.25 cm). The minimum inflorescence length was recorded in *Gliricidia sepium* (20.18 cm), which was on par with *Pongamia pinnata* (21.48 cm) and *Albizia lebbek* (20.64 cm). The Inflorescence length was significantly influenced by foliar application of leaf extracts and the maximum length of 23.40 cm was recorded in *Annona squamosa* and all other treatments *viz., Aegle marmellos* (21.71 cm), *Morina oleifera* (21.27 cm) and *Morinda tinctoria* (21.01 cm) were on par with each other. The interaction effect was not significant on length of the barnyard millet inflorescence. However, the treatment combination of *Peltophorum ferrugineum* + *Annona squamosa* recorded the maximum inflorescence length(25.6 4cm). The absolute control zero input recorded the inflorescence length of 20.45 cm and the recommended dose of inorganic fertilizers recorded the inflorescence length of 26.7 cm at harvest of the crop.

Inflorescence Weight

The incorporation of green leaf manures significantly influenced the inflorescence weight of the barnyard millet. The maximum weight was registered in *Peltophorum ferrugineum* with 22.0 g. It was followed by *Delonix regia* (19.98 g). The minimum weight (16.37 g) was recorded by *Gliricidia sepium*, it was on par with *Albizia lebbek* with 17.54 g. Inflorescence weight was significantly influenced by the foliar application of different leaf extracts. The maximum weight (20.43 g) was recorded in *Annona squamosa* and minimum weight (17.30 g) was registered in *Morinda tinctoria*. The other treatments were on par with each other. The interaction effect on inflorescence weight was significant. Among the 20 treatment combinations the maximum weight of 23.93g was registered by the treatment combination *Peltophorum ferrugineum* + *Annona squamosa*. It was followed by *Delonix regia* + *Annona squamosa* with 22.58 g and the minimum weight of 15.47 g was recorded by *Gliricidia sepium*+ *Moringa tinctoria*. The absolute control zero input recorded the inflorescence weight of 16.80 g and the recommended dose of inorganic fertilizers recorded the inflorescence weight of 25.27 g at harvest of the crop.

Table 13.1: Influence of *Vrikshayurveda* Treatments on Length of Inflorescence in Barnyard Millet

GLM Incorporation	*Length of Inflorescence (cm)*				
	Moringa	*Aegle*	*Morinda*	*Annona*	*Mean*
Albizia	19.07	20.94	20.28	22.29	**20.64**
Glyricidia	19.60	20.33	20.74	20.76	**20.18**
Pongamia	20.25	20.06	21.78	23.85	**21.48**
Delonix	23.60	23.74	21.23	24.41	**23.25**
Peltophorum	23.83	23.49	21.71	25.64	**23.66**
MEAN	**21.27**	**21.71**	**21.01**	**23.40**	
	M	***S***	***M at S***	***S at M***	
SEd	0.53	0.48	1.07	1.08	
CD (0.05)	1.48	1.03	NS	NS	
ZERO INPUT	20.45				
CPG	26.70				

Table 13.2: Influence of *Vrikshayurveda* Treatments on Inflorescence Weight of Barnyard Millet

GLM Incorporation	*Inflorescence Weight (g)*				
	Moringa	*Aegle*	*Morinda*	*Annona*	*Mean*
Albizia	17.54	17.19	16.72	18.71	**17.54**
Glyricidia	16.50	16.59	15.47	16.92	**16.37**
Pongamia	18.60	17.27	16.17	20.03	**18.01**
Delonix	20.14	20.00	16.91	22.58	**19.98**
Peltophorum	21.70	21.20	21.19	23.93	**22.00**
MEAN	**18.90**	**18.45**	**17.30**	**20.43**	
	M	***S***	***M at S***	***S at M***	
SEd	0.483	0.175	0.590	0.391	
CD (0.05)	1.341	0.373	1.513	0.835	
ZERO INPUT	16.80				
CPG	25.27				

Fodder Yield

The incorporation of green leaf manures did not have significant effect on fodder yield of barnyard millet crop. However, the highest fodder yield was obtained by *Peltophorum ferrugineum* with a mean fodder yield of 3834 kg ha^{-1}. A significant variation was found in fodder yield of barnyard millet crop due to foliar spraying of different leaf extracts *Annona squamosa* recorded higher fodder yield of 3926 kg ha^{-1} and the lower fodder yield of 3568 kg ha^{-1}was recorded in *Morinda tinctoria*. The other treatments, *Moringa oleifera* and *Aegle marmellos* were on par with each other. There was no interaction effect on fodder yield of barnyard millet. However, the combination *Peltophorum ferrugineum* + *Annona squamosa* recorded the highest fodder yield of 3986 kg ha^{-1}. The absolute control zero input recorded the fodder yield of 2791 kg ha^{-1}. When compared to Control *Peltophorum ferrugineum* + *Annona squamosa* showed a increased 29.9 per cent fodder yield and recommended dose of inorganic fertilizers recorded the fodder yield of 4202 kg ha^{-1} in barnyard millet at harvest of the crop. When compared to fertilizer application, *Peltophorum ferrugineum* + *Annona squamosa* showed a decreased 5.1 per cent fodder yield.

Table 13.3: Influence of *Vrikshayurveda* Treatments on Fodder Yield of Barnyard Millet

GLM Incorporation	*Fodder Yield (kg ha^{-1})*				
	Moringa	*Aegle*	*Morinda*	*Annona*	*Mean*
Albizia	3814	3688	3604	3901	**3751**
Glyricidia	3759	3751	3421	3896	**3706**
Pongamia	3764	3729	3580	3912	**3746**
Delonix	3867	3799	3480	3935	**3770**
Peltophorum	3835	3757	3758	3986	**3834**
MEAN	**3808**	**3745**	**3568**	**3926**	
SEd	41.2	30.8	79.3	75.6	
CD (0.05)	NS	66.1	NS	NS	
ZERO INPUT	2791				
CPG	4202				

Test Grain Weight

Neither incorporation of green leaf manures nor foliar spraying of leaf extracts did have significant effect on test grain weight of the barnyard millet. The incorporation of green leaf manure did not have significant effect on test grain weight of barnyard millet. However, numerically higher value 3.10 g were recorded by *Peltophorum ferrugineum* and *Delonix regia*. The test grain weight of barnyard millet was not influenced by foliar application of leaf extract. However, numerically higher values were recorded *Annona squamosa* and *Moringa oleifera* with the weight of 3.11 g. The interaction effect was not significant on test grain weight and numerically higher value was recorded by combination of *Peltophorum ferrugineum* + *Annona squamosa*

(3.15 g) at the harvest stage. The absolute control zero input recorded the test grain weight of 2.88 g and the recommended dose of inorganic fertilizers recorded the test grain weight of 3.24 g at harvest of the crop.

Table 13.4: Influence of *Vrikshayurveda* Treatments on Test Grain Weight of Barnyard Millet

GLM Incorporation	*Test Grain Weight (g)*				
	Moringa	*Aegle*	*Morinda*	*Annona*	*Mean*
Albizia	3.10	3.05	3.04	3.09	**3.07**
Glyricidia	3.09	3.09	3.02	3.1	**3.08**
Pongamia	3.11	3.10	3.01	3.13	**3.09**
Delonix	3.13	3.1	3.12	3.08	**3.10**
Peltophorum	3.09	3.11	3.06	3.15	3.10
MEAN	**3.11**	**3.09**	**3.05**	**3.11**	
	M	***S***	***M at S***	***S at M***	
SEd	0.11	0.05	0.06	0.01	
CD (0.05)	NS	NS	NS	NS	
ZERO INPUT	2.88				
CPG	3.24				

Table 13.5: Influence of *Vrikshayurveda* Treatments on Grain Yield of Barnyard Millet

GLM Incorporation	*Grain Yield (kg ha^{-1})*				
	Moringa	*Aegle*	*Morinda*	*Annona*	*Mean*
Albizia	1238	1175	1077	1326	**1204**
Glyricidia	1212	1062	950	1407	**1158**
Pongamia	1291	1089	1007	1475	**1216**
Delonix	1617	1326	1169	1760	**1468**
Peltophorum	1693	1615	1495	1960	**1691**
MEAN	**1410**	**1253**	**1139**	**1585**	
	M	***S***	***M at S***	***S at M***	
SEd	44.5	30.7	95.34	65.3	
CD (0.05)	95.0	65.4	212.5	131.2	
ZERO INPUT	861				
CPG	2151				

Grain Yield

The incorporation of green leaf manures had significant effect on grain yield of barnyard millet. The highest grain yield was recorded by *Peltophorum ferrugineum* with a mean yield of 1691 kg ha^{-1} and it was followed by *Delonix regia* with 1468 kg ha^{-1}. The lowest grain yield 1158 kg ha^{-1} was recorded by *Gliricidia sepium*. The grain yield of barnyard millet was significantly influenced by foliar spraying of different leaf extracts. *Annona squamosa* recorded higher grain yield of 1585 kg ha^{-1}, it was followed by *Moringa oleifera* and the lower grain yield was recorded in *Morinda tinctoria* (1139 kg ha^{-1}). The interaction effect was also found to be significant. The combination *Peltophorum ferrugineum* + *Annona squamosa* recorded the highest grain yield of 1960 kg ha^{-1}. It was followed by *Delonix regia* + *Annona squamosa* with a grain yield of 1760 kg ha^{-1}. The lowest grain yield of 950 kg ha^{-1} was recorded in *Gliricidia sepium* + *Morinda tinctoria* at harvest of the crop. The control zero input recorded the grain yield of 861 kg ha^{-1} only and the recommended dose of inorganic fertilizers recorded the grain yield of 2151 kg ha^{-1} in barnyard millet. When compared to Control *Peltophorum ferrugineum* + *Annona squamosa* showed 56.1 per cent increased in grain yield and compared to fertilizer application, it showed a decreased grain yield of 8.9 per cent.

Points for Discussion

In the investigation on inflorescence length and weight, the performance was higher in *Pheltophorum ferrugineum* and *Annona squamosa* combination and recorded 56.1 percent higher yield than the absolute control. This combination produced longest and heaviest inflorescence and also heavier grains. The better performance of this treatment *Pheltophorum ferrugineum* may also be attributed to the higher soil microbial population was evidenced from this study. The decomposition of green leaf manures strongly affects the structure and diversity of the microbial community. Increase in soil microbial population would help in root hair formation and thereby increase uptake capacity of roots by secreting growth hormones. The higher concentration of micronutrients in the foliar spray of *Annona squamosa* is also another factor responsible for higher yield in *Pheltophorum ferrugineum* + *Annona squamosa* combination. The foliar application of tree leaf extracts increased yield and nutrition of barnyard millet, due to the presence of secondary metabolites and plant growth regulators. A higher positive response of soil microbial population influenced by the application *Pheltophorum ferrugineum* as green leaf manures, higher concentration of micronutrients in *Annona squamosa* as foliar spray supported by low incidence of pest have been attributed for the better performance of *Pheltophorum ferrugineum* + *Annona squamosa* combination in terms of yield parameters and yield.

As usual, the recommended dose of inorganic fertilizers recorded the maximum inflorescence length, inflorescence weight and also the test grain weight and grain yield per hectare. This might be due to increased nutrient availability, better uptake and utilization of nutrients. The higher leaf area might have increased the photosynthates which could have lead to better source – sink relationship to accumulate more stored food. The absolute control, zero input recorded lower yield. This might be due to low level of nutrients in the soil. Hence the utilization would have been also on the lower side. The fodder yield showed positive response to the incorporation of green leaf manure and foliar application of leaf extracts. The maximum fodder yield recorded in the treatment combination *Pheltophorum ferrugineum* + *Annona squamosa*. This might be due to faster decomposition of green leaf manure and higher availability of nitrogen to the plants.

Finger Millet (Ragi)

Number of Fingers per Ear Head

Number of fingers was significantly influenced by green leaf manure incorporation and leaf extract spray at all growth stages of finger millet. Among the treatments, *Delonix regia* recorded maximum number of fingers per earhead (10.31). This was followed by *Albizia lebbek,* which was on par with *Leucaena leucocephala. Peltophorum ferrugineum* recorded less number of fingers per ear head of 8.13 among various green leaf manure incorporation. Among the treatments, *Aegle marmellos* produced more number of fingers per earhead of 9.68, which was on par with *Moringa oleifera* and *Annona squamosa*. Number of fingers per ear head was lowest in *Morinda tinctoria*. The interaction effect was significant and the treatment combination *Delonix regia + Moringa oleifera* which recorded more number of fingers per earhead (11.82) and it was on par with *Delonix regia + Aegle marmellos. Peltophorum ferrugineum + Annona squamosa* combination recorded less number of fingers per earhead (7.60). The absolute control (zero input) recorded 7.50 number of fingers per earhead. When compared to zero input, *Delonix regia + Moringa oleifera* showed a increased (4.32) number of fingers per ear head. While the recommended dose of inorganic fertilizers recorded 10.96 number of fingers per ear head. When compared to fertilizer application, it showed increased number of fingers per ear head of 0.86.

Length of Ear Head

Both green leaf manure incorporation and foliar spraying of leaf extract significantly influenced the length of earhead at harvest stage. *Delonix regia* produced longest ear head of 11.2 cm at harvest. It was on par with *Albizia lebbek* and *Gliricidia sepium*. The shortest ear head (10.18c m) was registered in *Peltophorum ferrugineum. Moringa oleifera* showed maximum length of ear head of 11.30 cm, which was on par with *Morinda tinctoria*. The treatment, *Annona squamosa* registered minimum length of earhead (10.51 cm) at harvest. The interaction effect was significant. *Delonix regia + Moringa oleifera* registered maximum length of earhead (11.82 cm), which was on par with *Delonix regia + Annona squamosa. Peltophorum ferrugineum + Annona squamosa* recorded minimum length of earhead (9.14 cm) and it was on par with *Peltophorum ferrugineum + Aegle marmellos*. The absolute control (zero input) recorded 9.73 cm of earhead length. When compared to zero input, *Delonix regia + Moringa oleifera* showed a increased 2.19 cm of earhead length. While the fertilizer application recorded 11.57 cm of ear head length. When it was compared to fertilizer application, *Delonix regia + Moringa oleifera* showed an increased 0.35 cm length of earhead.

Grain Yield

The green leaf manure incorporation and leaf extract spray significantly influenced the grain yield of finger millet. Among the green leaf manure incorporation, *Delonix regia* recorded higher grain yield of 542 kg ha^{-1}, which was followed by *Albizia lebbek*. The treatment *Peltophorum ferrugineum* recorded lower grain yield of 226 kg ha^{-1}. With respect to the leaf extract spray, *Moringa oleifera* recorded higher grain yield of 427 kg ha^{-1} and it was followed by *Aegle marmellos. Morinda tinctoria* recorded lower grain yield of 262 kg ha^{-1}. The interaction effect between green leaf manure incorporation and foliar spray of leaf extract showed that *Delonix regia* + *Moringa oleifera* recorded higher grain yield of 808 kg ha^{-1}. This was followed by *Delonix regia* + *Aegle marmellos*, which was on par with *Delonix regia* + *Annona squamosa*. Grain yield was lowest in *Peltophorum ferrugineum* + *Morinda tinctoria* and *Gliricidia sepium*. The recommended dose of fertilizers (CPG) recorded the grain yield of 1337.50 kg ha^{-1} in finger millet.

Straw Yield

Straw yield was significantly by green leaf manure incorporation and leaf extract spray. Straw yield was higher (4566 kg ha^{-1}) under and it was on par with *Leucaena leucocephala*. *Peltophorum ferrugineum* recorded lower straw yield of 2626 kg ha^{-1} among various green leaf manure application. Among the leaf extract spray, higher straw yield of 3892 kg ha^{-1} was recorded in *Aegle marmellos* and all other treatments were on par with each other. Lower straw yield of 3225 kg ha^{-1} was registered in *Annona squamosa*. The interaction effect between green leaf manure incorporation and foliar spray of leaf extract showed that recorded higher straw yield of 4808 kg ha^{-1}, which was comparable with the combination of *Leucaena leucocephala* + *Aegle marmellos* and all other treatments of *Delonix regia* combination. Straw yield was lowest (2125 kg ha^{-1}) in *Peltophorum ferrugineum* + *Morinda tinctoria*.

Test Grain Weight

Neither incorporation of green leaf manures nor foliar spraying of leaf extracts did have significant effect on test grain weight of the finger millet. However, numerically higher value of 3.05 g was recorded by *Delonix regia* and *Leucaena leucocephala* and *Moringa oleifera* and *Aegle marmellos*. The interaction effect was not significant on test grain weight. While numerically higher value was recorded by combination of *Delonix regia* + *Moringa oleifera* with value of 3.09 g. The control zero input recorded the test grain weight of 2.98 g and the recommended dose of inorganic fertilizers recorded the test grain weight of 3.07 g.

Points for Discussion

The yield is the output component which the crop is expected to give in returns to the inputs given to it. The following observations *i.e.*, number of ear heads per plant, ear head length, number of fingers per ear head, grain and straw yield were recorded in finger millet. It is evident from our study that the yield parameters were highest in the treatment with *Delonix regia* as green leaf manures and foliar spays of *Moringa oleifera*. This may be ascribed to the decomposition of organic leaf manures which provided nutrients and this could have improved the soil physical conditions which subsequently increased the availability of nutrient, owing to increase in microbial activity leading to more translocation of photosynthates and better source-sink relationship and photo assimilates stored in economic part. The presence of secondary metabolites in the leaf extracts would have enhanced the yield. The encouragement of vegetative growth due to application of green manure resulted in production of more tiller m^2 and test weight.

Continous N supply due to slow release from the incorporated green manure should have matched with the uptake pattern of crop resulting in better yield. The influence of *Moringa* leaf extract spray on higher yield in crops like groundnut, onion and other crops has been reported elsewhere. The recommended dose of inorganic fertilizers recorded the highest yield. This might be due to increased nutrient availability and better uptake and utilization of nutrient. The improvement in number of leaves might have increased the photosynthates which could have lead the source – sink relationship to accumulate more of the stored food. The zero input recorded lower yield. This might be due to low level of nutrients in the soil. Hence the utilization to would have been also on the lower side.

Maize

Cob Length

Neither incorporation of green leaf manures, nor application of leaf extract through foliar sprays did not have significant effect on cob length of maize. The incorporation of green leaf manures did not have significant effect on cob length of maize at harvest of the crop. The maximum cob length (13.25 cm) had been recorded in *Albizia lebbek*. It was followed by *Peltophorum ferrugineum* (12.50 cm) and minimum cob length (10.16 cm) was recorded in *Leucaena leucocephala* at harvest of the crop. The foliar spraying of leaf extract did not have significant effect on cob length of maize. *Aegle marmellos* recorded higher cob length of 12.70 cm and it was followed by *Annona squamosa* (12.07 cm) and minimum cob length was recorded in *Alangium salvifolium* (11.08 cm). The interaction effect was not significant. The combination *Albizia lebbek + Aegle marmellos* recorded the highest cob length (20.62 cm). It was followed by *Peltophorum ferrugineum + Annona squamosa* (14.13 cm) and the lowest cob length (9.40 cm) was recorded in *Leucaena leucocephala + Aegle marmellos* at harvest of the crop. The absolute control zero input produced cobs with a length of 9.40 cm and recommended dose of fertilizers recorded a cob length of 14.70 cm in maize at harvest of the crop. When compared to Control, *Albizia lebbek + Aegle marmellos* showed a increased cob

Table 13.6: Influence of *Vrikshayurveda* Treatments on Cob Length (cm) of Maize

GLM Incorporation	*Cob Length (cm)*				
	Alangium	*Aegle*	*Morinda*	*Annona*	*Mean*
Albizia	9.63	20.62	12.09	10.67	**13.25**
Gliricidia	10.46	13.12	12.56	12	**12.03**
Leucaena	10.71	9.4	9.41	11.14	**10.16**
Delonix	12.42	10.12	12.11	12.42	**11.76**
Peltophorum	12.22	10.28	13.44	14.13	**12.5**
MEAN	**11.08**	**12.70**	**11.92**	**12.07**	
	M	***S***	***M at S***	***S at M***	
SEd	1.64	1.28	2.97	2.86	
CD (0.05)	NS	NS	NS	NS	
CPG	14.7				
ZERO INPUT	9.4				

length of 11.20 cm and compared to recommended dose of fertilizers, it showed a decreased cob length of 5.90 cm.

Cob Girth

The incorporation of green leaf manures had significant effect on cob girth in maize at harvest of the crop. The maximum cob girth (13.48 cm) was recorded in *Albizia lebbek* which was on par with *Peltophorum ferrugineum*. The minimum cob girth (12.52 cm) was recorded in *Gliricidia sepium* at harvest of the crop. The foliar spraying of leaf extracts did not have significant effect on cob girth of maize. *Morinda tinctoria* recorded cob girth of 13.20 cm and *Annona squamosa* recorded the cob girth of 13.18 cm. The interaction effect was significant. The combination *Albizia lebbek* + *Annona squamosa* recorded the highest cob girth (14.45 cm) and it was on par with *Peltophorum ferrugineum* + *Aegle marmellos* and lowest cob girth (11.71 cm) was recorded in *Gliricidia sepium* + *Alangium salvifolium* at harvest of the crop. The absolute control zero input recorded cob girth of 10.60 cm and the recommended dose of fertilizers recorded cob girth of 14.70 cm in maize at harvest of the crop. When compared to Control, *Albizia lebbek* + *Annona squamosa* showed a increased cob girth (3.80 cm) and compared to fertilizer application, it showed a decreased cob girth (0.25 cm).

Table 13.7: Influence of *Vrikshayurveda* Treatments on Cob Girth (cm) of Maize

GLM Incorporation	*Cob Girth (cm)*				
	Alangium	*Aegle*	*Morinda*	*Annona*	*Mean*
Albizia	13.79	11.94	13.74	14.45	**13.4**
Gliricidia	11.71	13.81	12.33	12.23	**12.5**
Leucaena	13.05	13.17	13.05	12.62	**12.9**
Delonix	13.39	12.3	13.03	13.15	**12.9**
Peltophorum	12.07	14.25	13.87	13.48	**13.5**
MEAN	**12.8**	**13.0**	**13.2**	**13.1**	
	M	***S***	***M at S***	***S at M***	
SEd	0.30	0.24	0.56	0.54	
CD (0.05)	0.61	NS	1.12	1.08	
CPG	14.7				
ZERO INPUT	10.6				

Number of Grain Rows Cob^{-1}

Neither incorporation of green leaf manures nor foliar spraying of leaf extracts had a significant effect on number of grain rows per cob at the harvest of the crop. Numerically higher values were recorded with *Albizia lebbek* (14.0) and *Annona squamosa* (13.9). The interaction effect was significant. The combination *Albizia lebbek* + *Annona squamosa* recorded highest number of grain rows per cob (15.5). It was followed by *Albizia lebbek* + *Morinda tinctoria* which recorded 14.4 number of grain rows per cob and the lowest number of grain rows per cob (12.2) was recorded in *Albizia lebbek* + *Aegle marmellos* at harvest of the crop. The absolute control zero input recorded 10 number of grain rows per cob and the recommended dose of inorganic fertilizers recorded 15.0 number of grain rows per cob of at harvest of the crop. When compared to Control, *Albizia lebbek* + *Annona squamosa* showed a increased the number of grain rows per cob (4.4) and compared to fertilizer application, it showed a decreased number of grain rows per cob (0.6).

Table 13.8: Influence of *Vrikshayurveda* Treatments on Number of Grain Rows per Cob of Maize

GLM Incorporation	*Number of Grain Row per Cob*				
	Alangium	*Aegle*	*Morinda*	*Annona*	*Mean*
Albizia	14.1	12.2	14.4	15.5	**14.0**
Gliricidia	13.5	14.0	13.2	13.7	**13.6**
Leucaena	13.2	13.2	13.6	14.1	**13.5**
Delonix	13.3	14.1	13.0	14.2	**13.6**
Peltophorum	14.4	14.1	13.7	13.4	**13.9**
MEAN	**13.7**	**13.5**	**13.8**	**13.9**	
	M	***S***	***M at S***	***S at M***	
SEd	0.39	0.35	0.79	0.79	
CD (0.05)	NS	NS	1.57	1.56	
CPG	15				
ZERO INPUT	10				

Number of Grains per Row

The incorporation of green leaf manures had significant effect on number of grains per row in maize at harvest of the crop. The maximum number of grains per row (24.3) had been recorded in *Albizia lebbek*. It was followed by *Peltophorum ferrugineum* which recorded 23.0 grains per row. The minimum number of grains per row (18.2) was recorded in *Leucaena leucocephala* at harvest of the crop. The foliar spraying of leaf extracts had a significant effect on number of grains per row of maize. The maximum number of grains per row (23.9) was recorded in *Annona squamosa* and It was followed by *Morinda tinctoria*, which recorded 22.9 grains per row and minimum number of grains per row (20.5) was recorded in *Aegle marmellos* at harvest. The interaction effect was significant. The combination *Albizia lebbek* + *Annona squamosa* recorded more number of grains per row (29.5). It was followed by *Albizia lebbek* + *Morinda tinctoria* and *Peltophorum ferrugineum* + *Alangium salvifolium* with 25.8 grains per row and the less number of grains per rows (16.1) was recorded in *Leucaena leucocephala* + *Aegle marmellos* at harvest of the crop. The absolute control zero input recorded 10 grains per row and recommended dose of inorganic fertilizers recorded 31 grains per row in maize at harvest. When compared to Control, *Albizia lebbek* +

Annona squamosa had increased number of grains per row (19.5) while comparing to fertilizer application it had decreased number of grains row^{-1} (1.5).

Table 13.9: Influence of *Vrikshayurveda* Treatments on Number of Grains per Row of Maize

GLM Incorporation	*Number of Grains per Row*				
	Alangium	*Aegle*	*Morinda*	*Annona*	*Mean*
Albizia	23.7	18.5	25.8	29.5	**24.37**
Gliricidia	19.1	25.4	25.5	20.5	**22.62**
Leucaena	19.2	16.1	17.1	20.5	**18.22**
Delonix	20.2	22.7	25.1	23.0	**22.92**
Peltophorum	25.8	19.8	21.0	25.7	**23.07**
MEAN	**21.6**	**20.5**	**22.9**	**23.98**	
	M	***S***	***M at S***	***S at M***	
SEd	1.27	1.09	2.47	2.44	
CD (0.05)	2.59	2.16	4.93	4.84	
CPG	31				
ZERO INPUT	10				

Number of Grains per Cob

The incorporation of green leaf manures had significant effect on number of grains per cob in maize at harvest of the crop. The maximum number of 303.6 grains in a cob had been recorded in *Albizia lebbek*. It was followed by *Peltophorum ferrugineum* with 286.8 grains. The minimum of 222.2 grains cob^{-1} was recorded in *Leucaena leucocephala* at harvest of the crop. The foliar spraying of leaf extract had a significant effect on number of grains per cob of maize. The maximum number of 312.8 grains cob^{-1} was recorded in *Annona squamosa*. It was followed by *Alangium salvifolium* with 268.7 grains cob^{-1} and the minimum number of grains cob^{-1} (255.7) was recorded in *Aegle marmellos* at harvest of the maize. The interaction effect was significant and *Albizia lebbek* + *Annona squamosa* recorded the maximum number of grain per cob (417.6). It was followed by *Peltophorum ferrugineum* + *Annona squamosa* recorded 332.4 grains per cob and the lowest grains cob^{-1} (190.8) was recorded in *Leucaena leucocephala* + *Morinda tinctoria* at harvest of the crop. The absolute control zero input recorded 65.9 grains cob^{-1} and recommended dose of inorganic fertilizers recorded 437.4 grains cob^{-1} in maize at harvest of the crop. When compared to Control, *Albizia lebbek* + *Annona squamosa* showed a increased grains cob^{-1} (351.6) and when compared to fertilizer application, it showed a decreased grains cob^{-1} (19.8).

Cob Weight

The incorporation of green leaf manures had significant effect on cob weight of maize at harvest of the crop. *Albizia lebbek* had heaviest cobs (164.7 g) and it was followed by *Peltophorum ferrugineum* with cob weight of 152.4 g. The minimum cob

Table 13.10: Influence of *Vrikshayurveda* Treatments on Number of Grains per Cob of Maize

GLM Incorporation	*Number of Grains per Cob*				
	Alangium	*Aegle*	*Morinda*	*Annona*	*Mean*
Albizia	317.20	204.60	275.10	417.60	**303.62**
Gliricidia	246.00	327.40	319.20	247.90	**285.12**
Leucaena	231.50	211.70	190.80	255.00	**222.25**
Delonix	237.80	291.10	271.40	311.10	**277.85**
Peltophorum	311.00	243.80	260.00	332.40	**286.80**
MEAN	**268.70**	**255.72**	**263.30**	**312.80**	
	M	***S***	***M at S***	***S at M***	
SEd	22.36	16.72	40.30	38.72	
CD (0.05)	45.36	34.24	80.34	76.58	
CPG	437.46				
ZERO INPUT	65.96				

Table 13.11: Influence of *Vrikshayurveda* Treatments on Cob Weight (g/cob) of Maize

GLM Incorporation	*Cob Weight Average 5 Cobs/Treatment*				
	Alangium	*Aegle*	*Morinda*	*Annona*	*Mean*
Albizia	177.6	116.8	140	224.4	**164.7**
Gliricidia	171.2	125.8	132.2	159.8	**147.2**
Leucaena	143.2	115.1	114.4	133.4	**126.5**
Delonix	117.1	184.2	154.8	149.6	**151.4**
Peltophorum	124.6	166.4	161.1	157.6	**152.4**
MEAN	**146.7**	**141.5**	**140.5**	**164.9**	
	M	***S***	***M at S***	***S at M***	
SEd	12.06	8.54	20.48	19.11	
CD (0.05)	24.46	16.90	40.86	37.80	
CPG	251.07				
ZERO INPUT	48				

weight was recorded *Leucaena leucocephala* (126.5 g). The foliar spraying of leaf extract had a significant effect on cob weight of maize. *Annona squamosa* produced heaviest cobs (164.9 g) and it was followed by *Alangium salvifolium* which recorded the cob weight of 146.7 g. The minimum cob weight was recorded in *Morinda tinctoria* (140.5 g). The interaction effect was significant. The combination *Albizia lebbek* + *Annona squamosa* recorded the highest cob weight (224.4 g). It was followed by *Delonix regia* + *Aegle marmellos* with cob weight 184.2 g and the lowest cob weight (114.4 g) was recorded in *Leucaena leucocephala* + *Morinda tinctoria* at harvest of the crop. The absolute control zero input produced cob of 48.0 g weight and recommended dose of inorganic fertilizers recorded the cob weight of 250.0 g in maize at harvest of the crop. When compared to Control, *Albizia lebbek* + *Annona squamosa* showed a increased cob weight (176.4 g) and while compared to fertilizer application, it showed a decreased cob weight (26.6 g).

Test Weight

Neither incorporation of green leaf manures nor foliar spraying of leaf extracts did not have significant effect on test weight at the harvest of the crop. Numerically higher values were recorded by *Albizia lebbek* (29.3 g) and *Annona squamosa* (29.1 g). The interaction effect was also not significant on test grain weight. Numerically higher values were recorded by combination of *Albizia lebbek* + *Annona squamosa* recorded the test grain weight of 30.7 g at the harvest stage. The control zero input recorded the test grain weight of 26.1 g and the recommended dose of inorganic fertilizers recorded the test grain weight of 33.6 g in maize at harvest of the crop.

Table 13.12: Influence of *Vrikshayurveda* Treatments on Test Grain Weight (g 100^{-1} grains) of Maize

GLM Incorporation	*Test Grain Wt (g)*				
	Alangium	*Aegle*	*Morinda*	*Annona*	*Mean*
Albizia	28.50	29.50	28.75	30.75	**29.37**
Gliricidia	27.20	27.25	27.80	28.75	**27.75**
Leucaena	27.40	27.10	28.60	28.90	**28.00**
Delonix	28.45	26.20	29.55	29.80	**28.50**
Peltophorum	29.25	29.25	28.60	27.75	**28.71**
MEAN	**28.16**	**27.86**	**28.66**	**29.19**	
	M	***S***	***M at S***	***S at M***	
SEd	1.13	0.73	1.82	1.64	
CD (0.05)	NS	NS	NS	NS	
CPG (C_2)	33				
ZERO INPUT (C_1)	26				

Cob Yield

The incorporation of green leaf manures had significant effect on cob yield of maize at harvest of the crop. Cob yield was higher (4622 kg ha^{-1}) in *Albizia lebbek*. It was followed by *Peltophorum ferrugineum* (4252 kg ha^{-1}) and the lowest (3594 kg ha^{-1}) was recorded in *Leucaena leucocephala*. The foliar spraying of leaf extract had significant effect on cob yield of maize. *Annona squamosa* recorded higher cob yield of 4392 kg ha^{-1}. It was followed by *Morinda tinctoria* recorded the cob yield of 4131 kg ha^{-1} and minimum yield was recorded *Aegle marmellos* (3975 kg ha^{-1}). The interaction effect was also found to be significant. The combination *Albizia lebbek* + *Annona squamosa* recorded the highest cob yield (4895 kg ha^{-1}) and it was followed by *Delonix regia* +

Annona squamosa which recorded the cob yield of 4761 kg ha^{-1}. The lowest cob yield (3284 kg ha^{-1}) was recorded in *Leucaena leucocephala* + *Aegle marmellos* at harvest of the crop. The control zero input recorded the cob yield of 720 kg ha^{-1} and recommended dose of inorganic fertilizers recorded the cob yield of 8640 kg ha^{-1} in maize at harvest of the crop. When compared to Control, *Albizia lebbek* + *Annona squamosa* showed a increased cob yield (4115 kg ha^{-1}) and when compared to fertilizer application, it showed a decreased cob weight (3744 kg ha^{-1}).

Table 13.13: Influence of *Vrikshayurveda* Treatments on Cob Yield (kg ha^{-1}) of Maize

GLM Incorporation	*Cob Yield (Kg ha^{-1})*				
	Alangium	*Aegle*	*Morinda*	*Annona*	*Mean*
Albizia	4198.6	4693.4	4702.9	4895.8	4622.7
Gliricidia	3392.9	4434.5	4379.1	3692.0	3974.6
Leucaena	3432.2	3284.1	3431.8	4230.0	3594.5
Delonix	4637.4	3605.3	3809.4	4761.3	4203.3
Peltophorum	4427.8	3860.9	4335.4	4384.7	4252.2
MEAN	**4017.8**	**3975.6**	**4131.7**	**4392.8**	
	M	***S***	***M at S***	***S at M***	
SEd	255.2	178.7	471.7	357.6	
CD (0.05)	522.8	373.5	896.4	740.2	
CPG (C_2)	8640.00				
ZERO INPUT (C_1)	720.00				

Grain Yield

The incorporation of green leaf manures had significant effect on grain yield of maize at harvest of the crop. Grain yield was higher (1751 kg ha^{-1}) in *Albizia lebbek* and it was followed by *Peltophorum ferrugineum* (1607 kg ha^{-1}) and the lowest (1357 kg ha^{-1}) was recorded with *Leucaena leucocephala*. The foliar spraying of leaf extract had significant effect on grain yield of maize. *Annona squamosa* recorded higher grain yield of 1633 kg ha^{-1}. It was followed by *Morinda tinctoria* which recorded the grain yield of 1563 kg ha^{-1} and minimum grain yield was recorded *Aegle marmellos* (1516 kg ha^{-1}). The interaction effect was significant. The combination *Albizia lebbek* + *Annona squamosa* recorded the highest grain yield (1820 kg ha^{-1}). It was on par with *Albizia lebbek* + *Morinda tinctoria* which recorded the grain yield of 1795 kg ha^{-1}. The lowest grain yield (1252 kg ha^{-1}) was recorded in *Leucaena leucocephala* + *Aegle marmellos* at harvest of the crop. The absolute control zero input recorded the grain yield of 305 kg ha^{-1} and the recommended dose of inorganic fertilizers recorded the grain yield of 3900 kg ha^{-1} in maize at harvest of the crop. When compared to Control, *Albizia lebbek* + *Annona squamosa* showed a increased grain yield (1515 kg ha^{-1}) and when compared to fertilizer application it showed a decreased grain yield (2148 kg ha^{-1}).

Table 13.14: Influence of *Vrikshayurveda* Treatments on Grain Yield (kg ha^{-1}) of Maize

GLM Incorporation	Grain Yield (Kg ha^{-1})				
	Alangium	Aegle	Morinda	Annona	Mean
Albizia	1602.50	1790.00	1795.00	1820.00	**1751.87**
Gliricidia	1295.00	1691.25	1652.50	1372.50	**1502.81**
Leucaena	1310.00	1252.50	1295.00	1572.50	**1357.50**
Delonix	1770.00	1375.00	1437.50	1770.00	**1588.12**
Peltophorum	1690.00	1472.50	1636.00	1630.00	**1607.12**
MEAN	**1533.50**	**1516.25**	**1563.20**	**1633.00**	
	M	***S***	***M at S***	***S at M***	
SEd	153.7	83.1	232.8	175.8	
CD (0.05)	322.77	165.09	454.30	369.16	
CPG	3900.00				
ZERO INPUT	305.50				

Table 13.15: Influence of *Vrikshayurveda* Treatments on Stover Yield (kg ha^{-1}) of Maize

GLM Incorporation	Stover Yield (Kg ha^{-1})				
	Alangium	Aegle	Morinda	Annona	Mean
Albizia	5875	5500	7250	6750	**6343.75**
Gliricidia	4675	5875	5875	5375	**5450.00**
Leucaena	4125	5125	5375	4875	**4875.00**
Delonix	6250	4125	5250	6375	**5500.00**
Peltophorum	5875	4875	5750	6250	**5687.50**
MEAN	**5360**	**5100**	**5100**	**5925**	
	M	***S***	***M at S***	***S at M***	
SEd	535.0	460.5	1082	1076	
CD (0.05)	1120	990	2230	2220	
CPG	9000				
ZERO INPUT	630				

Stover Yield

The incorporation of green leaf manures had significant effect on stover yield of maize at harvest of the crop. Stover yield was higher (6343 kg ha^{-1}) in *Albizia lebbek*. It was on par with *Peltophorum ferrugineum* (5687 kg ha^{-1}) and the lowest (4875 kg ha^{-1}) was recorded in *Leucaena leucocephala*. The foliar spraying of leaf extract had significant effect on stover yield of maize. *Annona squamosa* recorded higher stover yield of 5925 kg ha^{-1}. It was on par with *Alangium salvifolium* recorded the stover yield of 5360 kg ha^{-1}. The minimum stover yield was recorded *Aegle marmellos* and *Morinda tinctoria* (5100 kg ha^{-1}). The interaction effect was found to be significant. *Albizia lebbek* + *Morinda tinctoria* recorded the highest stover yield (7250 kg ha^{-1}). It was closely and on par with *Albizia lebbek* + *Annona squamosa* which recorded the stover yield of 6750 kg ha^{-1}. The lowest stover yield (4125 kg ha^{-1}) was recorded in *Leucaena leucocephala* + *Alangium salvifolium* and *Delonix regia* + *Aegle marmellos* at harvest of the crop. The absolute control zero input recorded the stover yield of 630 kg $ha^{-1}$and the recommended dose of inorganic fertilizers recorded the stover yield of 9000 kg ha^{-1} in maize at harvest of the crop. When compared to Control, *Albizia lebbek* + *Morinda tinctoria* showed a increased stover yield (6120 kg ha^{-1}) and when compared to fertilizer application it showed a decreased stover yield (2250 kg ha^{-1}).

Points for Discussion

Yield is a function of cob length, cob girth, cob weight, number of grains per cob, number of grains per row, test grain weight and other related yield parameters. The positive changes in majority of the yield parameters will favourably influence the yield.

In the investigation on maize revealed that of cobs and grains, the performance was higher in *Albizia lebbek* + *Annona squamosa* combination. The combination had higher grain number and grain row, cob weight and also the test grain weight. The better performance of the combination *Albizia lebbek* + *Annona squamosa* may be attributed to the soil microbial population as evidenced from our studies.It is to imprint that decomposition of green leaf manures strongly affect the structure and diversity of the microbial community. Increasing soil microbes enhances root-hair formation and thereby increases root uptake capacity by secreting growth hormones and there by yield can be increased. The higher concentration of micronutrients in the foliar spray of *Annona squamosa* is also responsible for higher yield in *Albizia lebbek* + *Annona squamosa* combination. The foliar application of tree leaf extracts increased yield and nutrition of green gram, due to the presence of microelements and plant growth regulators.

Further the occurrence of pest incidence as well as weeds evidenced from this observation is also very low in the combination *Albizia lebbek* + *Annona squamosa*, which also contributed higher yield. Leaf and seed extract of *Annona squamosa* in water and also their powders were reported to show insecticidal, antifeedant and repellent activity against number of insects and pests. Higher positive influence of soil microbial population influenced by the application *Albizia lebbek* of green leaf manures, higher concentration of micronutrients in *Annona squamosa* as foliar spray in addition to low incidence of pest and weed infestation have been attributed for the better performance of *Albizia lebbek* + *Annona squamosa* combination in terms of yield parameters and yield. The recommended dose of inorganic fertilizers recorded the highest cob length, cob girth, cob weight, 100 grain weight, cob yield and grain yield per hectare. This might be due to increased nutrient availability and better uptake and utilization of nutrient. The more number of leaves might have increased the photosynthates which could have lead to better source – sink relationship to accumulate more of the stored food. The zero input recorded lower yield. This might be due to low level of nutrients in the soil. Hence the utilization to would have been also on the lower side.

Blackgram

Number of Pods per Plant

The incorporation of green leaf manures had significant effect on number of pods per plant in black gram at harvest of the crop. The maximum number of pods per plant (30) had been recorded in *Pongamia pinnata*. It was followed by *Albizia lebbek* recorded 21.5 pods per plant. The minimum number of pods per plant (18.0) was recorded in *Delonix regia* at harvest of the crop. The foliar spraying of leaf extracts had a significant effect on number of pods per plant of black gram. The maximum number of pods per plant (25.6) had been recorded in *Moringa oleifera*. It was followed by *Annona squamosa*) recorded 21.1 pods per plant and minimum number of pods per plant (19.3) was recorded in *Aegle marmellos* at harvest of the black gram. The interaction effect was significant. The combination *Pongamia pinnata* + *Moringa oleifera* recorded more number of pods per plant (29.5). It was followed by *Pongamia pinnata* + *Annona squamosa* with 33.3 pods per plant and the less number of pods per plant (14) was recorded in *Peltophorum ferrugineum* + *Morinda tinctoria* at harvest of the crop. The absolute control zero input had 15 pods per plant and the recommended dose of inorganic fertilizers produced 36.4 pods per plant in black gram at harvest of the crop. When compared to Control *Pongamia pinnata* + *Moringa oleifera* had increased number of pods per plant (20.2) and when compared to fertilizer application, it had lessor number of pods per plant (0.6).

Table 13.16: Influence of *Vrikshayurveda* Treatments on Number Pods Plant^{-1} of Black Gram

GLM Incorporation	*Number Pods Plant^{-1}*				
	Foliar Applications				*Mean*
	Moringa	*Aegle*	*Morinda*	*Annona*	
Albizia	19.9	17.9	26.4	21.7	**21.5**
Gliricidia	20.9	25.5	22.8	13.5	**20.7**
Pongamia	35.8	25.1	25.9	33.3	**30.0**
Delonix	29	13.5	15.5	14.1	**18.0**
Peltophorum	22.6	14.7	14	23	**18.6**
MEAN	**25.6**	**19.3**	**20.9**	**21.1**	
	M	***S***	***M at S***	***S at M***	
SEd	0.15	0.50	0.99	0.91	
CD (0.05)	0.47	1.06	1.94	2.12	
CPG	36.4				
ZERO INPUT	15.6				

Number of Clusters per Plant

The incorporation of green leaf manures had significant effect on number of clusters per plant in black gram at harvest of the crop. The maximum number of 11.9 clusters in a plant had been recorded in *Pongamia pinnata*. It was followed by *Albizia lebbek* with 7.3 clusters. The minimum of 6.2 clusters per plant was recorded in *Delonix regia* at harvest of the crop. The foliar spraying of leaf extracts had a significant effect on number of clusters per plant of black gram. The maximum number of 8.6 clusters per plant had been recorded in *Moringa oleifera*. It was followed by *Morinda tinctoria* with 7.8 clusters per plant and the minimum number of clusters per plant (7.5) was recorded in *Aegle marmellos* at harvest of the black gram. The interaction effect was significant and *Pongamia pinnata* + *Moringa oleifera* recorded the maximum number of clusters per plant (14.4). It was followed by *Pongamia pinnata* + *Annona squamosa which* recorded 12.4 clusters per plant and the lowest number of clusters per plant (5.0) was recorded in *Delonix regia* + *Aegle marmellos* at harvest of the crop. The absolute control zero input recorded 6.2 clusters per plant and the recommended dose of inorganic fertilizers recorded 14.2 clusters per plant in black gram at harvest of the crop. When compared to Control, *Pongamia pinnata* + *Moringa oleifera* showed a increased clusters per pod (8.2) and when compared to fertilizer application it showed a increased clusters per plant (0.2).

Table 13.17: Influence of *Vrikshayurveda* Treatments on Number of Clusters Plant^{-1} of Black Gram

GLM Incorporation	*Number of Clusters Plant^{-1}*				
	Foliar Applications				*Mean*
Basal Applications	*Moringa*	*Aegle*	*Morinda*	*Annona*	
Albizia	7.2	9.1	7.2	5.8	**7.3**
Gliricidia	6.3	6.1	5.4	9.0	**6.7**
Pongamia	14.4	10.8	10.1	12.4	**11.9**
Delonix	8.2	5.0	7.1	4.5	**6.2**
Peltophorum	7.0	6.3	9.4	6.0	**7.2**
MEAN	**8.6**	**7.5**	**7.8**	**7.5**	
	M	***S***	***M at S***	***S at M***	
SEd	0.36	0.32	0.72	0.72	
CD (0.05)	1.00	0.69	1.66	1.55	
CPG	14.2				
ZERO INPUT	6.2				

Pod Length

Incorporation of green leaf manures, application of leaf extract through foliar sprays had significant effect on pod length of black gram. The incorporation of green leaf manures had significant effect on pod length of black gram at harvest of the crop. The longest pod (4.43 cm) had been recorded in *Pongamia pinnata* and it was followed by *Albizia lebbek* (4.33 cm) and shortest pod length (4.17 cm) was recorded in *Delonix regia* at harvest of the crop. The foliar spraying of leaf extract had significant effect on pod length of black gram. *Annona squamosa* recorded higher pod length of 4.29 cm. It was followed by *Moringa oleifera* (4.27 cm) and minimum pod length was recorded in *Aegle marmellos* (4.15 cm). The interaction effect was significant. However the combination *Pongamia pinnata* + *Moringa oleifera* recorded maximum pod length (4.55 cm) and it was followed by *Pongamia pinnata* + *Morinda tinctoria* (4.50 cm) and minimum pod length (3.98 cm) was recorded in *Peltophorum ferrugineum* + *Aegle marmellos* at harvest of the crop. The absolute control zero input produced pod with a length of 3.85 cm and the recommended dose of inorganic fertilizers recorded pod length of 4.55 cm at harvest of the crop. When compared to Control, *Pongamia pinnata* + *Moringa oleifera* showed a increased pod length (0.7 cm) and when compared to fertilser application, it showed an equal pod length (4.55 cm).

Table 13.18: Influence of *Vrikshayurveda* Treatments on Pod Length (cm) of Black Gram

GLM Incorporation	*Pod Length (cm)*				
	Foliar Applications				*Mean*
	Moringa	*Aegle*	*Morinda*	*Annona*	
Albizia	4.10	4.40	4.50	4.33	**4.33**
Gliricidia	4.05	4.15	4.45	4.15	**4.20**
Pongamia	4.55	4.20	4.50	4.45	**4.43**
Delonix	4.35	4.00	4.40	3.93	**4.17**
Peltophorum	4.30	3.98	4.20	4.60	**4.27**
MEAN	**4.27**	**4.15**	**4.41**	**4.29**	
	M	***S***	***M at S***	***S at M***	
SEd	0.02	0.05	0.09	0.10	
CD (0.05)	0.06	0.10	0.20	0.22	
CPG	4.55				
ZERO INPUT	3.85				

Number of Seeds per Pod

Numerically higher values were recorded by *Pongamia pinnata* (4.9) and it was followed by *Albizia lebbek* (4.7). The lowest number of seeds per pod was recorded in *Peltophorum ferrugineum* (4.1). The foliar spraying of leaf extracts did not have significant effect on number of seeds per pod of black gram. *Moringa oleifera* (4.8) recorded maximum number of seeds per pod followed by *Morinda tinctoria* (4.5). The minimum number of seeds per pod was recorded at *Aegle marmellos* (4.4). The interaction effect was significant. The combination *Pongamia pinnata* + *Morinda tinctoria* recorded higher seeds per pod (5.2) and it was followed by *Pongamia pinnata* + *Moringa oleifera* and *Pongamia pinnata*+ *Aegle marmellos* which recorded 4.9 seeds per pod and the less number of seeds per pod (3.8) was recorded in *Peltophorum ferrugineum* +

Aegle marmellos at harvest of the crop. The absolute control zero input recorded 3.2 of seeds per pod and the recommended dose of inorganic fertilizers recorded 5.0 seeds per pod at harvest of the crop.

Table 13.19: Influence of *Vrikshayurveda* Treatments on Number of Seeds Pod^{-1} of Black Gram

GLM Incorporation	*Number of Seeds Pod^{-1}*				
	Foliar Applications				*Mean*
	Moringa	*Aegle*	*Morinda*	*Annona*	
Albizia	4.9	4.6	4.3	4.8	**4.7**
Gliricidia	4.9	4.4	4.5	4.2	**4.5**
Pongamia	4.9	4.9	5.2	4.5	**4.9**
Delonix	4.8	4.4	4.5	4.0	**4.4**
Peltophorum	4.4	3.8	3.8	4.4	**4.1**
MEAN	**4.8**	**4.4**	**4.5**	**4.4**	
	M	***S***	***M at S***	***S at M***	
SEd	0.13	0.11	0.24	0.23	
CD (0.05)	0.35	0.23	0.56	0.51	
CPG	5.0				
ZERO INPUT	3.2				

Test Grain Weight

Numerically higher values were recorded by *Peltophorum ferrugineum* (4.57 g) and *Moringa oleifera* 4.61 g. The interaction effect was not significant on test grain weight and numerically higher values were recorded by combination of *Gliricidia sepium* + *Moringa oleifera* recorded the test grain weight of 4.94 g at the harvest stage. The control zero input recorded the test grain weight of 4.28 g and the recommended dose of inorganic fertilizers recorded the test grain weight of 4.88 g at harvest of the crop.

Grain Yield

The incorporation of green leaf manures had significant effect on grain yield of black gram at harvest of the crop. Grain yield was higher (502.2 kg ha^{-1}) in *Pongamia pinnata*. It was followed by *Albizia lebbek* (434.7 kg ha^{-1}) and the lowest (291.1 kg ha^{-1}) was recorded in *Gliricidia sepium*. The foliar spraying of leaf extract had significant effect on grain yield of black gram. *Moringa oleifera* recorded higher grain yield of 412.6 kg ha^{-1}. It was followed by *Annona squamosa* which recorded the grain yield of 385.7 kg ha^{-1} and minimum yield was recorded *Aegle marmellos* (354.9 kg ha^{-1}). The interaction effect was also found to be significant. The combination *Pongamia pinnata* + *Moringa oleifera* recorded the highest grain yield (585.4 kg ha^{-1}). It was followed by *Pongamia pinnata* + *Annona squamosa* recorded the grain yield of 514.5 kg ha^{-1}. The lowest grain yield (286.4 kg ha^{-1}) was recorded in *Delonix regia* + *Aegle marmellos* at

Table 13.20: Influence of *Vrikshayurveda* Treatments on Test Grain Weight (g) of Black Gram

GLM Incorporation	*Test Grain Weight (g)*				
	Foliar Applications				*Mean*
	Moringa	*Aegle*	*Morinda*	*Annona*	
Albizia	4.72	4.65	4.05	4.50	**4.48**
Gliricidia	4.94	4.14	4.79	4.35	**4.55**
Pongamia	4.30	4.67	4.77	4.46	**4.55**
Delonix	4.53	3.59	4.58	4.43	**4.28**
Peltophorum	4.57	4.61	4.75	4.36	**4.57**
MEAN	**4.61**	**4.33**	**4.59**	**4.42**	
	M	***S***	***M at S***	***S at M***	
SEd	0.07	0.06	0.15	0.16	
CD (0.05)	NS	0.15	0.34	0.33	
CPG	4.88				
ZERO INPUT	4.28				

Table 13.21: Influence of *Vrikshayurveda* Treatments on Grain Yield (kg ha^{-1}) of Black Gram

GLM Incorporation	*Grain Yield (kg ha^{-1})*				
	Foliar Applications				*Mean*
	Moringa	*Aegle*	*Morinda*	*Annona*	
Albizia	408.75	407.08	431.25	492.08	**434.79**
Gliricidia	288.47	292.08	287.12	297.08	**291.19**
Pongamia	**585.42**	445.42	463.75	514.58	**502.29**
Delonix	396.25	286.42	299.58	282.92	**316.29**
Peltophorum	384.58	343.75	415.42	342.08	**371.46**
MEAN	**412.69**	**354.95**	**379.42**	**385.75**	
	M	***S***	***M at S***	***S at M***	
SEd	0.32	0.54	1.11	1.22	
CD (0.05)	0.89	1.16	2.42	2.61	
CPG	597.92				
ZERO INPUT	197.08				

harvest of the crop. The control zero input recorded the grain yield of 197.0 kg ha^{-1} only and the recommended dose of inorganic fertilizers recorded the grain yield of 597.9 kg ha^{-1} in black gram at harvest of the crop. When compared to Control, *Pongamia pinnata* + *Moringa oleifera* showed a increased grain yield (388.3 kg ha^{-1}) and when compared to fertilizer application it showed a decreased grain yield (12.5 kg ha^{-1}).

Stover Yield

The incorporation of green leaf manures had significant effect on stover yield of black gram at harvest of the crop. Stover yield was higher (961.56 kg ha^{-1}) in *Pongamia pinnata*. It was followed by *Albizia lebbek* (904.85 kg ha^{-1}) and the lowest (525.59 kg ha^{-1}) was recorded in *Gliricidia sepium*. The foliar spraying of leaf extracts had significant effect on stover yield of black gram. *Moringa oleifera* recorded higher stover yield of 770.6 kg ha^{-1} and it was followed by *Morinda tinctoria* recorded the stover yield of 756.2 kg ha^{-1}. The minimum stover yield was recorded at *Aegle marmellos* (685.9 kg ha^{-1}). The interaction effect was found to be significant. The combination, *Pongamia pinnata* + *Moringa oleifera* recorded the highest stover yield (1053.7 kg ha^{-1}) and it was followed by *Albizia lebbek* + *Annona squamosa which* recorded the stover yield of 1033.3 kg ha^{-1}. The lowest stover yield (513.4 kg ha^{-1}) was recorded in *Gliricidia sepium* + *Moringa oleifera* at harvest of the crop. The absolute control zero input recorded the stover yield of 315.3 kg ha^{-1} and the recommended dose of inorganic fertilizers recorded the stover yield of 1149.9 kg ha^{-1} in black gram at harvest of the crop. When compared to Control *Pongamia pinnata* + *Moringa oleifera* showed a increased stover yield (738.4 kg ha^{-1}) and when compared to fertilizer application, it showed decreased stover yield (96.1 kg ha^{-1}).

Table 13.22: Influence of *Vrikshayurveda* Treatments on Stover Yield (kg ha^{-1}) of Black Gram

GLM Incorporation	*Stover Yield (kg ha^{-1})*				
	Foliar Applications				*Mean*
	Moringa	*Aegle*	*Morinda*	*Annona*	
Albizia	846.11	838.59	901.31	1033.38	**904.85**
Gliricidia	513.48	543.27	516.82	528.81	**525.59**
Pongamia	1053.75	890.83	959.96	941.69	**961.56**
Delonix	713.25	538.46	539.25	480.96	**567.98**
Peltophorum	726.86	618.75	864.07	639.70	**712.34**
MEAN	**770.69**	**685.98**	**756.28**	**724.90**	
	M	***S***	***M at S***	***S at M***	
SEd	0.32	0.55	1.10	1.22	
CD (0.05)	0.90	1.17	2.42	2.61	
CPG	1149.92				
ZERO INPUT	315.33				

Points for Discussion

In general the yield of black gram in this experiment is low, which is affected to heavy rain during the flowering and harvesting stage of crop. In the present investigation on pods and grains, the performance was higher in *Pongamia pinnata* +

Moringa oleifera combination. The combination had higher grain number and number of pods in a plant, pod length, number of seeds in a pod and also the test grain weight. The better performance of the combination *Pongamia pinnata* + *Moringa oleifera* may be attributed to the higher soil microbial population as evidenced from this study. It is very well known that decomposition of green leaf manures strongly affect the structure and diversity of the microbial community and increase the soil microbes population and there by yield can be increased. The higher concentration of micronutrients in the foliar spray of *Moringa oleifera* is also responsible for higher yield in *Pongamia pinnata* + *Moringa oleifera* combination. The foliar application of tree leaf extracts increased yield and nutrition of green gram, due to the presence of microelements and plant growth regulators as observed in organically managed farms which records low productivity and yield losses during the early stage of farming but ensures an overall improvement in soil quality parameters during the course of time, indicating better soil health. It is economically feasible to practice organic farming when the farmers are able to get premium price for their produce and with the reduced cost of cultivation by not depending upon the purchased off-farm inputs. Low productivity in organic farming highlights needs in the current international and national research activities. Higher positive influence of soil microbial population influenced by the application *Pongamia pinnata* of green leaf manures, higher concentration of micronutrients in *Moringa oleifera* as foliar spray supported by low incidence of pest have been attributed for the better performance of *Pongamia pinnata* + *Moringa oleifera* combination in terms of yield parameters and yield of black gram in this study. The recommended dose of inorganic fertilizers recorded higher number of pods per plant, pod length, number of seeds per pod, test grain weight and grain yield per hectare. This might be due to increased nutrient availability and better uptake and utilization of nutrient. The more number of leaves might have increased the photosynthates which could have lead to better source – sink relationship to accumulate more of the stored food The zero input recorded lower yield. This might be due to low level of nutrients in the soil. Hence the utilization would have been also on the lower side.

Onion

Bulb Girth

The incorporation of green leaf manures had significant effect on bulb girth in aggregatum onion at harvest of the crop. The maximum bulb girth (13.81 cm) was recorded in *Gliricidia sepium* and minimum bulb girth (12.55 cm) was recorded in *Senna siamea* at harvest of the crop. The application of leaf extract through seed treatment and foliar sprays also had a significant effect on bulb girth of aggregatum onion at harvest of the crop. Among the foliar spraying treatments *Aegle marmelos* recorded in highest bulb girth (14.00 cm) and the lowest bulb girth (12.32 cm) was recorded in S_1 (*Alangium salvifloium*) at harvest of the crop. The interaction effect was significant and *Gliricidia sepium* + *Aegle marmelos* combination recorded the highest bulb girth (15.50 cm), which was on par with *Leucaena leucocephala* + *Alangium salvifolium* (14.87 cm) and the lowest bulb girth (11.25 cm) was recorded in *Delonix regia* + *Annona squamosa* at harvest of the crop. The absolute control recorded bulb girth of 7.25 cm and the recommended dose of inorganic fertilizers recorded bulb

Table 13.23: Influence of *Vrikshayurveda* Treatments on Number of Fingers Earhead^{-1} and Earhead Length of Finger Millet

GLM Incorporation	*Number of Fingers and Length of Ear head*									
	Number of Fingers					*Length of Ear Head*				
	Leaf Extract Spray					*Leaf Extract Spray*				
	Moringa	*Morinda*	*Annona*	*Aegle*	*Mean*	*Moringa*	*Morinda*	*Annona*	*Aegle*	*Mean*
Albizia	11.50	11.69	10.78	10.66	**11.16**	10.31	9.42	9.80	9.60	**9.78**
Gliricidia	11.40	11.61	10.16	11.25	**11.10**	8.25	7.97	8.43	9.16	**8.46**
Leucaena	10.75	11.31	10.67	10.76	**10.87**	9.09	9.25	10.32	10.33	**9.75**
Delanix	11.92	10.11	11.82	11.02	**11.22**	11.82	8.06	9.98	11.37	**10.31**
Peltophorum	10.91	10.74	9.14	9.94	**10.18**	8.27	8.72	7.60	7.94	**8.13**
Mean	**11.30**	**11.09**	**10.51**	**10.73**		**9.55**	**8.69**	**9.23**	**9.68**	
	SEd		***CD (0.05)***			***SEd***		***CD (0.05)***		
M	0.09		0.24			0.10		0.27		
S	0.17		0.37			0.33		0.71		
M at S	0.35		0.76			0.65		1.40		
S at M	0.39		0.83			0.75		1.59		
CPG		**10.96**					**11.57**			
Zero input		**7.50**					**9.73**			

girth of 16.23 cm in aggregatum onion at harvest of the crop. When compared to Control, *Gliricidia sepium + Aegle marmelos* showed a increased bulb girth of 8.25 cm and compared to recommended dose of fertilizers, it showed a decreased bulb girth of 0.73 cm.

Table 13.24: Influence of *Vrikshayurveda* Treatments on Test Grain Weight of Finger Millet

GLM Incorporation	*1000 Grain Weight (g)*				
	Leaf Extract Spray				
	Moringa	*Morinda*	*Annona*	*Aegle*	*Mean*
Albizia	3.02	3.02	3.01	3.01	**3.02**
Gliricidia	3.06	3.02	3.00	3.06	**3.04**
Leucaena	3.06	3.02	3.05	3.07	**3.05**
Delanix	3.09	3.00	3.03	3.08	**3.05**
Peltophorum	3.03	3.00	3.01	3.01	**3.01**
Mean	**3.05**	**3.01**	**3.02**	**3.05**	
	SEd		***CD (0.05)***		
M	0.020		NS		
S	0.017		NS		
M at S	0.039		NS		
S at M	0.037		NS		
CPG		**3.07**			
ZERO INPUT		**2.98**			

Bulb Weight

The incorporation of green leaf manures had significant effect on bulb weight in aggregatum onion at harvest of the crop. The heaviest bulb (50.28 g) was recorded in *Gliricidia sepium*, which was on par with *Senna siamea* with 49.57 g of bulb weight and the lightest bulb (46.99 g) was recorded in *Delonix regia* at harvest of the crop. The application of leaf extract through seed treatment and foliar sprays also had a significant effect on bulb weight of aggregatum onion at harvest of the crop. Among the foliar spraying treatments, *Aegle marmelos* recorded highest bulb weight of 48.94 g, which was on par with *Alangium salvifolium* recording 48.80 g of bulb weight and the lowest bulb weight (46.85 g) was recorded in *Annona squamosa* at harvest of the crop. The interaction effect was significant, among treatments *Gliricidia sepium + Aegle marmelos* combination recorded highest bulb weight (57.23 g) and lowest bulb weight (43.63 g) was recorded in *Albizia lebbek + Annona squamosa* at harvest of the crop. The absolute control recorded bulb weight of 33.25 g. When compared to Control, *Gliricidia sepium + Aegle marmelos* showed a increased bulb weight (23.98 g). The recommended dose of fertilizers recorded (61.25 g) bulb weight and when it was compared to *Gliricidia sepium + Aegle marmelos* the latter showed a decreased bulb weight (4.02 g).

Table 13.25: Influence of *Vrikshayurveda* Treatments on Grain and Fodder Yield of Finger Millet

GLM Incorporation	*Yield Characters*									
	Grain Yield (kg ha^{-1})					*Straw Yield (kg ha^{-1})*				
	Leaf Extract Spray					*Leaf Extract Spray*				
	Moringa	*Morinda*	*Annona*	*Aegle*	*Mean*	*Moringa*	*Morinda*	*Annona*	*Aegle*	*Mean*
Albizia	408	296	338	404	**361**	3208	3208	3417	3500	**3333**
Gliricidia	279	217	192	217	**226**	2458	3750	2583	3250	**3010**
Leucaena	354	296	283	400	**333**	3958	3583	3250	4708	**3875**
Delanix	808	313	488	558	**542**	4808	4750	4375	4333	**4566**
Peltophorum	288	192	196	229	**226**	2208	2125	2500	3667	**2626**
Mean	**427**	**262**	**299**	**362**		**3328**	**3483**	**3225**	**3892**	
	SEd		***CD (0.05)***			***SEd***		***CD (0.05)***		
M	38.23		106.16			263.62		731.40		
S	21.72		46.30			161.22		344.05		
M at S	56.85		137.76			408.78		980.81		
S at M	48.58		103.55			360.94		769.34		
CPG		**1337**					**4950**			
ZERO INPUT		**83**					**2125**			

Yield per Plot kg

The incorporation of green leaf manures had significant effect on yield per plot in aggregatum onion at harvest. The highest plot yield 13.02 kg was recorded in *Gliricidia sepium* and the lowest plot yield 11.50 kg was recorded in *Leucaena leucocephala* at harvest. The application of leaf extract through seed treatment and foliar sprays had a significant effect on yield per plot of aggregatum onion at harvest of the crop. Among the treatments *Aegle marmelos* recorded highest plot yield 12.73 kg and the lowest plot yield 11.04 kg was recorded in *Alangium salvifolium* at harvest. The interaction effect was significant, among treatments *Gliricidia sepium* + *Aegle marmelos* recorded the highest plot yield of 14.55 kg. It was on par with treatment *Senna siamea* + *Annona squamosa* which recorded 14.12 kg plots yield. The lowest plot yield 9.50 kg was recorded in *Senna siamea* + *Morinda tinctoria* at harvest.

The absolute control recorded 5.61 kg plot yield. When it was compared to absolute control *Gliricidia sepium* + *Aegle marmelos* showed an increased plot yield (8.91 kg). The recommended dose of inorganic fertilizers recorded 17.52 kg plot yield. When it was compared to RDF *Gliricidia sepium* + *Aegle marmelos* showed an decreased plot yield 2.97 kg in aggregatum onion at harvest.

Yield Hectare^{-1}

The incorporation of green leaf manures had significant effect on bulb yield per hectare in aggregatum onion at harvest of the crop. The highest bulb yield of 14.47 t ha^{-1} was recorded in *Gliricidia sepium* and the lowest bulb yield of 12.88 t ha^{-1} was recorded in *Leucaena leucocephala* at harvest. The application of leaf extract through seed treatment and foliar sprays had a significant effect on bulb yield per hectare of aggregatum onion at harvest. Among the treatments, *Aegle marmelos* recorded highest bulb yield per hectare of 14.23 t ha^{-1} and the lowest bulb yield per hectare of 11.26 t ha^{-1} was recorded in *Morinda tinctoria* in aggregatum onion. The interaction effect was significant. Among treatments *Gliricidia sepium* + *Aegle marmelos* recorded the highest bulb yield of 16.16 t ha^{-1}. This was on par with treatment *Senna siamea* + *Annona squamosa* which recorded 15.68 t ha^{-1} bulb yield and the lowest bulb yield of 11.27 t ha^{-1} was recorded in *Leucaena leucocephala* + *Annona squamosa* in aggregatum onion.

The absolute control recorded bulb yield of 4.75 t ha^{-1}. When it was compared to control *Gliricidia sepium* + *Aegle marmelos* showed an increased yield (11.41 t ha^{-1}). The recommended dose of inorganic fertilizers recorded bulb yield 18.65 t ha^{-1} and when it was compared to RDF *Gliricidia sepium* + *Aegle marmelos* showed an decreased yield (2.49 t ha^{-1}) in aggregatum onion at harvest.

Points for Discussion

The yield is the component which the crop is expected to give in returns to the inputs given to it. The yield is recorded the following observation *i.e.* bulb girth bulb weight, yield per plot and yield per hectare. It is understood from the study that yield parameters was highest in the treatment with *Gliricidia sepium* as green leaf manures and bulb soaking and foliar spays of *Aegles marmellos*. This might be due to the decomposition of organic leaf manures which provided nutrients and this could have improved the soil physical conditions with subsequently increased the availability of nutrient owing to increase in microbial activity leading to more translocation of photosynthates and better source-sink relationship and photo assimilates stored in bulbs. The presence of secondary metabolites in the leaf extracts would have enhanced the yield of onion. The recommended dose of inorganic fertilizers recorded the highest bulb girth, bulb weight, plot yield and total yield per hectare. This might be due to increased nutrient availability and better uptake and utilization of nutrient. The improvement in number of leaves might have increased the photosynthates which could have lead the source – sink relationship to accumulate more of the stored food in the bulb as observed in this study. The zero input recorded lower yield. This might be due to low level of nutrients in the soil. Hence the utilization to would have been also on the lower side.

Bhendi

Number of Fruits

The incorporation of green leaf manures had significant effect on number of fruits per plant in Bhendi. More number of fruits per plant (28.63) was recorded in *Albizia lebbek* and the lowest number of fruits per plant (27.03) was recorded in *Leucaena leucocephala*. The application of leaf extract through seed treatment and foliar sprays also had a significant effect on number of fruits per plant in Bhendi. Among the foliar sprayings (*Annona squamosa* recorded highest number of fruits per plant (28.07), which was on par with *Leucaena leucocephala* (28.05) and the lowest number of fruits per plant (26.69) was recorded in (*Ocimum sanctum*). The interaction effect was significant. Among the treatment combinations, *Albizia lebbek+ Annona squamosa* recorded the highest number of fruits per plant (30.73) and the least number of fruits per plant (24.95) was recorded in *Gliricidia sepium + Ocimum sanctum*. The number of fruits per

Table 13.26: Influence of *Vrikshayurveda* Treatments on Number of Fruits per Plant (Numbers) in Bhendi

GLM Incorporation	*Alangium*	*Annona*	*Aegles*	*Morinda*	*Ocimum*	*Mean*
Albizia	28.97	30.73	27.63	28.13	27.66	28.63
Delonix	27.68	28.47	27.88	27.85	27.49	27.87
Gliricidia	27.53	26.75	29.50	28.33	24.95	27.41
Leucaena	28.02	26.32	26.50	27.65	26.67	27.03
Mean	28.05	28.07	27.88	27.99	26.69	27.74
Control plots						
C_1						18.00
C_2						32.00
	M	***S***	***M x S***			
SEd	0.337	0.423	0.828			
CD (0.05)	0.824	0.861	1.742			

plant for absolute control was 18.0, when it was compared with *Albizia lebbek*+ *Annona squamosa*, it showed an increase in number of fruits per plant (12.73). The recommended dose of inorganic fertilizers recorded 32.0 numbers of fruits in Bhendi. When it was compared with *Albizia lebbek*+ *Annona squamosa*, it M_1S_2 showed a decrease in number of fruits per plant (1.27).

Fruit Yield per Plant

The incorporation of green leaf manures had significant effect on fruit yield per plant in Bhendi. The highest fruit yield per plant (371.6 g) was recorded in *Albizia lebbek* and lowest fruit yield per plant (325.2 g) was recorded in *Leucaena leucocephala*. The application of leaf extracts through seed treatment and foliar sprays also had a significant effect on fruit yield per plant of bhendi harvest. Among the treatments *Annona squamosa* recorded highest fruit yield per plant (364.7 g) and the lowest fruit yield per plant (303.1 g) was recorded in *Ocimum sanctum* at harvest of the crop. The interaction effect was significant, among the treatment combinations, *Albizia lebbek*+ *Annona squamosa* recorded the highest fruit yield per plant (490.3g) and the lowest fruit yield per plant (282.8 g) was recorded in *Delonix regia* + *Aegles marmellos* at harvest of the crop. The absolute control recorded fruit yield per plant of 255.3 g. When it was compared with *Albizia lebbek*+ *Annona squamosa*, it showed a increased fruit yield per plant (235.0 g).The recommended dose of inorganic fertilizers recorded 501 g of fruit yield per plant and when it was compared with *Albizia lebbek*+ *Annona squamosa* it showed a decrease in fruit yield per plant (10.6g).

Table 13.27: Influence of *Vrikshayurveda* Treatments on Fruit Yield per Plant (g) in Bhendi

GLM Incorporation	*Alangium*	*Annona*	*Aegles*	*Morinda*	*Ocimum*	*Mean*
Albizia	391.79	490.36	283.87	359.25	332.81	371.61
Delonix	390.80	310.94	282.85	374.30	295.28	330.83
Gliricidia	359.53					
		340.53				
			420.65			
				298.31		
					291.54	
						342.11
Leucaena	359.53	317.02	350.58	360.23	293.09	325.25
Mean	**361.87**	**364.71**	**334.49**	**348.02**	**303.18**	**342.45**
Control plots						
C_1						255.30
C_2						501.00
	M	***S***	***M x S***			
SEd	0.604	0.533	1.129			
CD (0.05)	1.479	1.086	2.433			

Yield per Hectare

The incorporation of green leaf manures had significant effect on fruit yield per

Table 13.28: Influence of *Vrikshayurveda* Treatments on Average Plot Yield (kg plot^{-1}) and Yield (t ha^{-1}) in Bhendi

GLM Incorporation	*Plot Yield (kg)*						*Fruit Yield (t ha^{-1})*					
	Alangium	*Annona*	*Aegles*	*Morinda*	*Ocimum*	*Mean*	*Alangium*	*Annona*	*Aegles*	*Morinda*	*Ocimum*	*Mean*
GLM incorporation												
Albizia	4.08	5.24	4.21	4.36	3.91	4.36	10.87	13.96	11.22	11.62	10.42	11.62
Delonix	4.21	4.35	4.52	4.13	4.01	4.24	11.24	11.61	12.06	11.01	10.68	11.32
Gliricidia	3.56	4.07	4.73	4.23	4.41	4.19	9.49	10.85	12.60	11.28	11.75	11.19
Leucaena	3.83	4.10	3.88	3.74	3.74	3.86	10.22	10.92	10.34	9.96	9.96	10.28
Mean	**3.92**	**4.44**	**4.33**	**4.11**	**4.01**	**4.16**	**10.46**	**11.84**	**11.56**	**10.97**	**10.70**	**11.10**
Control plots												
C_1						1.95						5.50
C_2						6.00						14.75
	M	***S***	***M x S***				***M***	***S***	***M x S***			
SEd	0.088	0.154	0.289				0.236	0.410	0.771			
CD(0.05)	0.216	0.313	0.599				0.577	0.836	1.599			

Table 13.29: Influence of *Vrikshayurveda* Treatments on Bulb Girth (cm) and Bulb Weight (g) in Aggregatum Onion

GLM Incorporation	*Bulb Girth*					*Bulb Weight*				
	Alangium	*Annona*	*Aegles*	*Morinda*	*Mean*	*Alangium*	*Annona*	*Aegles*	*Morinda*	*Mean*
Albizia	12.50	12.50	14.50	13.57	13.26	48.74	43.25	47.63	49.26	47.22
Senna	11.85	11.87	13.50	13.00	12.55	51.25	48.60	50.27	48.17	49.57
Gliricidia	12.75	13.50	15.50	13.50	13.81	49.28	49.88	57.23	49.60	50.28
Leucaena	14.87	12.50	12.50	13.07	13.23	47.99	48.25	47.80	44.30	47.08
Delonix	12.37	11.25	14.01	14.38	13.00	46.77	44.30	46.77	50.12	46.99
Mean	**12.87**	**12.32**	**14.00**	**13.50**	**13.17**	**48.80**	**46.85**	**48.94**	**48.29**	**48.22**
	M	***S***	***M X S***			***M***	***S***	***M X S***		
SED	0.074	0.082	0.176			1.17	0.78	1.92		
CD(0.05)	0.171	0.168	0.368			2.70	1.60	4.11		
Control plots										
C_1	7.25					33.25				
C_2	16.23					61.25				

Table 13.30: Influence of *Vrikshayurveda* Treatments on Average Plot Yield (kg) and Yield (t ha^{-1}) in Aggregatum Onion

GLM Incorporation	*Plot Yield (kg)*					*Average Yield (t ha^{-1})*				
	Alangium	*Annona*	*Aegles*	*Morinda*	*Mean*	*Alangium*	*Annona*	*Aegles*	*Morinda*	*Mean*
Albizia	12.15	11.15	13.15	10.15	11.65	13.50	12.38	14.61	11.27	12.94
Senna	13.25	14.12	10.15	9.50	11.75	14.72	15.68	11.27	10.55	13.05
Gliricidia	12.25	13.15	14.55	12.15	13.02	13.61	14.61	16.16	13.50	14.47
Leucaena	12.15	10.15	13.55	10.15	11.50	13.50	11.27	15.05	11.27	12.88
Delonix	12.25	13.45	12.25	13.25	12.80	13.61	14.94	13.61	14.72	14.22
Mean	**12.41**	**12.40**	**12.73**	**11.04**	**12.24**	**13.78**	**13.77**	**14.23**	**11.26**	**13.51**
	M	***S***	***M X S***			***M***	***S***	***M X S***		
SED	0.11	0.10	0.24			0.41	0.31	0.74		
CD(0.05)	0.26	0.22	0.50			0.96	0.64	1.57		
Control plots										
C_1	5.61					4.75				
C_2	17.52					18.65				

and foliar sprays had a significant effect on fruit yield per hectare of Bhendi at harvest. Among the treatments, *Annona squamosa* recorded highest fruit yield per hectare of 11.84 t ha^{-1} and the lowest fruit yield per hectare of 10.46 t ha^{-1} was recorded in *Alangium salvifolium*. The interaction effect was significant. Among the treatment combinations, *Albizia lebbek* + *Annona squamosa* recorded the highest fruit yield of 13.96 t ha^{-1}and the lowest fruit yield of 9.49 t ha^{-1} was recorded in *Gliricidia sepium* +*Alangium salvifolium*. The absolute control recorded fruit yield of 5.50 t ha^{-1}. When it was compared with *Albizia lebbek* + *Annona squamosa* there was increase in yield of 8.46 t ha^{-1}. The recommended dose of inorganic fertilizers recorded fruit yield of 14.75 t ha^{-1} and when it was compared with *Albizia lebbek* + *Annona squamosa* and it showed a decrease in yield (0.79 t ha^{-1}) in Bhendi at harvest.

Points for Discussion

The yield attributes *viz.*, number of flowers, number of fruits per plant, fruit girth, fruit length, fruit weight, fruit yield per plant are some important parameters which decide the yield of the crop. Factors like nutritional status of the soil and nutrient availability to the crop also influence the fruit yield and also quality. It is observed that, the plot received *Albizia lebbek* as green leaf manure with seed soaking and foliar spays of *Annona squamosa* registered high values in all the yield attributes whereas application of inorganic fertilizers at recommended dose recorded the highest values of yield attributes among all the treatments imposed. The high values registered in Vrikshyurveda farming may be due to the mineralization, solubilization and availability of all the nutrients to the crop for the entire growth period in addition to the improvement in soil physio-chemical properties that could have enhanced the growth and yield attributes. Moreover Increase in the availability of nutrients improved the synthesis and translocation of metabolities to various reproductive structures of the plant as observed in cluster bean. More foliage may help in better harvesting of light energy, higher photosynthesis activity, translocation and accumulation of photosynthates to different parts of the plant.

The recommended dose of inorganic fertilizers recorded the highest number of fruits per plant, fruit length, fruit weight which ultimately resulted in higher plot yield and total yield per hectare. The availability of nitrogen promoted the vegetative growth, the higher dose of nitrogen application lead to higher uptake of nutrients which would have helped to produce more vegetative growth and more number of flowers per plant by which the effective conversion of carbohydrate to reproductive (economic) part might have increased the yield. Application of phosphorus influenced the root growth in a positive manner which would have helped in better absorption and translocation of nutrients from source to sink of plants. Better root growth would have helped better synthesis of cytokinin which would have helped to divert photo assimilates to more yield attributes. These nutrients being important constituents of neucleotides, proteins, chlorophyll and enzymes, involved in various metabolic processes and have direct impact on vegetative and reproductive phases of plants

Bhendi crop generally produces higher biomass per unit area with in a short period. As the crop responds well to the application of organic manures and also fertilizers to realize higher yield of fruits. It is understood that application of green leaf manure with leaf extract spray, their interaction or application of inorganic fertilizers alone exhibited differences statistically for fruit yield over absolute control. However, among the treatment combinations, green leaf manures of *Albizia lebbek* with seed soaking of foliar sprays of *Annona squamosa* at 25th and 45th DAS recorded the highest yield. The yield obtained in the best treatment was 60.60 per cent higher when compared to absolute control. The increased yield can be ascribed to higher uptake of nitrogen, phosphorus and potassium promoted by higher CEC catalyzed by the application of green leaf manures. The increased yield in Vrikshyurveda farming could be attributed to better timing of the sprays at vegetative and flowering stages, which resulted in higher yield attributable and also yield and another possible reason attributed is the enhancement of the endogenous cytokinin levels of roots in leaf extract treated plants through seed soaking which are primarily responsible for some of the beneficial effects that have been reported to improve fruit size and yield.

Among all the treatments, the recommended dose of fertilizers registered the highest (14.75 t ha^{-1}) yield. This yield was 62.71 and 5.36 per cent higher than absolute control and one which registered high yield in Vrikshyurveda farming *i.e.* green leaf manuring of *Alibizia lebbek* with seed soaking and foliar sprays of *Annona squamosa*, respectively. The availability of nutrients at the immediate root zone might increase the nutrient availability which inturn resulted in better root development and absorption of nutrients which ultimately resulted in high yield. The basic reason of increasing in fruit yield owing in fertilizer application might be due to the fact that an adequate supply of P in the early vegetative growth of plant life, considered essential for the development of its reproductive process. Therefore, all the yield attributes were higher in case of fertilized plot which significantly contributed towards higher fruit yield.

The yield is the component which the crop is expected to give in returns to the inputs given to it. The yield is recorded the following observation *i.e.*, bulb girth bulb weight, yield per plot and yield per hectare. The present study revealed that above the yield parameters was highest in the treatment with *Gliricidia sepium* as green leaf manures and bulb soaking and foliar spays of *Aegles marmellos*. This might be due to the decomposition of organic leaf manures which provided nutrients and this could have improved the soil physical conditions with subsequently increased the availability of nutrient owing to increase in microbial activity leading to more translocation of photosynthates and better source-sink relationship and photo assimilates stored in bulbs. The presence of secondary metabolites in the leaf extracts would have enhanced the yield of crops (Swaminathan, 2004).

Chapter 14
Vrikshyurveda Farming Protocols for in Crops

Step 1

Collect fresh leaves of selected tree species, preferably N-fixing trees (NFTs) and incorporate leaves into the well prepared field, where cultivation is to be taken up, as green leaf manures. Since these leaves contain high nutrient content especially nitrogen and decomposition at a rate.

Step 2

The green leaf manures are to be applied at the rate of 5.0 t ha^{-1} at the time of field preparation and incorporated into the soil. Then the field left undisturbed for 45 days 45 days prior to sowing. time for facilitating decomposition of the applied green leaf manure. Further, if the field does not recive any rainfall one time irrigation of 20 mm water if required which will hasten the decomposition of leaves.

Step 3

Decomposition time: After incorporation of the fresh tree leaves a waiting period of 6-7 weeks will be allowed for the decomposition of the leaves and release of nutrients.

Step 4

After 45 days of waiting, thoroughly plough the field first and then harrow it and level with leveler. As per the requirement of crop to be grown the land configuaration should be perepared like ridges and furrows, beds and channels *etc* for sowing. Irrigation and drainage channels were provided according to the need and slope of the field.

Step 5

Seed treatment: the seeds were treated with *Vitex negundo* leaf extract or other selected tree leaf extracts at 5 per cent for 6-12 hours according to the size of the seed before sowing.

Step 6

The seeds are to be dibbled at the rate of single seed (for big size) or two seeds (smaller size) hill^{-1} by adopting optimum spacing between rows and plants.

First irrigation should be given immediately after sowing if there is no receipt of rains and later Life saving irrigation is to be given on 3rd or 4th day of sowing. Subsequent irrigations are given on need basis especially during the critical stages of the crop growth period.

Step 7

Gap filling and thinning, if needed can be done on 8th and 15th days after sowing respectively by leaving one healthy seedling per hill to maintain optimum population.

Step 8

For weed control, one or two handweeding may be given during 25-30 days of sowing and later during 40-45 days of sowing to avoid critical crop-weed competition. Avoid irrigating the field immedialtely after weeding as it will help in regorwth and proliferation of weeds.

Step 9

Foliar spray of leaf extracts for growth promotion

The leaf extract of each species will be prepared by grinding fresh leaves and distilled water at 1:1 proposition and the extracts to be filtered and this filtrate serve as stock solution. From the stock solution 5 per cent solution will be made for spraying to the field crops. Five spraying will be given to the crop to promote growth and development of crops. Leaf extract spray will be given once in two weeks. The foliar spray should coincide with the vegetative, flowering and maturity stages of crop growth.

Step 10

Foliar spray of lef extracts pest and disease management

Spraying neem lef extracts, neem seed kernel extracts, neem oil like products whenever the incidence was noticed.

Step 11

Stage of harvest

Observe the following symptoms, taking into consideration the average duration of the crop.

I. The sheath will turn yellow and dry at maturity.
II. The seeds become fairly hard and dry. At this stage the crop is ready for harvest.
III. Carry out harvest operations at a single stage for easy transportation.
IV. Dry the grains under the sun till the grains are dry.
V. Clean the seeds by winnowing.
VI. Collect and store the dry grains in gunnies.

The following pages tell about our experiences with different crops and knowledge gained by us in drawing the above protocol.

I) The First Systematic Study on Vrikshyurveda Farming in Green Gram

The recent awareness on harmful and ever lasting ill effects of agro chemical *viz.*, pesticides, fungicides, herbicides and fertilisers due to indiscriminate and unscientific usages in crop production have paved way for going back to human friendly, eco-friendly traditional agriculture. The importers of food from India also seek for produces produced in organic way. It is time now to reorient Indian System of Agriculture by looking back the traditional and natural ways of food production. *Vrikshayurveda* (Indian plant science) provides ample scope to produce quality food products by adopting traditional methods and by utilizing plant, including trees, extracts and by products Enormous and elaborate literatures are found in Indian plant science (*Vrikshayurveda*) on procedures and practices for raising of crops. The farmers in certain areas still practice, though not in a large scale, such procedures and practices. The plant products are naturally evolved ingredients on the biosphere; they not only have an edge over the synthetic alien molecules but also rate preference and acceptance from the viewpoint of environmental safety and eco-friendly. *Vrikshayurveda* (Indian plant science) is an ancient system exploited in Indian system of Agriculture. The three major ancient texts *inter alia* are 1) Varahamitra, 505 AD., Birhat Samhita, *Vrikshayurveda* Part I, Chapter 55 (edited by M. Ramakrishna Bhat, Motilal Banarasidas, Bangalore, India 1950) 2) Chaundarya, 1025 A.D. Lokopakaran, *Vrikshayurveda*, chapter 6 (edited by H. Shesa Iyengar, Government central Manuscripts library, Chennai, India1950) and 3) Sarangadhara, 1363 A.D. *Vrikshayurveda* (edited by S.K.Ramachandra Rao, Kalpatharu Research Academy, P.O. Box 1857, Bangalore, 1993, India). The most detailed of all the three seems to be that of Chaundarya, 1025 A.D. (Shamarao Jahadirgar *et al.*, 2003). The use of leaves of leguminous trees as green manures, because of its high nutrient content and faster decomposition rate, has been very well known. The leaves of *Albizia lebbek, Delonix regia, Leucaena, Azadirachta indica* and many other species are commonly used green leaf manure trees. The role of plant extracts in growth promotion, pest control and disease management is also very well documented. The leaf extracts from neem, Pungam, Bael, notchi and *Albizia amara* and dry fruits of kadukkai are few tree species having insecticidal and antifungal properties. Similarly some tree species (*Alangium salvifolium*, *Aegle marmellos* and *Annona squamosa* and neem) are found to have effects on growth and development of crops. Hence it is the right time we should think of researching and revisiting Indian system of Agriculture and exploit the information available in the ancient texts for present day modern agriculture so as to reduce the use and impact of agro chemicals in food production there by ensuring production of chem.-free foods.

The present study was designed with the following objectives i) To study the influence of different tree leaf extracts on food crops and its relation with yield. ii) To find out the suitability of tree products as an alternative for inorganic fertilizers to increase crop yields.

Materials and Methods

1. Site description: The experiments were set up at the central farm of Agricultural College and Research Institute, Madurai Tamil Nadu. The experiments were conducted during kharif season (June- September) of 2003 and 2004. The soil type is black soil
2. Test crop: Green gram cv pusa bold
3. Treatments: Trees species *viz., Albizia lebbek, Senna siamea, Gliricidia sepium, Leucaena leucocephala* and *Delonix regia* were selected for the study to provide basal nutrition to the test crop. The fresh leaves of these trees were incorporated into the soil 45 days prior to sowing at the rate of 10 t ha^{-1} at the time of field preparation and this formed main plot treatments. The Sub plot treatments included spraying of leaf extracts of tree species *viz., Alangium salvifolium, Annona squamosa, Aegle marmellos* and *Morinda tinctoria* at 5 per cent concentration, as foliar nutrition, during 30 and 45 Days after sowing (DAS). The leaf extracts of each species were prepared by grinding fresh leaves and water at 1:1 proposition and keeping it for 6 hours and then the extract was filtered and sprayed on crop.

Table 14.1: Composition of Nutrients in Leaves of Tree Species (Green Weight Basis)

Tree Species	*Nitrogen (per cent)*	*Phosphorus*	*Potassium*
Albizia lebbek	3.2	0.3	1.5
Senna siamea	2.5	0.27	1.35
Gliricidia sepium	3.0	0.5	3.0
Leucaena leucocephala	3.6	0	2.5
Delonix regia	3.1	0.3	2.0

4. Design and planting geometry: The experiments were set up in a Split Plot Design with two replications. Totally there were 40 plots (5 X 4 X 2) of 10 m^2 size and the test crop green gram was sown at 30 X 10 cm spacing

Table 14.2: Cultivation Details on Crops Grown

Sl.No.	*Details on Field Operation*	*First Crop-2003*	*First Crop-2004*
1.	Incorporation of fresh leaves of tree species	10.05.2003	10.05.2004
2.	Sowing of Green gram cv Pusa bold	25.06.2003	09.07.2004
3.	Spraying of leaf extracts	26.07.2003 and 9.08.2003	09.08.2004 and 24.08.2004
4.	Weeding (Manual)	11.07.2003	23.07.2004
5.	Pest control measure (Prophylactic) NSKE 5 per cent	29.07.2003	05.08.2004
6.	Harvesting	Five- 13th and 20th August and 3, 10 and 22 Sept. 2003	31th August;8, 15 and 23 Sept. 2003 and 5th October 2004

Note:

1) Before the start of the experiment the field was identified and kept fallow for 12 months to get the actual influence of GLM and sprays and avoid any residual effect of preceding crop.

2) The trials were conducted in the same filed successively for two years. Except cropping period the land remains fallow.

5. Data analysis: Observations on yield attribute, yield and total dry matter were recorded on two crops grown in successive years. The data were subjected to ANOVA for statistical significance after Panse and Sukhatme (2000) and the critical differences (CD) at p= 0.05 calculated using the appropriate t value.

Results and Discussion

The results of the experiments are presented in Tables 14.3–14.5.

(i) First Crop Inference

As regards pod weight among the leaf manure application a maximum pod weight was observed Vagai which was on a par with Gulmoher and Gliricidia. Among the growth promoters spray a maximum pod weight was registered with spraying of Bael leaf extract. The interaction effect was significant and the highest pod yield of 3.66 t ha^{-1} was registered in Gliricidia and Bael combination. As regards dry matter production among the leaf manure application a maximum DMP was observed with Leucaena which was on a par with Gliricidia. Among the growth promoter sprays a maximum DMP was registered with spraying of Bael leaf extract. The interaction effect was significant and the highest DMP of 8.75 t ha^{-1} was registered in Gulmoher + bael combination which was on a par with Gliricidia and Bael combination. As regards Grain yield, among the leaf manure application a maximum pod weight was observed with Gliricidia. Among the growth promoters spray a maximum grain yield was registered with spraying of Bael leaf extract. The interaction effect was significant and the highest pod yield of 1.94 t ha^{-1} was registered in Gliricidia and Bael combination.

Table 14.3: Influence on Pod Weight (t ha^{-1}) of Green Gram

GLM Incorporation	*Alangium*			*Annona*			*Bael*			*S4-Morinda*		
	I Year	*II Year*	*Mean*	*I Year*	*II Year*	*Mean*	*I Year*	*II Year*	*Mean*	*I Year*	*II Year*	*Mean*
Vagai	2.74	2.72	2.73	3.69	3.69	3.44	3.48	3.59	3.53	3.43	3.53	3.48
Siamea	2.80	2.78	2.79	2.60	2.60	2.80	2.97	3.36	3.16	2.95	3.31	3.13
Gliricidia	2.81	2.89	2.85	4.00	4.00	3.51	3.66	4.15	3.90	3.14	3.93	3.53
Leucaena	2.74	3.46	3.10	3.78	3.78	3.44	3.39	4.41	3.90	3.27	4.64	3.95
Gulmoher	2.83	4.28	3.55	3.79	3.79	3.54	3.09	3.95	3.52	3.43	4.56	3.99
Mean	**2.79**	**3.22**	**2.73**	**3.57**	**3.57**	**3.44**	**3.31**	**3.89**	**3.53**	**3.24**	**3.99**	**3.48**
CD= 0.05	**I year - M = 0.86; S = 0.26**				**II year- M = 0.86; S= NS; MXS =1.07**				**Pooled-M = 0.86; S= NS; MXS =1.07**			

Table 14.4: Influence on Mean Grain Yield (t ha^{-1})

GLM Incorporation	*Alangium*			*Annona*			*Bael*			*S4-Morinda*		
	I Year	*II Year*	*Mean*	*I Year*	*II Year*	*Mean*	*I Year*	*II Year*	*Mean*	*I Year*	*II Year*	*Mean*
Vagai	1.73	1.68	1.68	2.05	1.91	1.91	1.86	1.86	1.86	1.61	1.77	1.77
Siamea	1.64	1.59	1.59	1.48	1.60	1.60	2.31	2.03	2.03	1.80	1.75	1.75
Gliricidia	1.66	1.67	1.67	2.26	2.01	2.01	2.34	2.14	2.14	2.12	1.97	1.97
Leucaena	1.57	1.59	1.59	2.03	1.90	1.90	2.11	2.02	2.02	1.96	1.86	1.86
Gulmoher	2.14	1.88	1.88	2.44	2.08	2.08	2.03	1.85	1.85	2.48	2.12	2.12
Mean	**1.75**	**1.68**	**1.68**	**2.05**	**1.90**	**1.90**	**2.13**	**1.98**	**1.98**	**1.99**	**1.89**	**1.89**
CD = 0.05	**I year-M = NS; S = 0.15; MXS=0.42**				**II year- M = NS; S= NS; MXS = 0.72**				**Pooled- M = 0.62; S= NS; MXS =0.76**			

Table 14.5: Influence on Dry Matter Production (t ha^{-1})

GLM Incorporation	*Alangium*			*Annona*			*Bael*			*Morinda*		
	I Year	*II Year*	*Mean*	*I Year*	*II Year*	*Mean*	*I Year*	*II Year*	*Mean*	*I Year*	*II Year*	*Mean*
Vagai	6.15	5.75	5.95	7.15	6.00	6.58	9.15	5.65	7.40	8.10	5.45	6.78
Siamea	4.75	5.85	5.30	6.50	6.25	6.38	7.40	6.10	6.75	6.75	6.20	6.48
Gliricidia	6.65	5.75	6.20	7.75	6.25	7.00	8.00	7.25	7.63	8.05	7.00	7.53
Leucaena	8.30	6.75	7.53	8.20	6.85	7.53	8.75	7.25	7.00	8.00	7.00	7.50
Gulmoher	6.15	7.00	6.58	7.10	7.00	7.05	7.70	6.25	6.98	7.75	7.25	7.50
Mean	**6.40**	**6.22**	**6.31**	**7.34**	**6.47**	**6.91**	**8.20**	**6.50**	**7.35**	**7.73**	**6.58**	**7.16**
CD = 0.05	**I year-M = 2.68; S = 1.06; MXS =3.34**				**II year- M = 0.44; S = 0.35; MXS=NS**				**Pooled-M = 1.25; S= 0.71;MXS=1.62**			

(ii) Second Crop Inference

As regards pod weight, among the main plot treatments application of leaves of Gulmoher at 10 t ha^{-1} had a maximum mean pod weight of 4.14 t ha^{-1}. As regards spraying of leaf extracts Morinda had a maximum mean pod weight of 3.99 t ha^{-1} followed by Bael Leaf extract (3.89 t ha^{-1}) and there is no significant difference. Since the interaction is significant for M at S, Leucaena followed by Gliricidia is the best for soil application. As regards grain yield, among the main plot treatments application of leaves of Gulmoher at 10 t ha^{-1} had a maximum mean grain yield of 2.27 t ha^{-1} followed by Gliricidia application (2.09 t ha^{-1}) and there is no significant difference. As regards spraying of leaf extracts of bael had a maximum mean grain yield of 2.13 t ha^{-1}. Since the interaction is significant for M at S, the soil application of tree leaves showed significant differences. As regards DMP, among the main plot treatments application of leaves of Gulmoher at 10 t ha^{-1} had a maximum mean dry matter production of 7.13 t ha^{-1}. As regards spraying of leaf extracts Morinda had a maximum mean dry matter production of 6.58 t ha^{-1} and there is no statistical significance for interaction of treatments.

(iii) Overall Performance of Basal Nutrition and Foliar Sprays

Pod Weight

As regards pod weight, among the green leaf manures, Gulmoher at 10 t ha^{-1} had a maximum mean pod weight of 3.65 t ha^{-1}. As regards spraying of leaf extracts bael had a maximum mean pod weight of 3.99 t ha^{-1} followed by Bael Leaf extract (3.53 t ha^{-1}) and there is no significant difference. Since the interaction is significant Gulmoher + Morinda, Leucaena+ Morinda and Gliricidia + Bael are the best combinations and they are on a par (Table 14.3).

Grain Yield

Among the green leaf manures, Gulmoher at 10 t ha^{-1} had a maximum mean grain yield of 1.98 t ha^{-1} followed by Gliricidia application (1.95 t ha^{-1}) and were on a par. As regards spraying of leaf extracts of bael had a maximum mean grain yield of 1.98 t ha^{-1}. Since the interaction is significant a maximum mean yield of 2.14 t ha^{-1} was registered in Gliricidia + Bael combination (Table 14.4).

Dry Matter Production

Among the green leaf manures, Leucaena at 10 t ha^{-1} had a maximum mean dry matter production of 7.39 t ha^{-1}. As regards spraying of leaf extracts bael had a maximum mean dry matter production of 7.35 t ha^{-1} and since there is statistical significance for interaction of treatments, the highest DMP was registered by *Gliricidia* + Bael combination (7.63 t ha^{-1}). (Table 14.5).

Plants have been an essential part of human life. Since the start of civilization, during the Vedic period, great importance was given to plant as source of almost all the essentials in life, particularly for agriculture. Rig Veda and Ayurveda too described about the use of botanicals. Green plants appear to be the reservoirs of effective chemotherapy and would constitute inexhaustible sources of harmless chemical compounds. Botanicals do not possess quick knock down effect unlike synthetic

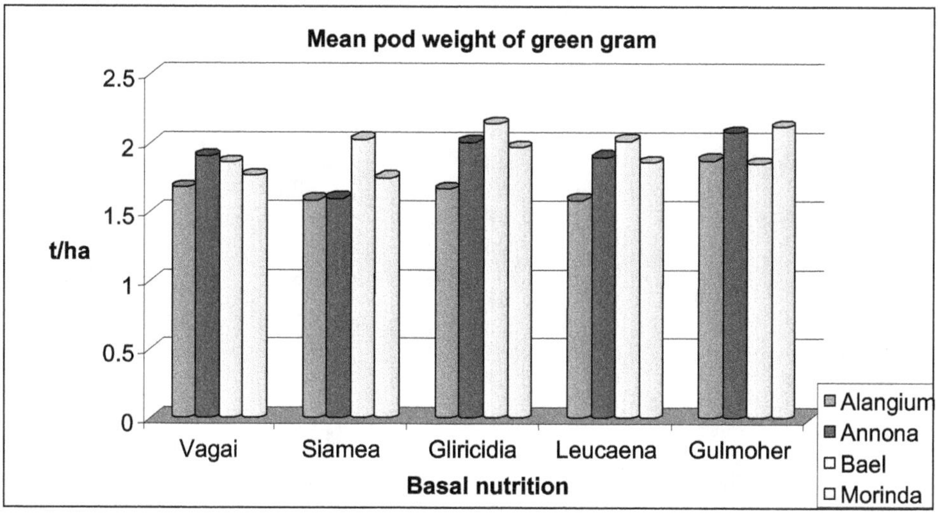

Figure 14.1: Response of Green Gram to Vrikshyurveda Farming.

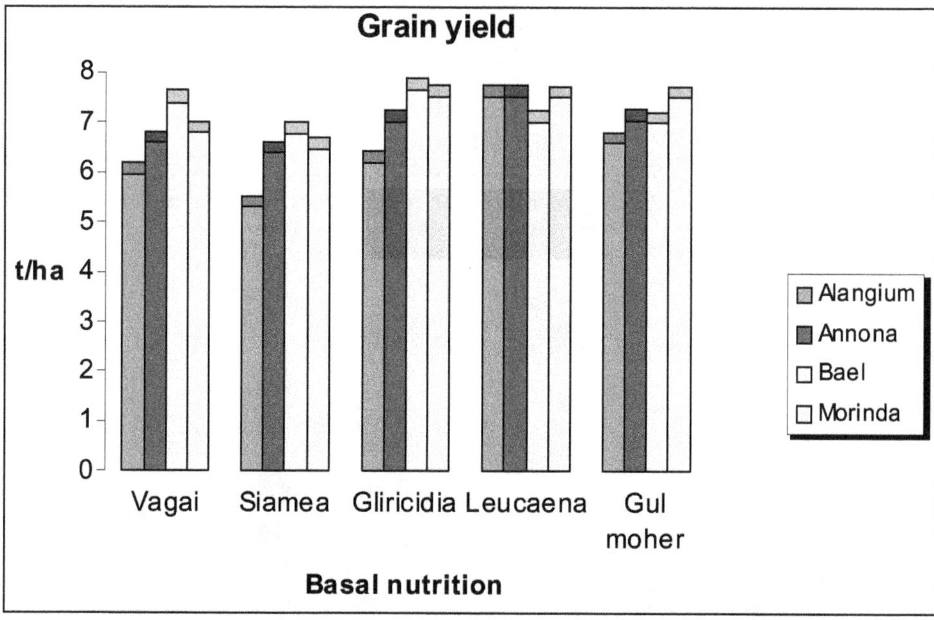

Figure 14.2: Response of Green Gram to Vrikshyurveda Farming.

biocides and fertilizers. They have no adverse effect on seed germinability cooking quality and milling of seeds. Many of the botanicals possess anti fungal and anti bacterial activity but few reports were available for anti fungal and anti bacterial activity. The leaf extracts from neem (Ramadevi *et al.*, 1988), Pungam (Naik and Dumbre 1985), Bael, notchi and *Albizia amara* (Anon.1991) and dry fruits of kadukkai are few tree species having insecticidal and antifungal properties. Similarly some

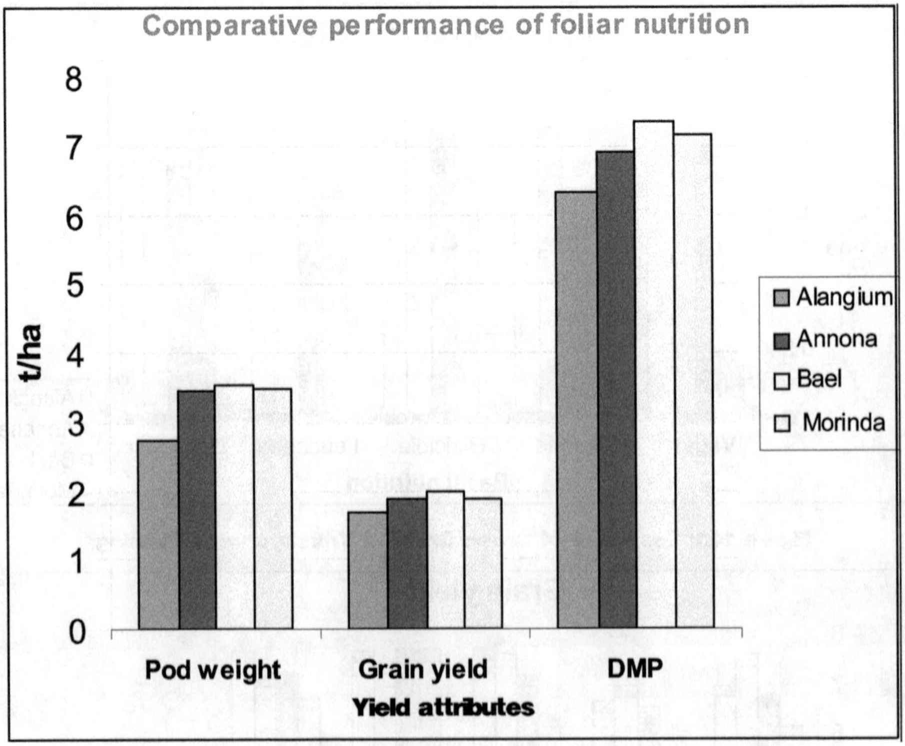

Figure 14.3: Comparative Performance of Trees Species for Foliar Nutrition.

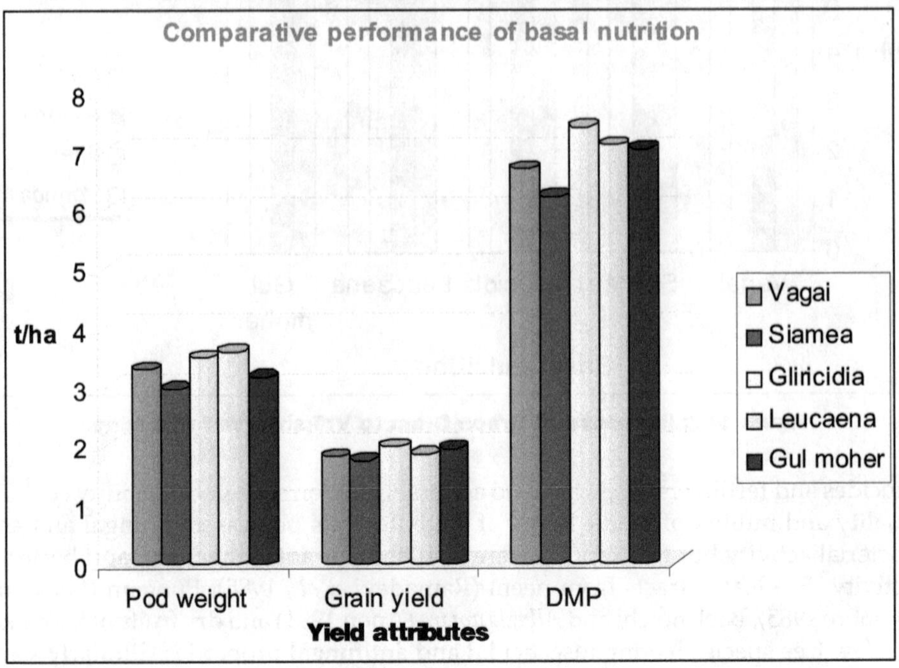

Figure 14.4: Comparative Performance of Trees Species for Basal Nutrition.

tree species (*Alangium salvifolium*, *Aegle marmellos* and *Annona squamosa* and neem) are found to have effects on growth and development of crops (Swaminathan, 2005). In the present investigation, the better performance of bael and Morinda in yield enhancement may be due to the presence of secondary metabolites (Figures 14.1 and 14.2). Spraying of leaf extracts of the tree species *Alangium salvifolium* and *Aegle marmellos* during 30 and 45 days after sowing could enhance food grain yield in cowpea (Swaminathan, 2004). The present study scientifically proves that the oldest Indian system of Agriculture is able to produce quality food grains without addition of modern day pesticides, herbicides and agro chemicals and which demands strengthening of research on revalidation of *Vrikshayurveda* for quality food production. This also has the added advantage of Promotion of tree groves in farmlands there by increasing the tree cover in the country.

Conclusion

The study revealed that application of leaves of *Gliricidia sepium* @ 10 t ha^{-1} 45 days before sowing of green gram and followed by that two sprayings of leaf extracts of *Aegle marmellos* @ 5 per cent during 30 and 45 days of sowing recorded an average yield of 2.14 t ha^{-1} and DMP of 7.63 t ha^{-1}. This was followed by application of leaves of *Delonix regia* @ 10 t ha^{-1} 45 days before sowing of green gram and sprayings of leaf extracts of *Morinda tinctoria* @ 5 per cent during 30 and 45 days is also with an average yield of 2.12 t ha^{-1} and DMP of 7.50 t ha^{-1}.

References

Panse, V.G. and Sukhatme, P.V. 2000. *Statistical Methods for Agricultural Workers*. Indian Council of Agricultural Research, New Delhi.

Swaminathan, C. 2004 Efficacy of tree origin botanicals on food grain yield in Cowpea *Vigna uniguculata* (l) Walp. Abstract paper presented in First World Congress of Agro forestry held during June 27 to June 2 at University of Florida, U.S.A.

Swaminathan, C. 2005. Eco-friendly production of quality food grains in ayurvedic way. In Natural Farming. Agricultural College and Research Institute, Madurai Tamil Nadu. 11-14.

II) Scientific Study on *Vrikshyurveda* Farming in Barnyard Millet (*Echinochloa frumentacea* L.)

Field experiment was conducted at Agricultural College and Research Institute, TNAU Madurai with an objective of studying the influence of green leaf manures and leaf extracts of selected trees on vegetative growth, flowering, quality, yield attributes and yield of barnyard millet and the best combination of green leaf manure and tree leaf extracts that ensures high yield in barnyard millet.

The experiment was carried out on a sandy clay loam soil at the Central Farm of Agricultural College and Research Institute, TNAU Madurai during 2011 - 2012 in a split plot design replicated twice. The experiment consisted of 20 treatment combinations; five different green leaf manures (*Albizia lebbek*, *Glyricidia sepium*, *Pongamia pinnata*, *Delonix regia*, *Peltophorum ferrugineum*) @ 2 t ha^{-1} in main plots and four different green leaf extracts (*Moringa oleifera*, *Aegle marmellos*, *Morinda tinctoria*,

Annona squamosa) @ 5 per cent in the form of four foliar sprays at 20, 40, 60 and 80 DAS in sub plots.

The results of the field study indicated that there were significant differences in growth attributes, yield attributes due to the application of green leaf manure and foliar sprays of tree leaf extracts. In the field study all the growth parameters and yield parameters were found to be high in the treatment plot of *Peltophorum ferrugineum* as green leaf manure with foliar sprayings of *Annona squamosa*.

The physiological parameters *viz.*, leaf chlorophyll content, dry matter production, harvest index were also found to be high in the treatment combination *Peltophorum ferrugineum* and *Annona squamosa*.

The highest yield of 1960 kg ha^{-1} was recorded in the plot that received *Peltophorum ferrugineum* as green leaf manure and foliar sprays of *Annona squamosa*, which is more than two times higher than the zero input management and almost 91.1 per cent of the yield realized from high input management (CPG management).

The quality characters like iron, zinc, fibre and protein content recorded high in the treatment combination *Peltophorum ferrugineum* and *Annona squamosa*.

Nutrients uptake (N, P and K) was increased when the plants were treated with *Peltophorum* ferrugineum as green leaf manure.

The study revealed that *Peltophorum ferrugineum* as green leaf manure and foliar sprays of *Annona squamosa* increased the net profit of 29376 and benefit cost ratio (2.5).

From the above study, the treatment combination of *Peltophorum ferrugineum* @ 2.0 t ha^{-1} as green leaf manure along with foliar sprays of *Annona squamosa* @ 5 per cent become a best treatment in growth and yield characters of barnyard millet.

Findings

The observations were recorded on germination percentage, growth parameters, physiological parameters, yield characters, quality characters, soil available nutrients and plant uptake in barnyard millet. The plant height, LAI and total dry matter production were recorded at three stages of crop growth at vegetative, flowering and harvesting stage indicated the the following.

1. The outcome of the investigation revealed that the highest growth parameters *viz.*, germination, plant height, LAI was recorded in the treatment with *Pheltophorum ferrugineum* as green leaf manure and foliar application of *Annona squamosa* leaf extract. In physiological parameters *viz.*, leaf chlorophyll content, dry matter production and harvest index were recorded, the highest in the treatment combination of *Pheltophorum ferrugineum* and *Annona squamosa*.
2. The yield contributing characters *viz.*, inflorescence length (25.64 cm), inflorescence weight (23.93 g) and test grain weight (3.15 g) per plant were highest in the treatment combination of *Pheltophorum ferrugineum* and *Annona squamosa*.

Albizia lebbek

Gliricidia sepium

Pongamia pinnata

Delonix regia

Peltophorum ferrugineum

Plate 14.1: Tree Speces Used for Green Leaf Manure Incorporation.

Moringa oleifera

Aegle marmellos

Morinda tinctoria

Annona squamosa

Plate 14.2: Tree Speces Used for Foliar Application.

Initial layout

Application of green leaf manure

Incorporation of green leaf manure

Field overview

BESR treatment plot

Standing crop

Plate 14.3: Field Preparation and Raising of Crops.

M_5S_4
(Peltophorum ferrugineum +Annona squamosa)

CPG

Zero input

Plate 14.4: General View of Inflorescence.

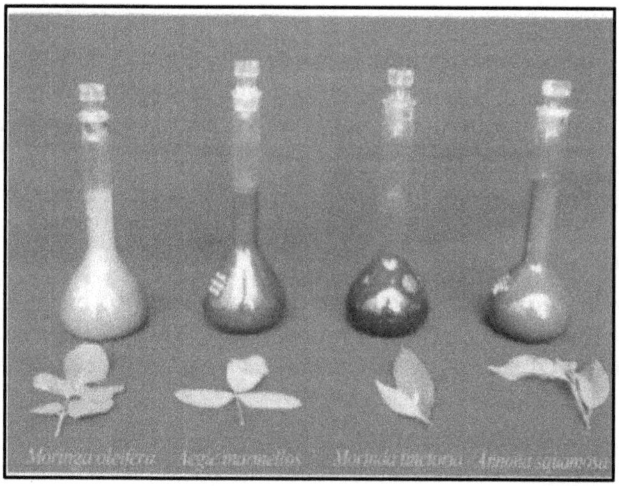

Plate 14.5: Leaf Extracts.

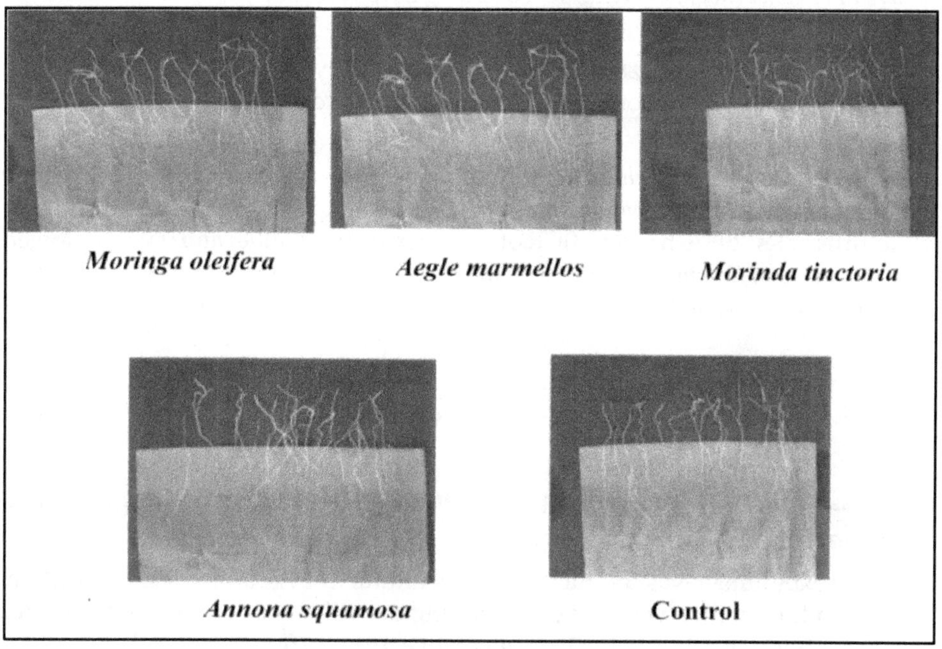

Plate 14.6: Roll Towel Technique.

3. The nutrient uptake *viz.* N (74.08 kg ha $^{-1}$), P (10.41 kg ha $^{-1}$) and K (64.03 kg ha $^{-1}$) was also high in the treatment combination of *Pheltophorum ferrugineum* and *Annona squamosa.*
4. The quality characters *viz.*, protein (7.82 g 100 g^{-1}), fibre (9.20 g 100 g^{-1}), iron (14.90 g 100 g^{-1}) and zinc (6.00 g 100 g^{-1}) content were highest in the treatment combination *Pheltophorum ferrugineum* and *Annona squamosa.*
5. The net return and benefit cost ratio were recorded the highest in treatment combination of *Pheltophorum ferrugineum* and *Annona squamosa.*

Conclusions

The growth, yield and physiological parameters of barnyard millet were significantly influenced by incorporation of *Pheltophorum ferrugineum* as green leaf manure, 45 DBI and foliar sprays of *Annona squamosa* leaf extracts at 20, 40, 60 and 80 DAS.

Application of *Pheltophorum ferrugineum* as green leaf manure, 45 DBS along with foliar sprays of *Annona squamosa* leaf extracts at 20, 40, 60 and 80 DAS recorded the highest grain yield of 1960 kg ha^{-1} and it can be recommended. It was followed by with application of *Delonix regia* as green leaf manure, 45 DBS along with foliar sprays of *Annona squamosa* at 20, 40, 60 and 80 DAS recorded the grain yield of 1760 kg ha^{-1}.

The absolute control zero input recorded grain yield of 861 kg ha^{-1} and the recommended dose of in inorganic fertilizers CPG recorded the grain yield of 2151 kg ha^{-1}.

Application of *Pheltophorum ferrugineum* as green leaf manure, 45 DBS along with foliar sprays of *Annona squamosa* at 20, 40, 60 and 80 DAS recorded the higher net returns of Rs. 29376 ha^{-1} and BCR of 1: 2.5 in barnyard millet and it was followed by application of *Delonix regia* as green leaf manure, 45 DBS along with foliar sprays of *Annona squamosa* at 20, 40, 60 and 80 DAS recorded the net returns of Rs. 22693 ha^{-1}.

The absolute control Zero input recorded the net loss of 2976 and benefit cost ratio of 1: 0.77 and the CPG recorded the net return of 9828 and benefit cost ratio of 1:1.31.

III) *Vrikshyurveda* Farming for Finger Millet (*Eleusine coracana* L.) Gaertn.

Field experiment was conducted at Agricultural College and Research Institute, TNAU Madurai with an objective of studying the influence of green leaf manures and leaf extracts of selected trees on vegetative growth, flowering, yield attributes and yield of finger millet and to select the best combination of green leaf manure and tree leaf extract that ensure high yield of this crop.

The experiment was carried out on a sandy clay loam soil at the Central Farm of Agricultural College and Research Institute, TNAU Madurai from 2012 to 2013 in a split plot design replicated twice. The experiment consisted of 20 treatments: Five different green leaf manures (*Albizia lebbek, Gliricidia sepium, Leucaena leucocephala, Delonix regia, Peltophorum ferrugineum*) @ 2 t ha^{-1} in main plots and four different green leaf extracts (*Moringa oleifera, Morinda tinctoria, Annona squamosa, Aegle marmellos*) @ 5 per cent in the form of foliar sprays at 36, 52, 68 and 85 DAS in sub plots. Before sowing, the seeds were treated with *Vitex negundo* leaf extract.

The results indicated that there were significant differences in growth attributes, yield attributes due to green leaf manure incorporation and foliar sprays of tree leaf extracts. All the growth parameters and yield parameters were found to be high in the plot that received leaves *Delonix regia* as green leaf manure with foliar spraying of *Moringa oleifera*.

The physiological parameters *viz.*, dry matter production, leaf chlorophyll content, harvest index were also found to be high in the treatment *Delonix regia* as green leaf manure and foliar sprays of *Moringa oleifera*.

The highest yield of 808 kg ha^{-1} was recorded in the plot that received *Delonix regia* as green leaf manure and foliar sprays of *Moringa oleifera*, which four times higher than the zero input management and almost 80 per cent of the yield realized from high input management (CPG management).

Nutrients uptake (N, P and K) was increased when the plants were treated with *Delonix regia* as green leaf manure.

The quality parameters of finger millet *viz.*, carbohydrate, crude protein, crude fibre, crude fat, calcium and iron content were found to be maximum in the treatment combination of *Delonix regia* as green leaf manure and foliar sprays of *Moringa oleifera*. But CPG treatment recorded low content for all the quality parameters tested.

The study revealed that *Delonix regia* as green leaf manure and foliar sprays of *Moringa oleifera* increased the net profit to a tune of Rs. 46152 and benefit cost ratio of (3.86).

From the above study it is inferred that application of *Delonix regia* as green leaf manure along with four foliar sprays of *Moringa oleifera* is found to be good in enhancing growth and yield characters of finger millet.

Findings

The observations were recorded on germination percentage, growth parameters, physiological parameters, yield characters, quality characters, soil available nutrients and plant uptake in finger millet. The plant height and LAI were recorded at three stages of crop growth at vegetative, flowering and harvesting stage found the following.

1. The outcome of the investigation revealed that the highest growth parameters *viz.*, germination, plant height, LAI was recorded in the treatment with *Delonix regia* as green leaf manure and foliar application of *Moringa oleifera* leaf extract. In physiological parameters *viz.*, leaf chlorophyll content, dry matter production and harvest index were recorded, the highest in the treatment combination of *Delonix regia* + *Moringa oleifera*.
2. The yield contributing characters *viz.*, inflorescence length, inflorescence weight and test grain weight per plant were highest in the treatment

combination of *Delonix regia* + *Moringa oleifera.*

3. In soil, nutrient uptake was more in the treatment of incorporation of *Delonix regia* as green leaf manure.
4. The quality characters *viz.*,carbohydrate, protein, fat, fibre, iron and calcium content were highest in the treatment combination of *Delonix regia* as green leaf manure + *Moringa oleifera* as leaf extract spray.
5. The net return and benefit cost ratio were recorded the highest in treatment combination of *Delonix regia* + *Moringa oleifera.*

Conclusions

The growth, physiological parameters and yield of finger millet were significantly influenced by incorporation of *Delonix regia* as green leaf manure and foliar spraying of *Moringa oleifera* leaf extract at 36, 52, 68 and 85 DAS.

Incorporation of *Delonix regia* as green leaf manure, 45 DBS along with foliar sprays of *Moringa oleifera* leaf extract at 36, 52, 68 and 85 DAS recorded the highest grain yield of 808 kg ha^{-1} and it can be recommended. It was followed by with application of *Delonix regia* as green leaf manure, along with foliar spraying of *Aegle marmellos* at 36, 52, 68 and 85 DAS recorded the grain yield of 558 kg ha^{-1}.

The absolute control zero input recorded grain yield of 83 kg ha^{-1} and the recommended dose of in inorganic fertilizers CPG recorded the grain yield of 1337 kg ha^{-1}.

Application of *Delonix regia* as green leaf manure, along with foliar sprays of *Moringa oleifera* at 36, 52, 68 and 85 DAS recorded the higher net returns of Rs. 46152 ha^{-1} and BCR of 1: 3.86 in finger millet and it was followed by application of *Delonix regia* as green leaf manure, along with foliar sprays of *Aegle marmellos* at 36, 52, 68 and 85 DAS recorded the net returns of Rs. 29535 ha^{-1}.The CPG recorded the net return of Rs. 31561 ha^{-1} and benefit cost ratio of 1:2.15.

IV) Influence of *Vrikshyurveda* Practices on Growth and Yield of Maize (*Zea mays* L.)

The experiment was conducted in field no.28 of Agronomy Research block at Central farm, Agricultural College and Research Institute, Madurai. The farm is geographically located at 9°54′ N latitude and 78°54′ E longitude at an elevation of 147 m above mean sea level. The field is located in the southern agro climatic zone of the Tamil Nadu. Treatments consisted of incorporation of five green leaf manures (*Albizia lebbek, Gliricidia sepium, Leucaena leucocephala, Delonix regia, Peltophorum ferrugineum*) 5 t ha^{-1} in main plots and foliar spraying of four green leaf extracts (*Alangium salvifolium, Aegle marmellos, Morinda tinctoria, Annona squamosa*) at 5 per cent in subplot. The experiment was carried out in split plot design with two replications. The maize was raised with a spacing 60 x 20 cm. The soil of the experiment field was sandy clay loam in textured. The green leaf manures collected and incorporated in the field @ 5 t ha^{-1} and allowed for 45 days for decomposition and the leaf extract of tree species were prepared by grinding the leaf with distilled water at

1:1 proportion on maize and the foliar spraying of leaf extracts were done on 25, 39, 53, 67 and 87 DAS. Two separate plots outside the experimental plot were maintained for comparison. They were zero input and recommended dose of fertilizer (RDF) management (130: 62.5: 60 NPK kg ha^{-1}).

Results and Discussion

The growth characteristics like plant height, LAI, dry matter production showed significant response to green leaf manures and foliar spraying of leaf extracts. The incorporation of green leaf manures had significant effect on growth of maize. The plant height was higher (191.9 cm) in M_1 (*Albizia lebbek*), which was followed by M_4 (*Delonix regia*) (185.2 cm) and the lowest (169.2 cm) was recorded in M_2 (*Gliricidia sepium*).

The foliar spraying of leaf extract had a significant effect on maize. Among the treatments S_4 (*Annona squamosa*) had tallest plants (187.6 cm) which was followed by S_3 (*Morinda tinctoria*) (182.0) and the shortest plant (175.6 cm) was recorded in S_2 (*Aegle marmellos*) at harvest stage. This might be due to consistent supply of nutrient to the maize plant by the decomposed green leaf manures of *Albizia lebbek*, which might have released the nutrient present in the leaves resulting in increased cell division and cell elongation, there by increased the plant height (Tripathi *et al.*, 2000). The same reason may also be attributed to higher plant height growth in crop production guide recommended dose of fertilizers. This might be also due to the inorganic fertilizers, which were readily available for growth of the plant through better fertilizer use efficiency (Thilakavathy, 1998). The zero input control recorded lowest plant height (109.7 cm) since there was no external application of organic or inorganic source of nutrients.

The Leaf Area Index (LAI) which determines the total assimilating area available for translocation to the sink showed differential response. With respect to LAI, among the treatments significantly higher in *Albizia lebbek* registered LAI of 5.19 which was followed by *Peltophorum ferrugineum* which recorded the LAI of 4.90. The minimum value of LAI was recorded in *Leucaena leucocephala* with 4.21 at harvest stage.

The foliar spraying of leaf extract had a significant effect on LAI in maize. Among the treatments *Annona squamosa* recorded higher LAI of 5.15. It was followed by *Morinda tinctoria* with a value of 4.77 and the lower LAI of 4.30 at harvest stage were observed with *Aegle marmellos*. This might be due to availability nutrients to the crop during different growth phases and could have improved the LAI (Patel *et al.*, 2003). The recommended dose of inorganic fertilizers recorded highest LAI. This might be due to the application of inorganic fertilizers and uptake by plants and utilization of nutrients, which has become constituent of nucleic acid, might have promoted cell division resulting in increased vegetative growth (Musmade and Knode, 1986). Similar observations were recorded Sirohi *et al.* (1998) in onion. The absolute control recorded least number of LAI, this might be due to low level of nutrients in the soil as evidenced from the initial analysis of the soil hence the utilization would have been also the lower side.

In respect of DMP the maximum production at harvest stage the higher DMP (7251 kg ha^{-1}) was recorded in *Albizia lebbek*, which was followed by *Peltophorum ferrugineum* with DMP of (6533 kg ha^{-1}) and the lowest DMP of (5582 kg ha^{-1}) was recorded in *Leucaena leucocephala*. Foliar spraying of *Annona squamosa* recorded the highest DMP of 6855 kg ha^{-1}. It was followed by *Morinda tinctoria* recorded the DMP of 6600 kg ha^{-1} and the lowest DMP of 5924 kg ha^{-1} was recorded in *Aegle marmellos* at harvest stage. This might be due to faster decomposition of the green leaf manures and subsequent availability and better uptake of nutrients, increased plant height, LAI which could have utilized the photosynthesis process and increased dry matter production in the crop. The presence of secondary metabolites in the leaf extracts might have increased the dry matter production. Similar observations were recorded by Swaminathan and Gururajan (2005) in green gram. The recommended dose of inorganic fertilizers recorded highest in all growth parameters. This might be due to nutrient availability and better uptake of nutrients and increased LAI and plant height with better photosynthesis activity and accumulation. As usual, the zero input recorded the lowest in all growth parameters. This might be due to the non-availability of the nutrients from the soil to the plant.

The incorporation of green leaf manures had significant effect on grain yield of maize at harvest of the crop. Grain yield was higher (1751 kg ha^{-1}) in *Albizia lebbek*. Followed by *Peltophorum ferrugineum*: 1607 kg ha^{-1} and the lowest (1357 kg ha^{-1}) were recorded with *Leucaena leucocephala*.

The foliar spraying of leaf extract had significant effect on grain yield of maize. *Annona squamosa* recorded higher grain yield of 1633.0 kg ha^{-1}. It was followed by *Morinda tinctoria* recorded the grain yield of 1563.2 kg ha^{-1} and minimum grain yield was recorded *Aegle marmellos* (1516.2 kg ha^{-1}), which five times higher than the zero input management and almost 50 per cent of the yield realized from high input management (CPG management).

Higher positive influence of soil microbial population influenced by the application *Albizia lebbek* of green leaf manures, higher concentration of micronutrients in *Annona squamosa* as foliar spray supported by low incidence of pest and weed infestation have been attributed for the better performance of *Albizia lebbek* + *Annona squamosa* combination in terms of yield. Increasing soil microbes enhances root-hair formation and thereby increases root uptake capacity by secreting growth hormones and there by yield can be increased (Tien *et al.*, 1979 and Vlassak and Reynders, 1981). The foliar application of tree leaf extracts increased yield and nutrition of green gram, due to the presence of microelements and plant growth regulators (Zodape *et al.*, 2010). The recommended dose of inorganic fertilizers recorded the highest grain yield per hectare. This might be due to increased nutrient availability and better uptake and utilization of nutrient. The more number of leaves might have increased the photosynthates which could have lead to better source – sink relationship to accumulate more of the stored food (Chelladurai, 2006). Similar observation was recorded by Thilakavathy and Ramasamy (1998) in aggregatum onion and Devi *et al.* (2003) in multiple onion. The zero input recorded lower yield. This might be due to low level of nutrients in the soil. Hence the utilization to would have been also on the lower side.

Plate 14.7: Field Preparation and Raising of Crops.

Plate 14.8: General View of Cobs and Grains.

Plate 14.9: Leaf Extracts.

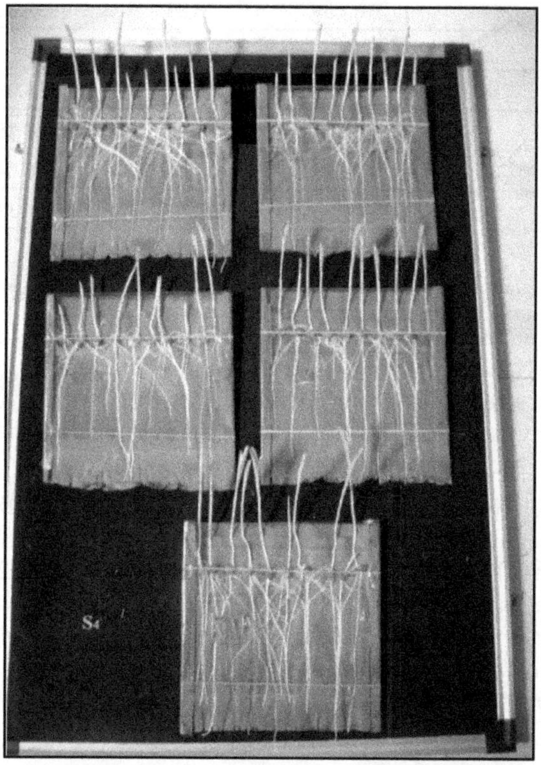

Plate 14.10: Lab Experiments: Germination Study (Inclinved Plate Method).

Table 14.6: Effect of *Vrikshyurveda* Practices on Plant Height, LAI, DMP and Yield of Maize

Treatments	*Plant Height (cm)*	*LAI*	*DMP (kg ha^{-1})*	*Yield (kg ha^{-1})*
GLM -Incorporation (M)				
Albizia lebbeck	191.9	5.19	7251	1751
Gliricidia sepium	169.2	4.56	6227	1502
Leucaena leucocephala	180.8	4.21	5582	1357
Delonix regia	185.2	4.69	6347	1588
Peltophorum ferrugineum	179.0	4.90	6533	1607
SEd	**4.70**	**0.34**	**127.40**	**153.70**
CD (p=0.05)	**9.53**	**0.70**	**353.72**	**322.77**
GLE Foliar Spraying (S)				
Alangium salvifolium	179.7	4.62	6173.48	1533
Aegle marmellos	175.6	4.30	5924.3	1516
Morinda tinctoria	182.0	4.77	6600.02	1563
Annona squamosa	187.6	5.15	6855.14	1633
SEd	**3.28**	**0.26**	**72.07**	**83.1**
CD (p=0.05)	**6.49**	**0.51**	**153.61**	**165.09**
Interaction M X S				
SEd	**7.90**	**0.61**	**188.96**	**232.8**
CD (p=0.05)	**15.77**	**1.21**	**458.18**	**454.30**
Zero input	109.78	4.56	835.84	305.0
CPG	211.96	5.51	11532	3900

V) *Vrikshyurveda* Farming in Black Gram (*Vigna mungo* L.)

Field experiment was conducted at Agricultural College and Research Institute, TNAU Madurai with an objective of studying the influence of green leaf manures and leaf extracts of selected trees on vegetative growth, flowering and yield attributes and yield of black gram and the best combination of green leaf manure and tree leaf extracts that ensure high yield in black gram.

The experiment was carried out on a sandy clay loam soil at the Central Farm of Agricultural College and Research Institute, TNAU Madurai from 2010 to 2011 in a split plot design replicated twice. The experiment consisted of 20 treatments: five different green leaf manures (*Albizia lebbek, Gliricidia sepium, Pongamia pinnata, Delonix regia, Peltophorum ferrugineum*) @ 2 t ha^{-1} in main plots and four different green leaf extracts (*Moringa oleifera, Aegle marmellos, Morinda tinctoria, Annona squamosa*) @ 5 per cent in the form of four foliar sprays at 30, 37, 44 and 51 DAS in sub plots.

Findings

The observations were recorded on germination percentage, growth parameters, physiological parameters, yield characters, nutrient and uptake of black gram. The plant height, LAI, total dry matter production and nutrient uptake were recorded at three stages of crop growth at vegetative, flowering and harvesting stage. The following is brief summary.

1. The outcome of the investigation revealed that the highest growth parameters *viz.*, germination percentage, plant height, LAI was recorded in the treatment with *Pongamia pinnata* as green leaf manure and seed soaking of notchi and foliar sprays of *Moringa oleifera.* In physiological parameters *viz.*, dry matter production, leaf chlorophyll content and harvest index were recorded the highest in the treatment with *Pongamia pinnata* as green leaf manure and foliar sprays of *Moringa oleifera.*
2. The yield contributing characters *viz.*, number of pods per plant, pod length, number of seeds per pod and test grain weight per hectare were highest in the treatment with *Pongamia pinnata* as green leaf manure and foliar sprays of *Moringa oleifera.*
3. The nutrient uptake *viz.* N, P and K was also highest in the treatment with *Pongamia pinnata* as green leaf manure along with foliar sprays of *Moringa oleifera.*
4. The net return and benefit cost ratio were recorded the highest in the treatment with *Pongamia pinnata* as green leaf manure and foliar sprays of *Moringa oleifera.*

Conclusions

The growth, yield and physiological parameters of black gram were significantly influenced by the *Pongamia pinnata* as green leaf manure incorporation of 45 days before sowing and foliar sprays of *Moringa oleifera* at 30, 37, 44 and 51 DAS.

Application of *Pongamia pinnata* as green leaf manure at 45 DAI along with foliar sprays of *Moringa oleifera* at 30, 37, 44 and 51 DAS recorded the highest grain yield of 585.42 kg ha^{-1} and it can be recommended. It was followed by with application of *Pongamia pinnata* as green leaf manure at 45 DAI along with foliar sprays of *Annona squamosa* at 30, 37, 44 and 51 DAS recorded the grain yield of 514.58 kg ha^{-1}.

The absolute control zero input recorded grain yield of 197.08 kg ha^{-1} and the recommended dose of in inorganic fertilizers CPG recorded the grain yield of 597.92 kg ha^{-1}.

Application of *Pongamia pinnata* as green leaf manure at 45 DAI along with four foliar sprays of *Moringa oleifera* at 30, 37, 44 and 51 DAS recorded the higher net returns of (Rs. 16160.83 ha^{-1}) and BCR of 1: 2.23 in black gram it was followed by application of *Pongamia pinnata* as green leaf manure at 45 DAI along with foliar sprays of *Annona squamosa* at 30, 37, 44 and 51 DAS recorded the net returns of Rs. 12619.17 ha^{-1}.

The absolute control Zero input (C1) had benefit cost ratio of 1: 0.78 and the CPG showed benefit cost ratio of 1: 1.14.

VI) *Vrikshyurveda* Farming in Bhendi cv. Arka Anamika

The experiment was laid out in split plot design with four different green leaf manures (*Albizia lebbek, Delonix regia, Gliricidia sepium* and *Leucaena leucocephala*) incorporation @ 5 t ha^{-1} in main plots and five different green leaf extracts (*Alangium salvifolium, Annona squamosa, Aegle marmelos, Morinda tinctoria* and *Ocimum sanctum*) @ 5 per cent in the form of seed soaking and two foliar sprays at 25 and 45 DAS in sub plots with three replications.

The results indicated that there were significant differences in the growth parameters *viz.*, germination percentage, plant height, number of leaves and number of branches per plant due to the application of green leaf manure and foliar sprays of tree leaf extracts. All the growth parameters were found to be high in the plot received *Albizia lebbek* as green leaf manure with seed soaking and foliar spraying of *Annona squamosa*.

Findings

The observations were recorded on germination percentage, growth parameters, physiological parameters, yield characters, quality characters, nutrient uptake and pest and disease incidence in bhendi. The plant height, number of leaves per plant, total dry matter production and nutrient uptake were recorded at four stages of crop growth at 30, 60, 90 and 120 DAS. The following is brief summary.

1. The out come of the investigation revealed that using *Albizia lebbek* as green leaf manure and *Annona squamosa* as seed treatment recorded the highest germination percentage.
2. Use of *Albizia lebbek* as green leaf manure and *Annona squamosa* as seed treatment along with foliar sprays were recorded the highest plant height, number of leaves and number of branches per plant.
3. The yield contributing characters *viz.*, number of flowers per plant, number of fruits per plant, fruit length, fruit girth, fruit weight, fruit yield per plant, number of seeds per fruit, seed weight per plant, average yield per plot and yield per hectares were higher in the treatment *Albizia lebbek* as green leaf manure and seed soaking along with foliar sprays of *Annona squamosa* leaf extract.
4. Incorporation of green leaf manure *Albizia lebbek* and seed soaking along with foliar sprays of *Annona squamosa* made the highest performance of physiological parameters *viz.*, dry matter production, crop growth rate and relative growth rate.
5. The quality parameters *viz.*, curde protein and vitamin-C were higher in the treatment combination of *Albizia lebbek* as green leaf manure and seed soaking along with foliar sprays of *Annona squamosa*.
6. The crude fibre content was the lowest in the treatment *Albizia lebbek* as green leaf manure and seed soaking along with foliar sprays of *Annona squamosa*.

Plate 14.11: Overview of the Field Experiment.

Plate 14.12: Best Treatment.

Plate 14.13: Best Treatment-M_1S_2 *(Albizia lebbeck + Annoa squamosa)*.

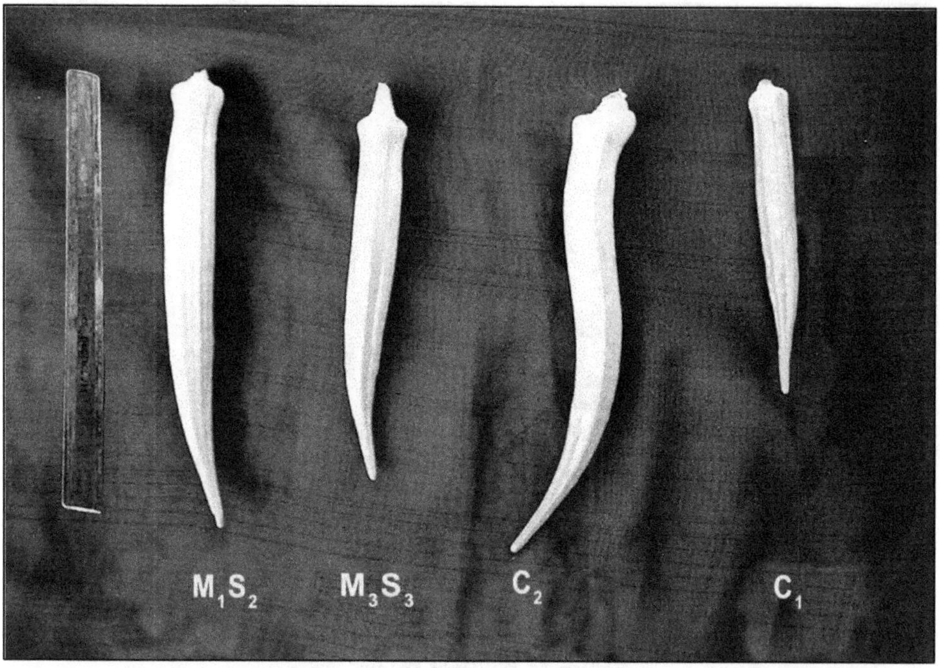

Plate 14.14: Influence of *Vrikshayurveda* Treatments on Fruit Length.

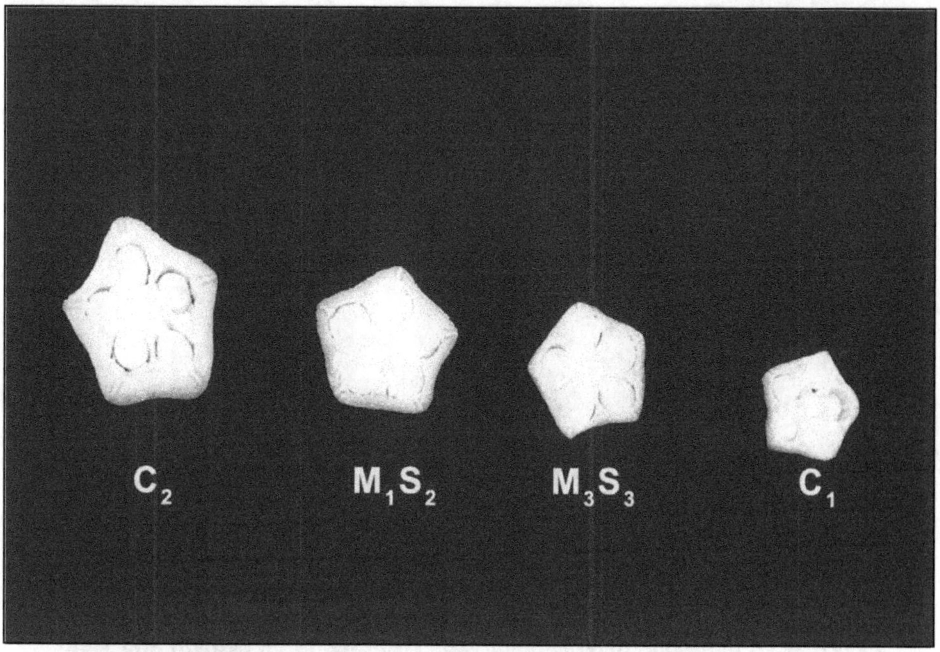

Plate 14.15: Influence of *Vrikshayurveda* Treatments on Fruit Girth.

Figure 14.16: Field Overview.

Figure 14.17: Influence of *Vrikshayurveda* Treatments on Growth and Yield.

7. The nutrient uptake *viz.* N, P and K was found to be the highest in the treatment with *Albizia lebbek* as green leaf manure and seed soaking along with foliar sprays of *Annona squamosa*.
8. The pest and disease incidence were recorded the lowest in the treatment with *Albizia lebbek* as green leaf manure and seed soaking along with foliar sprays *of Annona squamosa.*
9. The correlation studies indicated that number of flowers, number of fruits, fruit length and seed yield per plant correlated positively and significantly with fruit yield.
10. The highest net return and benefit cost ratio were recorded in the treatment with *Albizia lebbek* as green leaf manure and seed soaking along with foliar sprays of *Annona squamosa.*

VII) *Vrikshyurveda* Farming in Onion

Field experiment was undertaken at the College orchard, Department of Horticulture, Agricultural College and Research Institute, Madurai during 2005 - 2006. The experiment was laid out in split plot design with five different Green leaf manures incorporation @ 5 t ha^{-1} (*Albizia lebbek, Senna siamea, Gliricidia sepium, Leucaena leucocephala* and *Delonix regia*) in main plots and four different Green leaf extracts @ 5 per cent in the form of bulb soaking and two foliar sprays (*Alangium salvifolium, Annona squamosa, Aegle marmelos* and *Morinda tinctoria*) in sub plots with two replications and twenty treatment. The study revealed that *Gliricidia sepium* as green leaf manures and bulb soaking along with foliar sprays of *Aegle marmelos* showed a significant increase in the sprouting level, plant height, number of leaves and root length.

Findings

The observations were recorded on sprouting percentage, growth parameters, physiological parameters, yield characters, quality characters, nutrient uptake and Pest and Disease incidence of aggregatum onion. The plant height, number of leaves per plant, root length, total dry matter production and nutrient uptake were recorded at three stages of crop growth at 30, 60 and 90 DAS. The following is brief summary.

1. The out come of the investigation revealed that the highest sprouting percentage were recorded in the treatment with *Gliricidia sepium* as green leaf manure and bulb soaking of *Aegle marmelos*.
2. The plant height, number of leaves per plant and root length were recorded the highest in the treatment with *Gliricidia sepium* as green leaf manure and bulb soaking along with foliar sprays of *Aegle marmelos*.
3. The yield characters *viz.*, bulb girth, bulb weight, average yield per plot and yield per hectares were recorded the highest in the treatment with *Gliricidia sepium* as green leaf manure and bulb soaking along with foliar sprays of *Aegle marmelos*.
4. The physiological parameter *viz.*, dry matter production, crop growth rate, relative growth rate, shoot to bulb ratio and harvest index were recorded

the highest in the treatment with *Gliricidia sepium* as green leaf manure and bulb soaking along with foliar sprays of *Aegle marmelos*

5. The quality parameter *viz.* total soluble solids, ascorbic acid, pyruvic acid and sulphur content were recorded the highest in the treatment with *Gliricidia sepium* as green leaf manure and bulb soaking along with foliar sprays of *Aegle marmelos.*
6. The total sugar was recorded the highest in the treatment with *Leucaena leucocephala* as green leaf manure and bulb soaking along with foliar sprays of *Aegle marmelos.*
7. The nutrient uptake *viz.*, N, P and K was recorded the highest in the treatment with *Gliricidia sepium* as green leaf manure and bulb soaking along with foliar sprays of *Aegle marmelos.*
8. The pest and disease incidence was recorded the highest in the treatment with *Gliricidia sepium* as green leaf manure and bulb soaking along with foliar sprays *Aegle marmelos.*
9. The pesticide residual analysis was recorded the below detectable limit in the Vrikshyurveda farming.
10. The net return and benefit cost ratio were recorded the highest in the treatment with *Gliricidia sepium* as green leaf manure and bulb soaking along with foliar sprays *Aegle marmelos.*

References

Bheemaiah, G. and. Subramaniyam, V. R. 2004. Effect of green leaf manuring on productivity of sunflower (*Helianthus annus*) under guava (*Psidium guajava*)-based Agri- Horticultural system. *Indian J. Agri. Sci.*, **74 (10):** 515-520.

Chelladurai, M. 2006. Studies on performance of (*Allium cepa* var. *aggregatum*) cv. Co 5 Under '*Vrikshyurveda* farming' M.Sc. (Ag.) Thesis, Tamil Nadu Agricultural University, Department of Horticulture, Agricultural College and Research Institute, Madurai, TNAU.

DES, 2006. Department of Economics and Statistics season and crop report of Tamil Nadu, Govt. of Tamil Nadu Chennai – 600 006.

Devi, M. N., T. K. Singh and L. C. Devi. 2003. Efficacy of certain botanicals insecticides against cotton aphid, *Aphis gossypii* Glorer on brinjal. *Medicinal and Aromatic plants abste.* **25(4):** 819.

Gazi, A.V., A. Kyricou, M. Kotsou, K.E. Lasaridi 2007. *Global NEST J.* **9(1)**: 35-41. http://www.agricoop.nic.in

Musmade, B. N and B.K. Knode. 1986. Effect of *Azospirillum brasilense* and *A. lipoferum* inoculation under various levels of nitrogen on growth and yield of onion. Nat. Seminar on Microbial Eco. Tamil Nadu. Agricultural University, Coimbatore on 16-18th Feb.p.47.

Patel, J. J., B. M. Patel, B. T. Patel and R. G. Patil. 2003. Study on use of *Gliricidia sepium* leaves for leaf manuring in cluster bean-pearl millet rotation under dryland condition. *Agri. Sci. Digest.*,**23(1):** 10-13.

Rather, S. A., Ahmed, N. and Chatto, M. S. 2003. Response of onion to microbial inoculation and chemical nitrogen. *J. Horti. Sci.*, **33(3 and 8):** 270-271.

Sirohi, H. S, K. Rohal and Satish. 1998. A note on the affect of different level of nitrogen and plant spacing on growth, yield and quality of kharif onion (*Allium cepa*. L) *Har. J. Hort. Sci.*, **21(1-2):** 113-114.

Swaminathan, C. 2004. Ayurvedic farming in: natural farming published by national service scheme of Agricultural College and Research Institute, Madurai.

Swaminathan, C. and Gururajan, B. 2005. Revisiting and researching an Indian systems of agriculture through *vrikshayurveda*. 93rd Indian science congress" integrated rural development and science and technology. Acharya N.G. Ranga Agri.Univ. Rajendranagar. Andhra Pradhesh. Jan. 3-7 pp.35-39.

Thilakavathy, S. 1998. Studies on effect of inorganic and biofertlizer treatments on yield and quality parameters of country onion (*Allium cepa* var.*aggegatum*). cv. CO-4. M.Sc. (Horti) Thesis, Tamil Nadu Agricultural University, Dept. of Horti, Agri College. Madurai.

Thilakavathy, S. and N. Ramaswamy.1999. Effect of inorganic and biofertilizer on yield and quality parameter of multiplier or country onion (*Allium cepa* var.*aggegatum*). *Veg. Sci.*, **26(1):** 97-98.

Tien, T. M., M. H. Gaskins and D. H. Hubbell. 1979. Plant growth substances produced by *Azospirillum brasilense* and its effect on the growth of pearl millet. *Appl. Environ. Microbiol.*, **37:** 1016-1024.

Tripathi, S., A.Tripathi, D. C. Kari and S. Paroha. 2000. Effect of *Dalbergia sissoo* extracts rhizohium and nitrogen on germination, growth and yield of *Vigna radiata*. *Allelo J.*, **7(2):** 255-264.

Zodape, S. T., Soumita Mukhopadhyay, K. Eswaran, M. P. Reddy and J. Chikara 2010. Enhanced yield and nutritional quality in green gram (*Phaseolus radiate* L.) treated with seaweed (*Kappahycus alvarezii*) extract. *J. of Scien. and Indu. Res.*, **69(2):** 468-471.

Chapter 15
Description of Trees Used for Vrikshayurvedic Farming

I. *Albizia lebbek*

Scientific Classification

Kingdom:	Plantae
Order:	Fabales
Family:	Fabaceae
Genus:	*Albizia*
Binomial name	***Albizia lebbek*** (L.) Benth.

Albizia lebbek is a species of *Albizia*, native to Indomalaya, New Guinea and Northern Australia and widely cultivated and naturalised in other tropical and subtropical regions. English names for it include **lebbeck, lebbek tree, flea tree, frywood, koko** and **woman's tongues tree**. The latter name is a play on the sound the seeds make as they rattle inside the pods. Being one of the most widespread and common species of *Albizia* worldwide, it is often simply called **"siris"** though this name may refer to any locally common member of the genus.

It is a tree growing to a height of 18–30 m tall with a trunk 50 cm to 1 m in diameter. The leaves are bipinnate, 7.5–15 cm long, with one to four pairs of pinnae, each pinna with 6–18 leaflets. The flowers are white, with numerous 2.5–3.8 cm long stamens and very fragrant. The fruit is a pod 15–30 cm long and 2.5-5.0 cm broad, containing six to twelve seeds.

Uses

Its uses include environmental management, forage, medicine and wood. It is cultivated as a shade tree in North and South America. In India and Pakistan, the tree is used to produce timber. Wood from *Albizia lebbek* has a density of 0.55-0.66 g/cm^3 or higher. Even where it is not native, some indigenous herbivores are liable to utilize lebbeck as a food resource. For example, the greater rhea (*Rhea americana*) has been observed feeding on it in the cerrado of Brazil.

II. *Gliricidia sepium*

Scientific Classification

Kingdom:	Plantae
Order:	Fabales
Family:	Fabaceae
Genus:	*Gliricidia*
Binomial name	***Gliricidia sepium*** (Jacq.) Kunth ex Walp.

Gliricidia sepium, often simply referred to as **Gliricidia** (common names: *Mata Ratón; Cacao de nance, Cachanance*, it is commonly known as *Pinon Cubano* in the Dominican Republic; *Madreado* in Honduras; *Kakawate* in the Philippines; *Madre Cacao* or *Madre de Cacao* in the Philippines and Guatemala; and *Madero negro* in

Nicaragua), is a medium size leguminous tree belonging to the family Fabaceae. It is considered as the second most important multi-purpose legume tree, surpassed only by *Leucaena leucocephala*.

Description

Gliricidia sepium is a medium-sized tree and can grow to from 10 to 12 meters high. The bark is smooth and its color can range from a whitish gray to deep red-brown. It has composite leaves that can be 30 cm long. Each leaf is composed of leaflets that are about 2 to 7 cm long and 1 to 3 cm wide. The flowers are located on the end of branches that have no leaves. These flowers have a bright pink to lilac color that is tinged with white. A pale yellow spot is usually at the flower's base. The tree's fruit is a pod which is about 10 to 15 cm in length. It is green when unripe and becomes yellow-brown when it reaches maturity. The pod produces 4 to 10 round brown seeds. The tree grows well in acidic soils with a pH of 4.5-6.2. The tree is found on volcanic soils in its native range in Central America and Mexico. However, it can also grow on sandy, clay and limestone soils.[2][3]

Uses

Gliricidia sepium in Agoo, La Union (foreground tree, with numerous other examples behind it), used as a landscaping/reforesting tool. In this case the main means of preventing erosion and a quick way of providing shade.

The tree is used in many tropical and sub-tropical countries for various purposes such as live fencing, fodder, coffee shade, firewood, green manure and rat poison. Live fences can be grown from 1.5 m to 2.0 m stakes of *Gliricidia sepium* in just a month. Gliricidia can be intercropped with maize. Its effect is that of a potent fertilizer.

G. sepium is also used for its medicinal and insect repellent properties. Farmers in Latin America often wash their livestock with a paste made of crushed *G. sepium* leaves to ward off torsalos. In the Philippines, the extract obtained from its leaves is used to remove external parasites.[2]

According to World Agroforestry Centre, this species is becoming an important part of farming practices in Africa. *G. sepium* has a combination of desirable properties. Because it fixes nitrogen in the soil, it boosts crop yields significantly without the expense of chemical fertilizers. In addition, it tolerates being cut back to crop height year after year. The trees go into a dormant state when they are cut back, so the root system is not competing straight away for the nutrients and the crop is free to become established. The trees only really start to come out of the dormant phase when the crop is already tall.

III. *Leucaena leucocephala*

Scientific Classification

Kingdom:	Plantae
Order:	Fabales
Family:	Fabaceae
Subfamily:	Mimosoideae
Tribe:	Mimoseae
Genus:	***Leucaena***
Species:	***L. leucocephala***
Binomial name	***Leucaena leucocephala*** (Lam.) de Wit[1]

Leucaena leucocephala is a small, fast-growing mimosoid tree native to southern Mexico and northern Central America (Belize and Guatemala), but is now naturalized throughout the tropics. Common names include **white leadtree, jumbay** and **white popinac.** The specific name is derived from the Greek words, meaning "white", meaning "head", referring to its flowers. *L. leucocephala* is used for a variety of purposes, such as firewood, fiber and livestock fodder.

Forage and Fodder

The legume provides an excellent source of high-protein cattle fodder.

Green Manure and Biomass Production

Leucaena leucocephala has been considered for biomass production, as its reported yield of foliage corresponds to a dried mass of 2,000–20,000 kg ha year^{-1} and that of wood 30–40 m^3 ha year^{-1}, with up to twice those amounts in favourable climates. It is also efficient in nitrogen fixation, at more than 500 kg ha year^{-1}. It has a very fast growth rate, young trees reach a height of more than 20 ft in two to three years.

IV. *Delonix regia*

Scientific Classification

Kingdom:	Plantae
Order:	Fabales
Family:	Fabaceae
Subfamily:	Caesalpinioideae
Genus:	*Delonix*
Species:	***D. regia***
Binomial name	***Delonix regia*** (Boj. ex Hook.) Raf.

Delonix regia is a species of flowering plant in the family Fabaceae, subfamily Caesalpinioideae. It is noted for its fern-like leaves and flamboyant display of flowers. In many tropical parts of the world it is grown as an ornamental tree and in English it is given the name **Royal Poinciana** or **Flamboyant**. It is also one of several trees known as Flame tree.

In India it is known as **Gulmohar** in Hindi. It is also known there as *Krishnachura* or *Krusnachuda* (Bengali/Oriya: crown of the Krishna) and *Krishnasura* (in Assamese and Bengali). In Kerala, it is known as *Kaalvaripoo*. In Vietnam, it is known as Phýng v) (means "Phoenix's Tail) (Vietnamese), Malinche and Tabachine. In Khmer, the tree and the flower is known collectively as "Peacock" or (tree) or (flower). In Guatemala, Antigua Guatemala, it is known as *llama del bosque*, in Paraguay as *chivato* and in Cuba as *flamboyán* (taken from the French *flamboyant*).

It is widely cultivated elsewhere. In addition to its ornamental value, it is also a useful shade tree in tropical conditions, because it usually grows to a modest height (mostly 5 meters, but it can reach an maximum height of 12 meters) but spreads widely and its dense foliage provides full shade. In areas with a marked dry season, it sheds its leaves during the drought, but in other areas it is virtually evergreen. Flowers appear in corymbs along and at the ends of branches. Pods are green and flaccid when young and turn dark-brown and woody.

V. *Peltophorum pterocarpum*

Scientific Classification

Kingdom:	Plantae
Order:	Fabales
Family:	Fabaceae
Genus:	*Peltophorum*
Species:	***P. pterocarpum***
Binomial name	***Peltophorum pterocarpum*** (DC.) K. Heyne

Peltophorum pterocarpum (Copperpod, Golden Flamboyant, Yellow Flamboyant, Yellow Flame Tree, Yellow Poinciana, Konda chinta in Telugu, "Perunkonrai" in Tamil, "Radhachura" in Bengali) is a species of *Peltophorum*, native to tropical southeastern Asia and a popularly ornamental tree grown around the world.

Description

It is a deciduous tree growing to 15–25 m (rarely up to 50 m) tall, with a trunk diameter of up to 1 m belonging to Family Leguminosae and sub-family Caesalpiniaceae. The leaves are bipinnate, 30–60 cm long, with 16-20 pinnae, each pinna with 20-40 oval leaflets 8–25 mm long and 4–10 mm broad. The flowers are yellow, 2.5–4 cm diameter, produced in large compound racemes up to 20 cm long. The fruit is a pod 5–10 cm long and 2.5 cm broad, red at first, ripening black and containing one to four seeds. Trees begin to flower after about four years.[1][2]

Uses

The tree is widely grown in tropical regions as an ornamental tree, particularly in India, Nigeria, Pakistan and Florida and Hawaii in the United States. The trees have been planted alternately in India as a common scheme for avenue trees in India alternately with *Delonix regia* (Gulmohur) to give a striking yellow and red effect in summer, as has been done on Hughes road in Mumbai.

The wood has a wide variety of uses, including cabinet-making and the foliage is used as a fodder crop. The brownish colour called *sogan* typical of batik cloth from inland Java is produced from *P. pterocarpum*, which is known there as *soga*.

VI. *Pongamia pinnata*

Scientific Classification

Kingdom:	Plantae
Order:	Fabales
Family:	Fabaceae
Genus:	***Pongamia*** Ventenat

Pongamia is a genus of legume in the Fabaceae family. Recently it was proposed that the genus *Pongamia* be rejected in favor of the genus *Millettia* and many species have been reclassified. Due to recent interest in biofuels, Pongamia is often the generic name given for *Millettia pinnata*, a tree being explored for producing biodiesel.

Species

Species include:

- *Pongamia velutina*

Species Formerly in the Genus

- *Pongamia ovalifolia* is a synonym for *Millettia peguensis*
- *Pongamia pinnata* or *Pongamia glabra* are synonyms for *Millettia pinnata*
- *Pongamia uliginosa* is a synonym for *Derris trifoliata*

Description

Calyx cup-shaped somewhat truncated and 5-toothed. Corolla papilionaceous glabrous. Stamens at first diadelphous at the base and apex and monadelphons about the middle, afterwards usually entirely diadelphous. Legume more or less compressed, more or less oval, with a short recurved point, 1-celled (neither contracted nor with partitions between the seeds), 1-2 seeded: valves concave on the inside, not separating naturally. Trees or twining shrubs. Leaves unequally pinnated: leaflets opposite.

VII. *Moringa oleifera*

Scientific Classification

Kingdom:	Plantae
Order:	Brassicales
Family:	Moringaceae
Genus:	***Moringa***
Species:	***M. oleifera***
Binomial name	***Moringa oleifera*** Lam.

Moringa oleifera is the most widely cultivated species of the genus *Moringa*, which is the only genus in the family Moringaceae. English common names include: **moringa, drumstick tree** (from the appearance of the long, slender, triangular seed-pods), **horseradish tree** (from the taste of the roots, which resembles horseradish), **ben oil tree**, or **benzoil tree** (from the oil which is derived from the seeds). It is a fast-growing, drought-resistant tree, native to the southern foothills of the Himalayas in northwestern India and widely cultivated in tropical and subtropical areas where its young seed pods and leaves are used as vegetables. It can also be used for water purification and hand washing and is sometimes used in herbal medicine.

Description

M. oleifera is a fast-growing, deciduous tree. It can reach a height of 10–12 m (32-40 ft) and the trunk can reach a diameter of 45 cm (1.5 ft). The bark has a whitish-grey colour and is surrounded by thick cork. Young shoots have purplish or greenish-white, hairy bark. The tree has an open crown of drooping, fragile branches and the leaves build up a feathery foliage of tripinnate leaves.

The flowers are fragrant and bisexual, surrounded by five unequal, thinly veined, yellowish-white petals. The flowers are about 1.0-1.5 cm (1/2")long and 2.0 cm (3/4")broad. They grow on slender, hairy stalks in spreading or drooping later flower clusters which have a length of 10–25 cm.

Flowering begins within the first six months after planting. In seasonally cool regions, flowering only occurs once a year between April and June. In more constant seasonal temperatures and with constant rainfall, flowering can happen twice or even all year-round.

The fruit is a hanging, three-sided brown capsule of 20–45 cm size which holds dark brown, globular seeds with a diameter around 1 cm. The seeds have three whitish papery wings and are dispersed by wind and water.

In cultivation, it is often cut back annually to 1–2 m (3-6 ft) and allowed to regrow so the pods and leaves remain within arm's reach.

VIII. *Aegle marmelos*

Scientific Classification

Kingdom:	Plantae
Order:	Sapindales
Family:	Rutaceae
Subfamily:	Aurantioideae
Tribe:	Clauseneae
Genus:	*Aegle* Corrêa
Species:	***A. marmellos***
Binomial name	***Aegle marmelos*** (L.) Corrêa

Aegle marmelos, commonly known as **bael, Bengal quince, golden apple, stone apple, wood apple, bili**, is a species of tree native to India. It is present throughout

Southeast Asia as a naturalized species. The tree is considered to be sacred by Hindus. Its fruits are used in traditional medicine and as a food throughout its range.

Vernacular Names

The tree is called "belada mara" and the religious tree "bilva" or "bilpathre" in Kannada, "vilvam" in Tamil. The fruits are known as *belada hannu* (edible variety), *bilva* (sacred variety) in Kannada, "bela" in Oriya, "bael" in Bengali and *bilva and maredu* in Telugu. It is a native of India and is found widely in Asia, in countries like Burma, Thailand, Sri Lanka, etc. It is called as Sivadruma by the Hindus and is considered as a sacred herb. It is widely found in Indian Siva temples since the herb is considered sacred to Lord Siva, the lord of health. The leaves of the plant are being offered to Gods as part of prayers. The fruits can be eaten either freshly from trees or after drying them.

All parts of the herb (leaves, fruits, roots) are used for medicinal purposes. The herb is widely helpful to kapha and vata dosas. It is not suited to pita dosa.

Uses

The fruit is eaten fresh or dried. If fresh, the juice is strained and sweetened to make a drink similar to lemonade. It can be made into sharbat (Hindi) or *Bela pana* (or *bel pana*, a refreshing drink made of the pulp with water, sugar and lime juice, mixed, left to stand a few hours, strained and put on ice. One large bael fruit may yield five or six liters of sharbat.

The leaves and small shoots are eaten as salad greens.

The Tamil Siddhars call the plant *koovilam* and use the fragrant leaves for medicinal purposes, including dyspepsia and sinusitis.In the system of Ayurveda this drug finds several and frequent therapeutic uses in different forms and recipes. They are prescribed in a number of diseases such as gastro intestinal diseases, piles, oedema, jaundice, vomiting, obesity, pediatric disorders, gynecological disorders, urinary complaints and as a rejuvenative. Besides the wide medicinal utility the plant and its certain parts (leaves and fruits) are of religious importance since the tree is regarded as one of the sacred trees of Indian heritage.

Aegeline (N-[2-hydroxy-2(4-methoxyphenyl) ethyl]-3-phenyl-2-propenamide) is a known constituent of the bael leaf and consumed as a dietary supplement for a variety of purposes.

IX. *Morinda tinctoria*

Scientific Classification

Kingdom:	Plantae
Order:	Gentianales
Family:	Rubiaceae
Genus:	*Morinda*
Species:	***M. tinctoria***
Binomial name	***Morinda tinctoria***

X. *Alangium salviifolium*

Scientific Classification

Kingdom:	Plantae
Order:	Cornales
Family:	Cornaceae (Alangiaceae)
Genus:	*Alangium*
Species:	*A. salvifolium*
Binomial name	***Alangium salvifolium***

Alangium salviifolium is a small tree native to India. In Ayurveda the roots and the fruits are used for the treatment of rheumatism and hemorrhoid. Externally, it is used for the treatment of bites by rabbits, rats and dogs.

XI. *Annona squamosa*

Scientific Classification

Kingdom:	Plantae
Order:	Magnoliales
Family:	Annonaceae
Genus:	*Annona*
Species:	***A. squamosa***
Binomial name	***Annona squamosa*** L.

Annona squamosa is a small, well-branched tree or shrub from the family Annonaceae that bears edible fruits called sugar-apples. It tolerates a tropical lowland climate better than its relatives *Annona reticulata* and *Annona cherimola* (whose fruits often share the same name) helping make it the most widely cultivated of these species.

Description

Annona squamosa is a small, semi-(or late) deciduous, much branched shrub or small tree 3 metres (9.8 ft) to 8 metres (26 ft) tall very similar to soursop (*Annona muricata*) with a broad, open crown or irregularly spreading branches and a short trunk short, not buttressed at base. The fruit of *A. squamosa* (sugar-apple) has delicious whitish pulp and is popular in tropical markets.

Distribution

Annona squamosa is native to the tropical Americas and West Indies, but the exact origin is unknown. It is now the most widely cultivated of all the species of *Annona*, being grown for its fruit throughout the tropics and warmer subtropics, such as Indonesia, Thailand and Taiwan; it was introduced to southern Asia before 1590. It is naturalized as far north as southern Florida in the United States and as south as Bahia in Brazil and is an invasive species in some areas.

Uses

Heat-extracted oil from the seeds has been employed against agricultural pests. High concentrations are potent for 2 days and weaken steadily, all activity being lost after 8 days. See also Annonin.In Mexico, the leaves are rubbed on floors and put in hens' nests to repel lice.

For uses of the fruit, see sugar-apple.

Chemical Constituents

The diterpenoid alkaloid atisine is the most abundant alkaloid in the root. Other constituents of *Annona squamosa* include oxophoebine, reticuline, atidine, histisine, hetisine, hetidine, heterophyllisine, heterophylline, heterlophylline, isoatisine, dihydroatisine, hetisinone benzoyl heteratisine and citronella oil.

Bayer AG has patented the extraction process and molecular identity of the annonaceous acetogenin annonin, as well as its use as a biopesticide. Other bioactive acetogenins have been isolated from the seeds, bark and leaves of *Annona squamosa*.

XII. *Senna siamea*

Scientific Classification

Kingdom:	Plantae
Order:	Fabales
Family:	Fabaceae
Subfamily:	Caesalpinioideae
Tribe:	Cassieae
Subtribe:	Cassiinae
Genus:	*Senna*
Species:	***S. siamea***
Binomial name	***Senna siamea*** (Lam.) Irwin et Barneby

Senna siamea also known as **Kassod Tree**, **Cassod Tree** and (incorrectly) as **Cassia tree**, is a legume in the subfamily Caesalpinioideae. It is native to South and Southeast Asia, although its exact origin is unknown. It is a medium-size, evergreen tree growing up to 18 m with beautiful yellow flowers. It is often used as shade tree in cocoa, coffee and tea plantations. Leaves are alternate, pinnately compound, with slender, green-reddish, tinged axis and 6 to 12 pairs of leaflets on short stalks, rounded at both ends.

Use

This plant has medicinal value and it contains a compound named Barakol. The leaves, tender pods and seeds are edible, but they must be previously boiled and the water discarded. Other uses include as fodder plant, in intercropping systems, windbreaks and shelter belts.

XIII. *Ocimum sanctum*

Scientific Classification

Kingdom:	Plantae
Order:	Lamiales
Family:	Lamiaceae
Genus:	*Ocimum*
Species:	***O. tenuiflorum***
Binomial name	***Ocimum tenuiflorum*** L.

Ocimum sanctum, also known as ***Ocimum tenuiflorum***, **holy basil**, or ***tulasî***, is an aromatic plant in the family Lamiaceae which is native to the Indian Subcontinent and widespread as a cultivated plant throughout the Southeast Asian tropics. It is an erect, many branched subshrub, 30–60 cm tall with hairy stems and simple opposite green or purple leaves that are strongly scented. Leaves have petioles and are ovate, up to 5 cm long, usually slightly toothed. The flowers are purplish in elongate racemes in close whorls. Tulasi is cultivated for religious and medicinal purposes and for its essential oil. It is widely known across the Indian Subcontinent as a medicinal plant and an herbal tea, commonly used in Ayurveda and has an important role in worshipping God.

Uses

Tulasi has been used for thousands of years in Ayurveda for its diverse healing properties. It is mentioned in the Charaka Samhita, an ancient Ayurvedic text. *Tulsi* is considered to be an adaptogen, balancing different processes in the body and helpful for adapting to stress. Marked by its strong aroma and astringent taste, it is regarded in Ayurveda as a kind of "elixir of life" and believed to promote longevity. *Tulasi* extracts are used in ayurvedic remedies for a variety of ailments. Traditionally, *tulasi* is taken in many forms: as herbal tea, dried powder, fresh leaf or mixed with *ghee*. Essential oil extracted from Karpoora *tulasi* is mostly used for medicinal purposes and in herbal cosmetics and is widely used in skin preparations and for fever, colds and infections. It is an Insect repellent.For centuries, the dried leaves have been mixed with stored grains to repel insects and as a mosquito repellent.

Examples for Usufructs of Trees

Basal Nutrition: Tree Leaves as Green Manures

Tree species as Green Leaf Manure (fresh leaves will be incorporated into the soil) will be used. The leaves of these trees have higher nitrogen content which is released after decomposition and available for uptake by crops. The green leaf manures will be applied at the rate of 5 t ha^{-1} at the time of field preparation and the manures will be incorporated into the soil. Then the land will be allowed to be fallow for 45 days time for facilitating decomposition of the applied green leaf manure.

Sl.No.	Tree Species	Family	Major Nutrient Present
1.	*Albizia lebbeck* (L.) Benth.	Leguminaceae	Nitrogen
2.	*Gliricidia sepium* (Jacq.)	Leguminaceae	Nitrogen
3.	*Leucaena leucocephala* (Lam.)	Leguminaceae	Nitrogen
4.	*Delonix regia* Raf.	Leguminaceae	Nitrogen
5.	*Peltophorum ferrugineum* (DC.) Hayne.	Leguminaceae	Nitrogen
6.	*Sesbania grandiflora*	Leguminaceae	Nitrogen

The leaf extracts of trees having secondary metabolites which are capable of promoting growth through physiological and biochemical changes in the plant system.

Sl.No.	Tree Species	Family	Secondary Metabolite Present
1.	*Alstonia scholaris*	Apocyanaceae	Alkaloid
2.	*Tabernaemontana coranaria*	Apocyanaceae	Alkaloid
3.	*Erythrina indica*	Papilionaceae	Alkaloid
4.	*Dalbergia latifolia*	Papilionaceae	Alkaloid
5.	*Dalbergia sissoo*	Papilionaceae	Alkaloid
6.	*Sesbania grandiflora*	Papilionaceae	Alkaloid

The leaf extracts for each species will be prepared by grinding fresh leaves and water at 1:1 proposition and the extract will be filtered and this will serve as stock solution.

From the stock solution 5 per cent solution will be made for spraying to the field crops.

Three or Four sprayings will be given to the crop as per the life period of the crop to promote growth and development in food crops. The first spraying will be done on during 25-35 days of sowing and there after once in two weeks. The spraying will coincide with the vegetative, flowering and maturity stages of crop growth.

Plant Protection

The leaf extracts of tree species having the bio-pesticide properties will be used to control the pests and diseases incidence of the grown crop. The trees listed in the table will be utilised.

Sl.No.	Tree Species	Family	Secondary Metabolite Present
1.	*Azadirachta indica*	Milaca	Azadirachtin
2.	*Vitex negundo*	Meliaceae	Alkaloids
3.	*Anthocephalus cadamba*	Rubiaceae	Alkaloid
4.	*Adina cordifolia*	Rubiaceae	Alkaloid
5.	*Morinda tinctoria*	Rubiaceae	Alkaloid
6.	*Pongamia pinnata*	Leguminosae	Fixed oil-Karanjin

References

"*Gliricidia sepium*". *Treating Livestock with Medicinal Plants: Beneficial or Toxic?*. Cornell University. Retrieved 2008-09-26.

"*Gliricidia sepium*". *Tropical Forages*. Australian Centre for International Agricultural Research. Retrieved 2008-09-26.

"*Leucaena leucocephala* (Lam.) de Wit". *Germplasm Resources Information Network*. United States Department of Agriculture. 1995-03-24. Retrieved 2010-01-18.

"*Leucaena leucocephala* (tree)". *Global Invasive Species Database*. Invasive Species Specialist Group. Retrieved 2010-01-18.

"*Leucaena leucocephala*". *AgroForestryTree Database*. World Agroforestry Centre. Retrieved 2010-01-18.

"PLANTS Profile for *Leucaena leucocephala* (white leadtree)". *PLANTS Database*. United States Department of Agriculture. Retrieved 2009-09-19.

"The Plant List: A Working List of All Plant Species". Retrieved 13 January 2015.

"Tulasi". Tamilnadu.com. 28 January 2013.

Ali, S.I. (1973): *Albizia lebbek* (L.) Benth. *In: Flora of Pakistan* (Vol. 36: Mimosaceae). University of Karachi, Karachi. HTML fulltext

Blatter, Ethelbert; Millard, Walter Samuel (1977) [1937]. *Some beautiful Indian Trees* (2 ed.). Mumbai: Bombay natural History Society. p. 165.

Boning, Charles (2006). *Florida's Best Fruiting Plants: Native and Exotic Trees, Shrubs and Vines*. Sarasota, Florida: Pineapple Press, Inc. p. 35.

Brijesh, S; Daswani P; Tetali P; Antia N; Birdi T (2009). "Studies on the antidiarrhoeal activity of Aegle marmelos unripe fruit: Validating its traditional usage". *BMC Complementary and Alternative Medicine* **9** (47): 47. doi:10.1186/1472-6882-9-47. Retrieved 28 February 2013.

Brown, Sandra (1997): Appendix 1 - List of wood densities for tree species from tropical America, Africa and Asia. *In:* Estimating Biomass and Biomass Change of Tropical Forests: a Primer. *FAO Forestry Papers* **134**. ISBN 92-5-103955-0 HTML fulltext

Chatterjee, Gautam (2001). *Sacred Hindu Symbols*. Abhinav Publications. p. 93. ISBN 978-81-7017-397-7.

Cowen, D. V. (1984). *Flowering Trees and Shrubs in India, Sixth Edition*. Bombay: THACKER and Co. Ltd. p. 1.

Don Burke (1 November 2005). *The complete Burke's backyard: the ultimate book of fact sheets*. Murdoch Books. p. 269. ISBN 978-1-74045-739-2. Retrieved 9 March 2011.

Duke, James A. (2008): Dr. Duke's Phytochemical and Ethnobotanical Databases - *Albizia lebbek*. Retrieved 2008-FEB-23.

Gajalakshmi S, Divya R *et al.* (2011). "Pharmacological activities of *Annona squamosa*: A review". *International Journal of Pharmaceutical Sciences Review and Research* **10** (2): 24–29

Golshahi H., Ghasemi E., Mehranzade E. (2011). "Antibacterial activity of *Ocimum sanctum* extract against E. coli, S. aureus and P. aeruginosa". *Clinical Biochemistry. Conference: 12th Iranian Congress of Biochemistry, ICB and 4th International Congress of Biochemistry and Molecular Biology* **44** (13 SUPPL. 1): S35.

Hughes, Colin E. (1998). *Monograph of Leucaena (Leguminosae-Mimosoideae)*. Systematic botany monographs v. 55. ISBN 0-912861-55-X.

International Legume Database and Information Service (ILDIS) (2005): *Albizia lebbek* (L.) Benth.. Version 10.01, November 2005. Retrieved 2008-MAR-30.

Lowry, J.B.; Prinsen, J.H. and Burrows, D.M. (1994): 2.5 *Albizia lebbeck* - a Promising Forage Tree for Semiarid Regions. *In:* Gutteridge, Ross C. and Shelton, H. Max (eds.): *Forage Tree Legumes in Tropical Agriculture.* CAB International. HTML fulltext

McCann, Charles. (1966). *100 Beautiful Trees of India.* (3/ed) Taraporevala, Mumbai. Pg 259.

Morton, Julia (1987). *"Annona squamosa". Fruits of warm climates.* Department of Horticulture and Landscape Architecture, Purdue University. p. 69. Archived from the original on 5 April 2008. Retrieved 2008-04-17.

National Research Council (2006-10-27). "Moringa". *Lost Crops of Africa: Volume II: Vegetables.* Lost Crops of Africa **2**. National Academies Press. ISBN 978-0-309-10333-6. Retrieved 2008-07-15.

Natural Resources Conservation Service (NRCS). "PLANTS Profile, *Annona squamosa* L.". *The PLANTS Database.* United States Department of Agriculture,. Retrieved 2008-04-17.

Olson, M. E. (2010). Flora of North America Editorial Committee, ed. *Moringaceae:* Drumstick Family. Flora of North America North of Mexico **7**. New York and Oxford. pp. 167–169/

Pattnaik, S; Subramanyam VR; Kole C. (1996). "Antibacterial and antifungal activity of ten essential oils *in vitro*". *Microbios,* **86** (349): 237–246. PMID 8893526.

Puy, D. *et al.* (1998). *Delonix regia.* 2006. *IUCN Red List of Threatened Species.* IUCN 2006. www.iucnredlist.org. Retrieved on 11 May 2006. Listed as Vulnerable (VU B1+2c v2.3).

Raamachandran, J. *Herbs of Siddha Medicines,* The First 3D Book on Herbs, pp.16.

Ramachandran, C.; Peter, K. V.; Gopalakrishnan, P. K. (1980). "Drumstick (*Moringa oleifera*): A multipurpose Indian vegetable". *Economic Botany* **34** (3): 276. doi:10.1007/BF02858648.

Rani Batish, Daizy (2007). *Ecological Basis of Agroforestry.* CRC Press. p. 44. ISBN 1-4200-4327-7. Retrieved 2008-09-26.

Rätsch, Christian (2004): *Enzyklopädie der psychoaktiven Pflanzen, Botanik, Ethnopharmakologie und Anwendungen* (7th ed.). AT Verlag. ISBN 3-85502-570-3

Sharma, BR; Rattan RK; Sharma P (1981). "Marmeline, an alkaloid and other components of unripe fruits of *Aegle marmelos.*". 20 **11**. pp. 2606–2607.

Chapter 16

Healthy Foods for Happy Living

The health is a state of complete physical, mental and social well being and not merely the absence of disease or infirmityaccording to the World Health Organization.

The essential requisites of health would include,

- ☆ Achievement of optimal growth and development, reflecting the full expression of one's genetic potential.
- ☆ Maintenance of the structural integrity and functional efficiency of body tissues necessary for an active and productive life.
- ☆ Mental well-being.
- ☆ Ability to withstand the inevitable process of ageing with minimal disability and functional impairment and
- ☆ Ability to combat disease, such a
  a. Resisting infections (immunocompetence)
  b. Preventing the onset (and retarding the progress) of degenerative diseases and cancer and
  c. Resisting the effect of environmental toxins and pollutants.

God has given us a lot of food stuff by nature to choose best foods according to the needs of human beings by age, gender, socio-economic status, cultural practices and physiological conditions of individual. Foods are classified into energy yielding foods, body building foods and protective foods. Human beings need energy for every walks of lifeincluding sleeping for respiration and basal metabolism. Foods rich in carbohydrates (CHO) and fat provide energy required for individual. Protein also provides energy, if carbohydrates are not available to give energy to our body. One gram of carbohydrates, protein and fat will provide 4.1 kcal, 4kcal and 9 kcal of energy respectively.

Carbohydrates can give the quick energy to the body. It is stored as energy reserve in liver and muscle (300g) and blood (10g) which will give a half a day moderate energy when an emergency called for. A minimum of 100g of glucose is required a day for protein–sparing action and ketogenic effect. Glucose is the only source of energy for central nervous system. Brain demands minute–by-minute supply of glucose from blood. Fats are not only dense source of energy but also associated with the fat soluble vitamins A, D, E and K. The fat layer under the skin provides thermal insulation by controlling body temperature, adipose fat surrounding the organs act as a protection pad against shocks to sensitive organs like kidney.

Cereals, millets, sugars, jaggery, roots and tubers were rich in carbohydrates and fats. Body building foods are rich in proteins which are very much essential for growth and maintenance of human body. Only protein can supply body building materials as amino acids. The function of proteins not stops by providing body building materials to growth but also forms immune proteins, hormones, enzymes for digestion, absorption, nucleoprotein to control and govern the synthesis of all body protein and blood proteins. Protein rich foods are milk, pulses, oilseeds nuts of medium biological value. Protective foods are rich in proteins, vitamins and minerals, vitamins are organic substances present in small amounts in many foods. Vitamins are not providing energy as CHO, protein and fats but perform specific metabolic functions of growth, maintenance and reproduction or to prevent an associated deficiency disease.

The sources of fat soluble vitamins are vitamin A (vision), yellow and orange colored fruits and vegetables, egg yolk, liver, kidney of fish and dairy products. Vitamin D (calcium absorption) egg yolk, liver of fish and oil. Vitamin E (anti sterilitic vitamin) – nuts and oil seeds, vegetables, fruits; vitamin K (coagulation vitamin) sea weed, soybeans, broccoli, egg yolk, liver etc. The water soluble vitamins are vitamin C and B. Vitamin C has wound healing property, rich in fruits vegetables, cereal and pulses sprouts. B vitamins as vital control agents, many of these B group vitamins serve as co-enzymes when they collaborate with key cell enzymes in performing in energy metabolism and tissue building jobs, B vitamins are rich in whole grains, milk products. Water soluble vitamins, are not accumulated in the body but are readily excreted while fat soluble vitamins are stored in the body. For this reason, excessive intake of fat soluble vitamins vitamin A and D can prove toxic.

A large number of minerals and trace metals are present in the body. Some form part of structural component and some acts as catalytic agents in many body reactions. Calcium, magnesium and phosphorus are components of bones and skeleton and iron is a component of blood. Zinc, molybdenum, copper, manganese and magnesium are either structural part or activate a large number of enzyme systems. Iodine is a part of hormonethyroxin. Sodium, potassium are important elements present in fluids within and outside the cells along with ions like chloride, bicarbonate and carbonate keep water and acid base balance in the body. On an average man excretes 20-30 g of mineral salts daily consisting of Na, K, Mg, Ca, Cl, sulphate and phosphates and there must be made good by an adequate intake of these mineral salts through our food.

Each and every nutrients is essential for proper function of our body from head to foot. The requirements of nutrients varies with the age of the individual, gender, physical activity (sedentary, moderate,heavy work) and physiological state *viz.*, pregnancy, lactation etc. hence, either infant or oldage, male or female, at any physical activity it is essential to have a balanced diet for an individual.

Based on the nutritive value foods are classified into 11 food groups. They are cereals and millets, pulses, nuts and oilseeds, vegetables, fruits, milk and milk products, eggs, meat, fish and other animal foods, fat and oils, sugar and other carbohydrate foods, condiments and spices.

A balanced diet is one which contains the various groups of foodstuffs such as energy yielding foods, body building foods and protective foods in the correct proportions so that an individual is assured of obtaining the minimum requirements of all the nutrients. A vast variety of foods are available to choose foods according to individual needs, food habits and socio economic status, seasonal variations, etc. for example, a preschool child, spending moderate cost and vegetarian group can take extra milk instead of flesh foods and egg of non-vegetarian diet to get adequate or on par with non-vegetarian nutrition through diet. A balanced diet should provide around 60-70 per cent of total calories from carbohydrates, 10-12 per cent from protein and 20-25 per cent from fat.

In addition a balanced diet should provide bioactive phytochemicals such as dietary fibre, antioxidants and nutraceuticals which have positive health benefits. Carbohydrate foods include not only the sugars/saccharides, but also the dietary fibre, indigestible (not digested by the enzymes of stomach and small intestine) carbohydrates present in plant foods. They are divided into insoluble and soluble dietary fibre. The insoluble dietary fibre are cellulose, lignin and soluble dietary fibre are pectins, gums, β-glucans, mucilages and major portions of hemicelluloses. Dietary fibre has physical properties such as water holding capacity, cation binding capacity and viscosity and the physiological properties include hypoglycemic and hypolipidemic effects. The suggested intake of total dietary fibre is 25-40g/day or 12-14g/1000kcal energy. The important dietary fibre sources are whole grains of cereals and pulses, spices and condiments, green leafy vegetables, other vegetables like *avarai,* beans, plantain stem etc. nuts like soya and fruits.

As we, the human body is totally made by food we have to be very careful while consuming food or feeding others. The consumer should aware of the healthy and unhealthy foods. Generally quantity of food is taken into account for a given cost. But quality of food is much important as with quantity. Good sanitary and hygienic practices are must while preparing and consuming foods, unhygienic environment may lead to affect the availability of nutrients thus affects the health. Nutritional scientists agree that at all ages of life, well planned plant based diets that incorporate the principles of adequacy, balance and moderation can be nutritious and healthful.

Concern

It is obvious to anyone concerned with agriculture that weds often exert harmful interference. The use of inorganic chemicals and fertilizers in the last three decades have increased manifold and has ended with problems like human and livestock health hazard, deterioration of soil health, poor quality foods and environment pollution. There is, therefore a need to develop alternate strategies to support food production. Further to keep out the inorganic chemicals, fertilizers and other agrochemicals, to certain extent and residual toxicity of chemicals in the food that we eat daily we need to think of alternate strategies in farming. The ideas and concepts discussed above is an eye opener to go beyond in the years to come through practicing chemicals free farming.

Future Research Focus

The increasing need for less cost, eco friendly and safer and more selective bio formulations, for total health care in crop production, should accelerate the interest and research on the natural tree products as crop health tonics. In many cases, natural products offer improved selectivity better toxicological and environmental safety and increased efficiency. The new tool of molecular biology and biotechnology will allow industry to amplify the production of such developed crop health tonics/bio formulations in agriculture so that the commercial biosynthesis will become economically feasible.

- ✰ Assessing the performance of ruling and local varieties of barnyard millet under *Vrikshyurveda* farming is required to develop management protocol with low cost and cheap natural resources.
- ✰ Identifying the principal compound present in the leaf extracts of tree species needs to be done.
- ✰ Exploring possibility reducing the duration of *in situ* decomposition of 45 days through external application of microbial agents.
- ✰ Post harvest technology on thus produced grains needs to be done.
- ✰ Assessing the performance of ruling and local varieties of finger millet under *Vrikshyurveda* farming under rainfed condition is required to develop management protocol with low cost and cheap natural resources.
- ✰ Identifying the principal compound present in the leaf extracts of tree species needs to be done.
- ✰ Exploring possibility reducing the duration of *in situ* decomposition of 45 days through external application of microbial agents as accelerating agents.

Index

Q

R

S